Methods in Quaternary Ecology

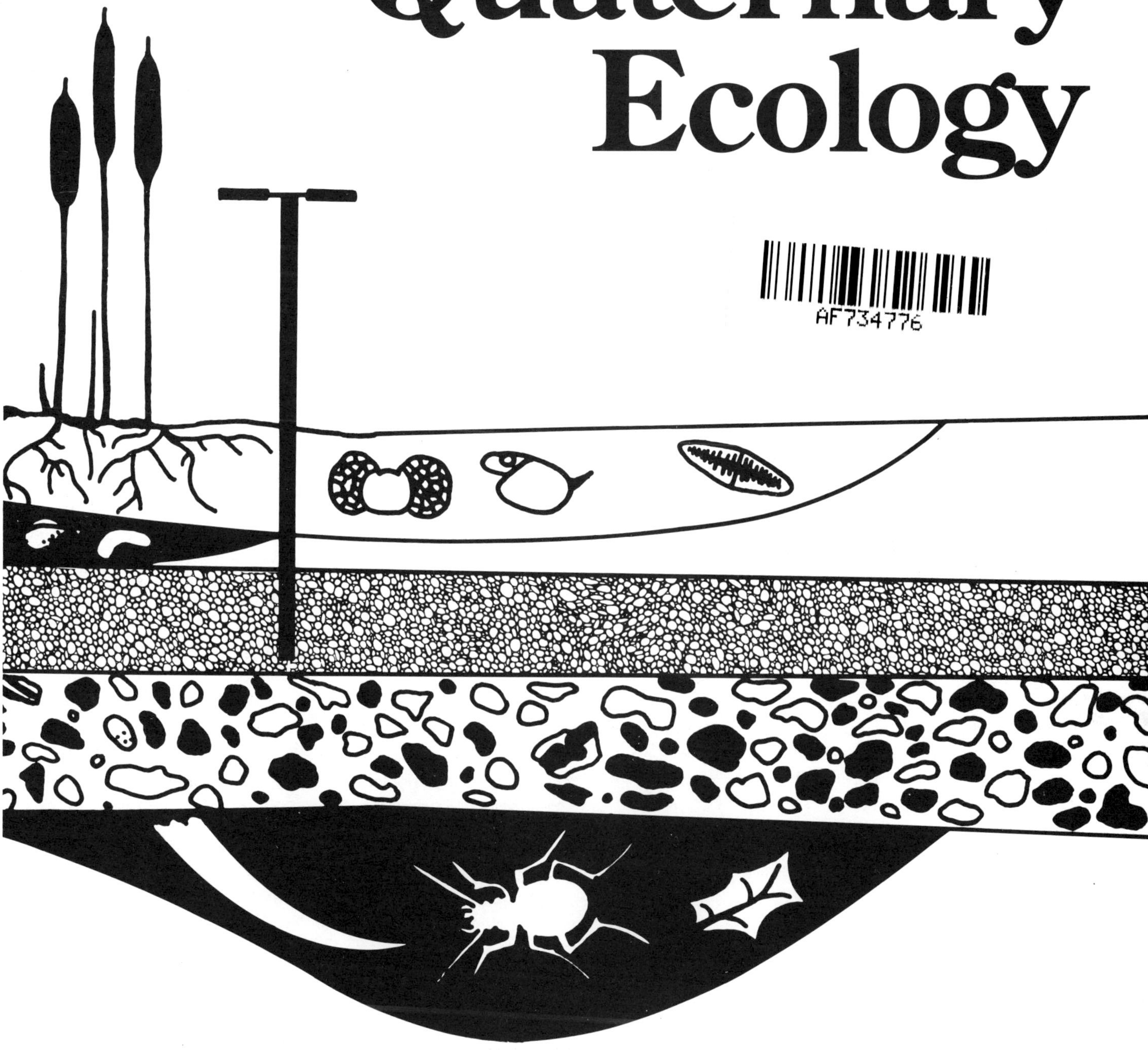

Geoscience canada

Reprint Series 5

Edited by B.G. Warner

Canadian Cataloguing in Publication Data

Main entry under title:

Methods in Quaternary ecology

(Geoscience Canada reprint series ; 5)
Includes bibliographic references.
ISBN 0-919216-42-0

1. Geology, Stratigraphic—Quaternary. 2. Ecology.
I. Warner, Barry G. II. Geological Association of Canada.
III. Series.

QE741.M48 1990 551.7′9 C90–090376–7

Cover Design: Peter Russell, Waterloo

Additional copies may be obtained by writing to:

Geological Association of Canada
Publications
Department of Earth Sciences
Memorial University of Newfoundland
St. John's, Newfoundland A1B 3X5
Canada

 ISBN 0-919216-42-0

Typesetting:
Geological Association of Canada
and Typeline, Mississauga, Ontario

Printing:
Love Printing Service Ltd.
Stittsville, Ontario

CONTENTS

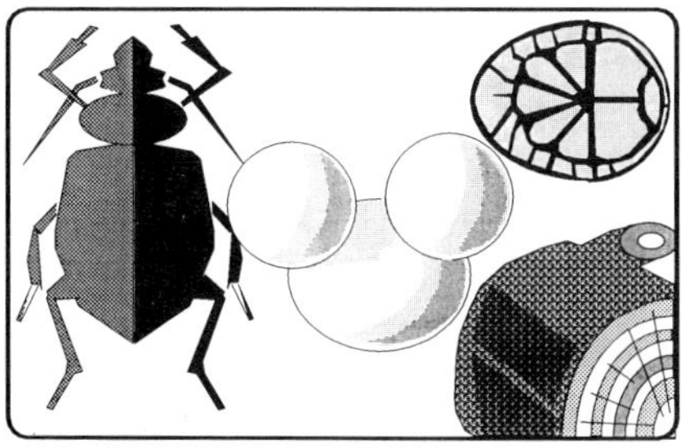

Methods in Quaternary Ecology. Introduction

Barry G. Warner
Department of Earth Sciences
and Quaternary Sciences Institute
University of Waterloo
Waterloo, Ontario N2L 3G1

This volume contains a series of papers devoted to various terrestrial and freshwater biological indicators used in the reconstruction of Quaternary environments. A volume by a group of international collaborators would make a valuable companion volume to this one, as large parts of the present Canadian landscape were covered by inland seas or were affected by marine waters at various times during the Quaternary. This volume highlights the experience of Quaternary workers concerned with non-marine environments.

Quaternary fossil studies have a long tradition in Canada, beginning with publications by Dawson in 1857 and Bell in 1861 on fossil plants and molluscs from Champlain Sea deposits in the vicinity of Montreal. Subsequently, Canadians have maintained a strong committment to Quaternary fossil analysis, most importantly during the last 20 years. It is intended that these papers will bring to light the accomplishments and progress of Canadian research, and research from other regions also.

The unique Canadian landscape and its Quaternary history have contributed much to the understanding of Quaternary processes and events, some not clearly evident in other geographic regions of the world. For instance, pollen studies in eastern Canada have been involved in demonstrating the importance of forest pathogens on shaping past forest communities. Consequently, forest pathogens have been accepted as a plausible explanation for unusual trends in pollen records in parts of Europe. Newly discovered fossil packrat middens in British Columbia provide new sources of paleoenvironmental information. Hopefully, Canadian fossil middens will play as major a role in Quaternary paleoenvironmental reconstructions as middens have in the southwest United States. A somewhat novel area of research being conducted in Canada is freshwater sponge analysis. Systematic analysis of fossil freshwater sponges is not being undertaken outside of North America. Canadians have taken a leading role in demonstrating the use of freshwater ostracodes as paleoenvironmental indicators. Large ecological data bases on the distribution of modern faunas have been compiled and specific paleoecological calibration indices developed for interpretation of fossil faunas. These and many other topics contained in the papers of this volume are rarely covered in other collected works.

Two groups of organisms important in Quaternary ecology are not included in this volume: caddisflies (Trichoptera) and chironomids (Diptera). Readers are referred to the papers by Williams (1988, 1989) on the former group, and to Walker's (1987) paper on the latter group. Gray's (1988) catalogue should be of interest to Quaternary ecologists wishing to study the origin and evolution of the various freshwater groups through the complete geological record as well as the Quaternary record. All of these papers appeared during the time that the "Methods in Quaternary Ecology" series was being published.

The papers in this volume demonstrate the close interdependant need of information collected by paleoecologists and neoecologists. The degree to which paleoecological processes in the Quaternary can be inferred is directly proportional to the size of the taxonomic and ecological data base amassed by ecologists. Similarly, paleoecologists can only provide ecologists with the refined temporal resolution of ecosystem dynamics if given the tools to do so.

West (1985) and Halstead (1988) have reminded us of the role and importance of Quaternary research as a science. Fossil studies give us the means by which to assess natural processes and rates of change through geological time. Processes viewed from Quaternary time are equally applicable to earlier geological time periods, nor is there any reason to expect such processes to cease on into the future. Thus, when dealing with problems of global change and environmental impact, Quaternary paleoecological methods play a central role in providing invaluable foresight into the rate and magnitude of natural ecosystem behaviour, and of ecosystem response to such unnatural factors as pollution, acid rain, and "greenhouse effects". Paleoecological data can be of great assistance to resource managers and governmental agencies in formulating use and conservation policies. Furthermore, as Quaternary deposits are readily accessible and represent major economic resources, fossil studies may aid in evaluating their true economic value.

These papers exemplify the multidisciplinary nature of Quaternary ecology. The future of Quaternary research depends upon close collaborative efforts between respective specialists and upon the teaching of multidisciplinary courses. These points were emphasized by the Report of the Committee investigating the Quaternary Geosciences in Canada (Bird, 1987). It is hoped that the papers in this volume will serve as an introduction to students and non-specialists actively engaged in Quaternary research.

ACKNOWLEDGEMENTS

This volume would not have been possible without the willingness and co-operation of the authors to summarize research in their respective specialties. The assistance of those who agreed to critique the manuscripts contributed much to the quality of the final contributions. Peter Russell's artistic talents with the cover design and graphic blocks at the top of each paper add an invaluable aesthetic flare to overall presentation of the subject matter. In her usual expert and conscientious manner, Monica Easton provided editorial assistance that made production of this volume at all stages a real delight. The Natural Sciences and Engineering Research Council of Canada is gratefully acknowledged for financial assistance to the editor. To all, a most appreciative thank you.

REFERENCES

Berglund, B.E., 1986, Handbook of Holocene Palaeoecology and Palaeohydrology: John Wiley and Sons, Chichester, 869 p.

Bird, J.B., 1987, ed., Quaternary Geosciences in Canada: Their status, national need and future direction: Geological Survey of Canada, Paper 87-18, p. 1-41.

Gray, J., 1988, The evolution of the freshwater ecosystem: The fossil record: Palaeogeography, Palaeoclimatology, Palaeoecology, v. 62, p. 1-214.

Halstead, B., 1988, The fall and rise of Quaternary geology: Modern Geology, v. 13, p. 1-12.

Rutter, N.W., 1985, ed., Dating Methods of Pleistocene Deposits and Their Problems: Geological Association of Canada, Geoscience Canada Reprint Series 2, 87 p.

Walker, I.R., 1987, Chironomidae (Diptera) in paleoecology: Quaternary Science Reviews, v. 6, p. 29-40.

West, R.G., 1985, The future of Quaternary research: The Geologists' Association, Proceedings, v. 96, p. 193-197.

Williams, N.E., 1988, The use of caddisflies (Trichoptera) in palaeoecology: Palaeogeography, Palaeoclimatology, Palaeoecology, v. 62, p. 493-500.

Williams, N.E., 1989, Factors affecting the interpretation of caddisfly assemblages from Quaternary sediments: Journal of Paleolimnology, v. 1, p. 241-248.

March 1990

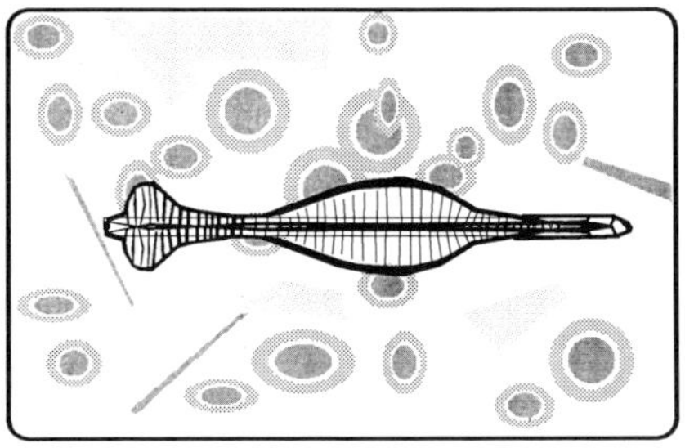

Freshwater Algae

John P. Smol
Department of Biology
Queen's University
Kingston, Ontario K7L 3N6

SUMMARY
Morphological and biogeochemical remains of past algal populations can be recovered from most sedimentary deposits. The majority of these fossils are diatom valves, but many algal groups are preserved in some form. All algal taxa have specific ecological requirements, and therefore their stratigraphic distributions in sedimentary profiles can be used to infer past environmental conditions. This paper briefly reviews some of the ways that freshwater algal assemblages have been used in paleoecological studies, with an emphasis on North American research.

INTRODUCTION

The word "algae" has no real taxonomic meaning, except to circumscribe a diverse grouping of plants that share only a few characteristics. A simple, technical definition is that algae are non-vascular plants that contain chlorophyll *a* as their primary photosynthetic pigment and have simple reproductive structures (Trainor, 1978). Even these criteria are not met by all organisms generally considered to be algae (*e.g.*, some dinoflagellates are not photosynthetic). Nonetheless, the above is a useful working definition for a group of organisms whose classification includes upwards of 15 divisions (phyla) containing over 20,000 species.

One of the first tasks in any algal study is to outline the classification system one will be following — there are several (Round, 1973). Table 1 lists the class system I will be using. The Rhodophyceae (red algae) and Phaeophyceae (brown algae) are almost exclusively marine and are not considered here.

HISTORICAL DEVELOPMENT

The study of algae, namely algology or phycology, was underway, at least in a minor way, since the time of Theophrastus (*ca.* 300 B.C.) — considered by many to be the father of botany. He divided the plant kingdom into three groups: trees, shrubs, and herbs. The latter category included the algae known at that time. Studies of microscopic forms, which incidentally include almost all of the freshwater taxa, only began after van Leeuwenhoek introduced the seventeenth-century scientific community to the microscope, and undoubtedly some of his "animalcules" were actually algae (Whitford, 1968).

Algae have a long fossil record, dating back to the first life forms of the Precambrian. Morphological fossils, which phycologists would immediately recognize as algae, were amongst the first living organisms, with blue-green algal forms recovered from two billion year old stromatolites (Schopf, 1975). Other sedimentary deposits, spanning many geologic time periods, also contain diverse algal assemblages (Round, 1981). Despite this rich record, my review is restricted to the discussion of algae from Quaternary freshwater deposits.

The fact that siliceous algae, such as diatoms, were well preserved in sediments was observed by the early phycologists. For example, Christian G. Ehrenberg (1854) made extensive use of diatomaceous deposits in his pioneering work with diatom systematics in Europe and Asia. Ehrenberg was also the first to write on North American diatoms, although he never visited this continent.

Patrick's (1984) history of diatom research in the United States is germane to this review, as little work was being done in Canada prior to the 1960s. The earliest account that I am aware of is the nineteenth-century study by Hinde (1877) on the Scarborough beds, where he noted three fossil diatom taxa, followed by Dawson's (1893) report of diatoms from what he believed were interglacial deposits near Rolling River, Manitoba. A few species lists of fossils identified from diatomaceous deposits in southern Canada also were compiled (Boyer, 1926).

Interestingly, a re-examination of the same interglacial beds where Hinde had noted diatoms and *Chara* almost a century earlier rekindled interest in the use of diatoms as paleoecological markers in Canada. Terasmae's (1960) palynological study of the Don Formation included a brief discussion of 20 diatom taxa. Later, Duthie and Mannada Rani (1967) re-investigated the site, and recorded over 200 diatom species; they concluded that the beds represented a freshwater river estuary.

Lasalle (1966) listed the diatoms he recorded in his primarily palynological studies in the St. Lawrence lowlands, and showed how fossil pollen and diatoms could be used together to reconstruct the paleoenvironments of the Champlain Sea. Shortly thereafter, Stockner (1971) assessed the applicability of diatoms as indicators of water quality, and Sreenivasa and Duthie (1973) published the postglacial diatom history of Sunfish Lake (Ontario). A large number of Canadian studies were soon to follow.

Table 1 The dominant freshwater algal classes (Lee, 1980) and their morphological fossils. These algal groups are also represented by fossil pigments.

CLASS	COMMON NAME	MORPHOLOGICAL FOSSIL
Cyanophyceae	blue-green algae or blue-green bacteria	vegetative and reproductive cells
Chlorophyceae	green algae	vegetative and reproductive cells
Euglenophyceae	euglenoids	vegetative and reproductive cells
Charophyceae	charophytes, stone worts, brittle worts	vegetative and reproductive cells
Dinophyceae	dinoflagellates	vegetative and reproductive cells
Cryptophyceae	cryptophytes	none known
Bacillariophyceae	diatoms	valves/frustules
Chrysophyceae	chrysophytes	statospores, mallomonadacean scales and bristles/spines

MORPHOLOGICAL AND BIOGEOCHEMICAL FOSSILS

The main requirement for any freshwater fossil is that it be insoluble in water, and that it be relatively resistant to bacterial decay and chemical solution. Despite these stringent pre-conditions, a lake's sediment contains a surprisingly large suite of indicators, with virtually every algal group represented (Table 1). I distinguish two main categories of algal fossils: morphological fossils, which maintain some morphological integrity in sedimentary deposits, and biogeochemical fossils, which are preserved in sedimentary profiles by chemical markers.

Morphological fossils are those that can be studied visually, although high resolution microscopy is usually required. These include the "hard parts" of an alga's structure, many of which are taxonomically diagnostic. By far, the most widely studied are the siliceous valves of diatoms (Figures 1-3). Diatoms represent one of the most widely distributed and ecologically diverse group of algae, whose systematics is based on the size, shape, and sculpturing of their cell walls or frustules (one frustule is composed of two interlocking halves or valves). Because of their siliceous nature, diatom valves are usually well preserved in sedimentary deposits, where they can reach concentrations on the order of 10^9 per cm^3. Barber and Haworth (1981) provide a useful summary of the terminology used in diatom taxonomy, and Battarbee (1986) reviews the methodology and applicability of diatoms to paleoecological studies.

Chrysophycean algae similarly are represented in the fossil record by siliceous remains. All chrysophytes produce siliceous resting cysts known as statospores or stomatocysts (Figures 4 and 5), the morphologies of which are often species-specific, and, like diatom valves, appear to have all the necessary characteristics to be useful paleoindicators. However, in contrast to diatom valves, the taxonomy of the cysts remains poorly documented, and in fact only about 5% of cyst morphotypes have been linked to known chrysophyte taxa. Several artificial classification systems have been proposed (*e.g.*, Nygaard, 1956) and used (*e.g.*, Carney and Sandgren, 1983), although rarely can paleolimnologists identify many fossil cyst morphs. These taxonomic shortcomings are being addressed by the International Statospore Working Group (ISWG), which has proposed a series of guidelines for describing statospores, even if their taxonomic identities are unknown (Cronberg and Sandgren, 1986).

Several common chrysophyte genera (such as those in the Synurophyceae) are further characterized by an external armour of overlapping siliceous scales (Figure 6) and bristles or spines. Unlike the stomatocysts, these species-specific scales have been studied extensively with the electron microscope, and presently the entire systematics of the Synurophyceae is based on the shape and sculpturing of the scales (Asmund and Kristiansen, 1986). Chrysophyte scales are usually well preserved in lake sediments, where they can be identified and enumerated alongside diatoms using the same preparations (Munch, 1980; Smol, 1980). Cronberg (1986a, b) and Smol (1988a) have reviewed the methodology and applicability of chrysophyte cysts and scales to paleolimnological analyses.

Most other algal groups also preserve morphological remains (Figure 7), although research on these has lagged behind the siliceous fossils. For example, vegetative remains of green algae, such as intact desmids or entire colonies of *Pediastrum*, are not uncommon in pollen preparations (Cronberg, 1986a; Frederick, 1981). Similarly, the resting stages and vegetative remains of other Chlorophyceae (Van Geel *et al.*, 1981; Livingstone, 1984), blue-green algae (*ibid.*), dinoflagellates (Norris and McAndrews, 1970), and charophytes (Burne *et al.*, 1980) are occasionally reported, although these indicators still represent a largely unstudied and probably underutilized paleoecological tool.

Past algal populations also can be inferred from biogeochemical fossils, most notably photosynthetic pigments and their derivatives (Brown, 1969). Chlorophylls have been useful in this regard, but carotenoids (*i.e.*, carotenes and their oxidized derivatives the xanthophylls) are more taxonomically specific, and can be used to trace the wax and wane of algal and photosynthetic bacterial groups. Fossil pigments are especially important in the reconstruction of populations that do not usually form morphological fossils. For example, cryptophytes can be traced by their specific carotenoid alloxanthin, blue-green algal populations by myxoxanthophyll, and so forth. Certain pigments are restricted to even lower taxonomic categories, such as oscillaxanthin (Brown and Colman, 1963), which is only found in some *Oscillatoria* and *Spirulina* species.

From a historical perspective, it is interesting to note that the pioneering Canadian research with pigments (Vallentyne, 1955, 1956; Brown, 1962) predated much of the early work done on morphological fossils.

FIELD AND LABORATORY TECHNIQUES

Most sedimentary deposits contain at least some algal fossils. Field collection techniques are identical to those used for other paleoecological work. In general, sediment cores from lakes or peatlands are retrieved using either gravity or piston corers (techniques reviewed by Aaby and Digerfeldt, 1986). The sediment is then extruded and sectioned at defined intervals. In some circumstances, fine resolution sampling is required (*e.g.*, the algae in a single varve). The sediment is then either impregnated with a resin and thin sectioned (Tippett, 1964), or a thin rind of sediment is adhered to a length of transparent tape from a longitudinal section of a frozen core (Simola, 1977). For certain specialized investigations, such as the study of sections (Terasmae, 1960), a systematic sequence of samples can be collected using a trowel or spatula.

Preparation procedures depend on the type of paleoindicators to be studied. Siliceous microfossils must first be separated from the sediment matrix by treatment with strong oxidizing acids, which digest any associated organic matter, followed by several rinses in distilled water (Battarbee, 1986). The resulting concentrate is then mounted on glass slides in a medium of high refractive index, such as Hyrax or Naphrax. Analyses can be prepared quantitatively by using evaporation trays (Battarbee, 1973) or by introducing a known quantity of exotic markers to the sediment, such as glass microspheres (Battarbee and Kneen, 1982). For non-siliceous microfossils, most procedures are similar to palynological preparations (Cronberg, 1986a; van Geel, 1986).

Morphological remains are usually identified and enumerated using high resolution optics (Figures 1, 4 and 7), although certain groups may require some scanning (Figures 3 and 5) or transmission (Figure 6) electron microscopy to confirm taxonomic identifications. Although the morphological taxonomy of most algal groups is based on systems that originated in the early nineteenth century, the single most important quality assurance/quality control problem in morphology-based paleoecology is consistent systematics.

Procedures for extracting fossil pigments and their derivatives are less standardized and more rigorous, and the validity of certain techniques continues to be an area of considerable debate. Isomers and the various pigment degradation products must all be considered (Brown, 1969). In general, analyses include the use of solvents, such as acetone and alcohol, to extract the pigments from the sediments. The extract is then separated into its various pigment components using chromatographic techniques. Spectrophotometry is often used to aid in identifying the pigments and their derivatives. In contrast to morphological microfossils, more stringent laboratory procedures (*e.g.*, working under dim lights, storage of sediments under a nitrogen atmosphere, *etc.*) are required to minimize degradative processes and to avoid the introduction of artifacts.

APPLICATIONS

Many algal taxa have specific ecological requirements; therefore at least some information on past ecological conditions can be gleaned from a careful analysis of the algal assemblage the environment once supported.

Because of the large number of environmental factors that can potentially affect the size and composition of an algal community, the autecologies of many taxa are still poorly documented — but this shortcoming is being pursued aggressively and continued progress should be expected (Davis and Smol, 1986). An especially popular and successful approach has been the use of "calibration sets", *viz.* the study of algal microfossils in the recent sediments (*e.g.*, top cm, usually representing the last few years of sediment accumulation) in well studied lakes, and the correlation of these assemblages with modern environmental conditions. A variety of statistical techniques can then be applied to the data set, and transfer functions relating algal composition to environmental variables (such as pH; see papers in Smol *et al.*, 1986) can be constructed. Other measures of community structure, such as species diversity, have been used, but these approaches should be used with caution (Smol, 1981).

Below, I briefly discuss some of the ways that algal assemblages have been used to infer past environmental conditions. Most of the examples deal with diatoms, but other algal groups are discussed where data are available.

Habitat.
The life forms of most algae, and especially diatoms, are sufficiently well known that taxa can be categorized as having optimal growth either in the open water (phytoplankton) or attached in the littoral zone (periphyton). Often, the latter can be characterized further by the type of associated substrate, namely attached to plants (epiphytic), rocks (epilithic), sand grains (epipsammic), or living on mud (epipelic). Certain taxa are also characteristic of flowing waters (rheophilic; Patrick and Reimer, 1966), snow and ice (cryophilic; Lichti-Federovich, 1980), tree bark (corticolous; Cox and Hightower, 1972), and subaerial and aerial habitats, such as sandstone cliffs (Camburn, 1982) or even the hanging gardens at national monuments (Johansen *et al.*, 1983)! The paleoecologist can use the known life strategies of algae to reconstruct the relative size and composition of these different communities (*e.g.*, shallow water *versus* deep water forms), from which other environmental information can be deduced (*e.g.*, water level fluctuations).

Ritchie and Koivo (1975) related habitat preferences of diatoms to water level changes in the Lake Agassiz basin throughout postglacial time. They interpreted the abundance of planktonic diatoms in the early sediments as representing deposits of a large glacial lake. These communities were replaced by benthic floras as the lake gradually receded to the shallow pond, still present today. Björck and Keister (1983) also included diatom analyses to aid their interpretations of the Emerson Phase of Lake Agassiz. Manny *et al.* (1978) recorded fluctuations in the relative frequencies of planktonic and littoral taxa in a diatom stratigraphy from Wintergreen Lake (Michigan). They postulated that water levels have fluctuated in the past, which altered the size of the lake's littoral zone.

Osborne and Moss (1977) and Moss (1978) noted changes in the stratigraphic distribution of epiphytic diatoms in some Norfolk (UK) ponds. They believed that the decline in epiphytes indicated the elimination of macrophytes by the striking increase in planktonic forms and the resulting reduction in light transparency following eutrophication. Meanwhile, Jones *et al.* (1978) and Smol and Dickman (1981) attributed the decline in shallow water taxa to increased turbidity in their study lakes' littoral zones following road construction; the latter example was from a twentieth-century highway built in Algonquin Park (Ontario), the former was from a Welsh road built by the Romans in A.D. 74!

Hickman and Klarer (1981) recorded a decline in planktonic diatoms and a compensatory increase in epipelic forms about 6,600 years ago in Lake Isle (Alberta). This coincided with deposition of Mt. Mazama ash on the lake.

Florin and Wright (1969) identified a diatom assemblage characteristic of terrestrial habitats, along with wood debris and fungal spores, in the basal sediments of Kirchner Marsh (Minnesota). These data were interpreted as evidence for the persistence of stagnant glacial ice. The earliest sediments probably reflect the floor of a forest that once grew on the stranded ice block. Subsequently, the ice melted and a kettle lake formed, and planktonic diatoms began to dominate the assemblage. Haworth (1976) also found terrestrial diatoms in the late-glacial sediments of two Scottish lakes. She postulated that these diatoms were indicative of solifluction processes in the drainage basin.

Temperature.
Algae are generally considered to be relatively poor indicators of climate, although temperature optima for some taxa have been suggested. Diatoms have been included in many studies to reinforce paleoclimate interpretations, especially in African lakes (*e.g.*, Richardson and Richardson, 1972; Stoffers and Hecky, 1978; Stager, 1984). Preliminary research on light and nutrient resource partitioning amongst African diatoms, and the relationships these variables have with climate, also provides interesting paleoecological possibilities (Kilham *et al.*, 1986).

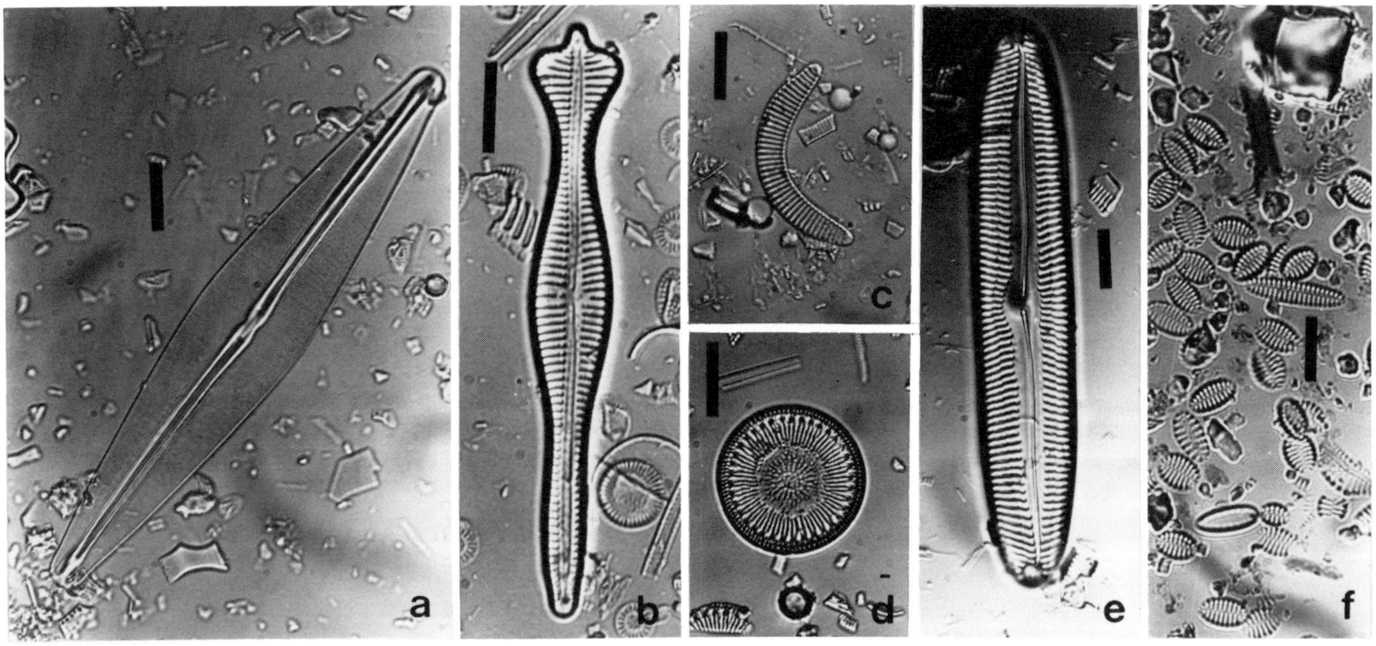

1

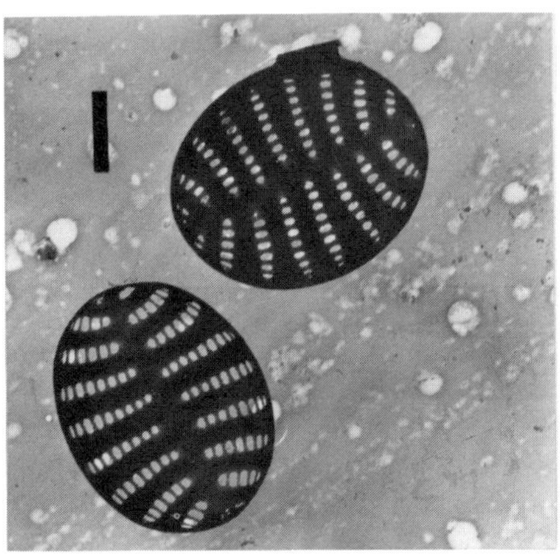

2

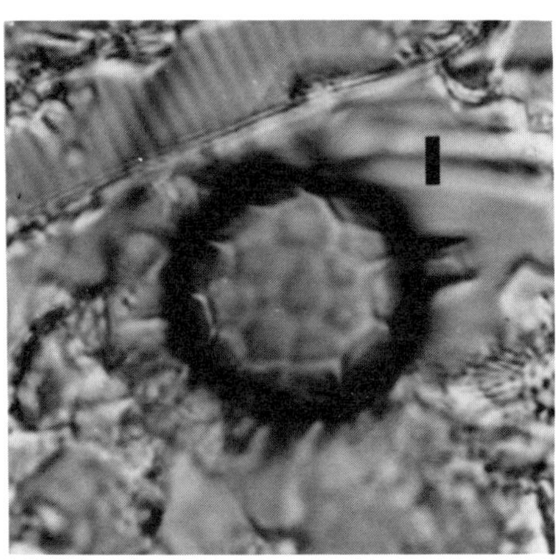

4

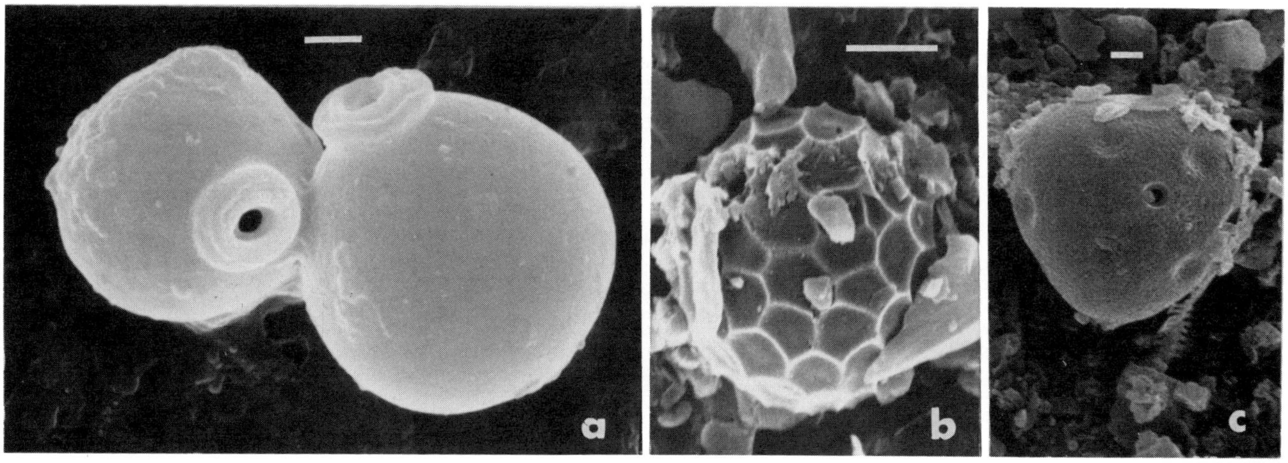

5

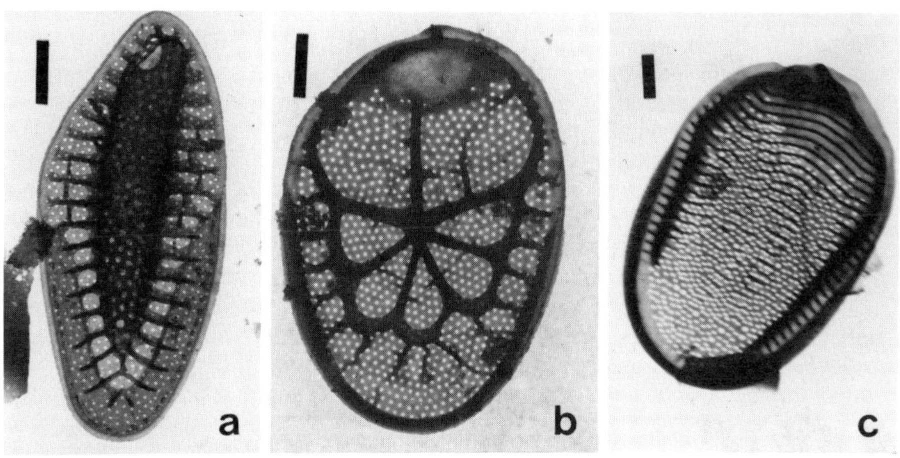

6

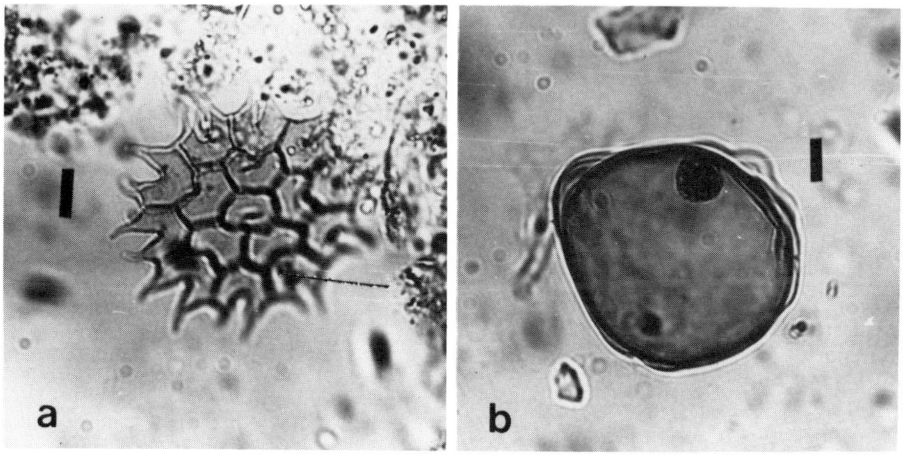

7

Figure 1 *Light micrographs of fossil diatom valves. Scale bar = 10 μm.*
(a) Frustulia rhomboides *(Ehr.) DeT. — a common benthic diatom from shallow, acidic waters.*
(b) Gomphonema acuminatum *Ehr. — found in circumneutral, shallow waters.*
(c) Semiorbis hemicyclus *(Ehr.) Patr. — a shallow water, acidobiontic diatom.*
(d) Cyclotella comta *(Ehr.) Kütz. — a common planktonic diatom in non-acidic waters.*
(e) Pinnularia viridis *(Nitz.) Ehr. — a benthic diatom found in circumneutral waters.*
(f) *Small* Fragilaria *species from the sediments of a small lake from Northern Greenland. These diatoms are commonly found in the early post-glacial sediments of temperate lakes.*

Figure 2 *Transmission electron micrograph of some of the small* Fragilaria *diatoms shown in Figure 1 (f). The detail of the valve striation pattern can be seen. Scale bar = 2 μm.*

Figure 3 *Scanning electron micrographs of fossil diatom valves. Scale bar = 2 μm.*
(a) Anomoeoneis serians *(Bréb.* ex *Kütz.) Cl. — a shallow water diatom from acidic waters.*
(b) Cyclotella stelligera *Cl.* et *Grun. — a common planktonic diatom in many Canadian waters.*

Figure 4 *Light micrograph of a fossil chrysophycean stomatocyst. Scale bar = 2 μm.*

Figure 5 **(a-c)** *Scanning electron micrographs of chrysophycean stomatocysts. Scale bar = 2 μm.*

Figure 6 *Transmission electron micrographs of chrysophycean scales. Scale bar = 0.5 μm.*
(a) Synura petersenii *Korsh. — a widespread species common in many water bodies.*
(b) Mallomonas hindonii *Nicholls — a chrysophyte characteristic of very acidic waters.*
(c) Mallomonas transsylvanica *Peterfi* et *Momeu — a relatively rare chrysophyte in most North American waters.*

Figure 7 *Light micrographs of non-siliceous algal microfossils. Scale bar = 20 μm.*
(a) Pediastrum *colony — these planktonic algae have been used to infer past changes in a lake's eutrophication.*
(b) *dinoflagellate cyst.*

In certain environments, the defined habitat requirements of algal taxa may be used to infer changes in past climate. Smol (1983, 1988b) suggested that in polar regions, such as the high arctic, algal microfossils may provide proxy data for paleoclimates because temperature changes are intimately related to the extent of snow and ice cover on high latitude lakes. Therefore, temperature determines the area of a lake that can be colonized by photosynthetic organisms. During cold summers, a portion of the lake's pelagic region may remain permanently covered by a raft of snow and ice, and therefore only shallow water, littoral species will thrive. As temperatures ameliorate, a larger portion of the lake is made available for algal growth, and progressively deeper water floras develop.

Similar reasoning might explain the abundance of shallow water diatoms in the late-glacial and early postglacial sediments of most lakes (*e.g.*, Sreenivasa and Duthie, 1973; Smol and Boucherle, 1985). Planktonic communities may have been late in developing because the cold temperatures characteristic of the late-glacial truncated the growing season in the pelagic region. Only later, as temperatures ameliorated and lake ice was less extensive, were deeper water algae more competitive.

Seasonality.
Seasonal changes in algal populations have been the subject of intense study since the time of the first limnologists. Knowledge of the marked seasonality of certain algal groups has been exploited by paleoecologists to demonstrate that laminations in some lake sediments are annual deposits. For example, Tippett (1964), in a classic paper on varves in Ontario lakes, used the known seasonal distributions of diatoms and chrysophytes as evidence for the annual nature of the laminations. Other studies that examined algal fossils in varves, and related them to seasonal events, include Simola (1977), Battarbee (1981), and Peglar *et al.* (1984).

Conductivity/salinity.
Water salinity is one of the most important factors influencing the distribution of algae. Much of the available data are for diatom species. Most diatoms can be characterized ecologically according to their salinity tolerances (*i.e.*, halobian spectrum; defined by Kolbe (1927), and subsequently modifed by Hustedt (1957)). Recent work on calibration sets indicates that diatoms may be characterized further by even more specific salinity variables, such as assemblages common to sodium carbonate—bicarbonate, sodium chloride, or calcium magnesium carbonate—bicarbonate waters (Gasse, 1986). The biostratigraphy of other algal remains, such as charophyte oospores (Burne *et al.*, 1980; Chivas *et al.*, 1986), also may be important paleosalinity indicators, and certainly warrant further work.

Alhonen (1971) used the known salinity tolerances for diatoms in Finnish sediment cores to document the stages in the development of the Baltic Sea. Similarly, Kjemperud (1981) traced the post-glacial isolation from the sea of eight Norwegian basins by recording the shift from marine, to brackish, and finally to freshwater diatoms in his sediment cores. Coincident with the shift to a freshwater flora, he also noted the presence of green algae, such as *Pediastrum* and *Botryococcus* colonies — further evidence that the water salinity had declined markedly.

Osborne and Moss (1977) and Buzer (1981) interpreted the presence of brackish and marine diatoms in their lacustrine sediments as evidence for marine transgressions. In the latter study, changes in dinoflagellate cysts were noted. Tuchman *et al.* (1984) recorded salinification in a Michigan lake based on a diatom profile, but this was due to seepage from a salt storage facility adjacent to the lake.

Paleosalinity interpretations may also be used to infer past climates and water levels, as these variables are closely related, especially in closed basins. For example, Bradbury (1971) delineated 15 diatom zones in a 46 m core from Lake Texcoco (Mexico). He suggested that past changes in water level resulted in drastic salinity changes, which were reflected by the diatom assemblages. These data indicated that Lake Texcoco had fluctuated from such extremes as a deep, freshwater lake with low salinity, to a shallow, brackish marsh. Stoermer *et al.* (1971) believed that lake level fluctuations were related to salinity changes during the ontogeny of Devil's Lake's in North Dakota.

Nutrients/eutrophication.
Many algal taxa are relatively stenotypic with respect to their trophic requirements. Although a large number of nutrients may potentially limit algal growth, the vast majority of temperate lakes are phosphorus limited. Not surprising, most paleoecological studies dealing with eutrophication have centered on phosphorus dynamics.

A lake may be enriched with nutrients from a variety of sources, ranging from the steady, albeit usually slow release of nutrients from the terrestrial system (natural eutrophication), to the greatly accelerated nutrient input resulting from human activities (cultural eutrophication). The latter includes direct sources, such as sewage input, as well as more diffuse sources, such as agricultural runoff.

Virtually every type of eutrophication process has been studied in the paleoecological record. Diatoms are the most widely used indicators; however, in contrast to most of the previous examples, other algal markers have been used. This is at least partially because other groups, such as blue-green or green algae, are usually more competitive in eutrophic waters. Most of the work on the blue-green algae is via pigments (Engstrom *et al.*, 1985), whilst the green algae are usually studied by their morphological fossils (Cronberg, 1986a). Chrysophycean scales (Munch, 1980; Smol, 1980) and stomatocysts (Carney and Sandgren, 1983; Smol, 1985; Rybak, 1986) are also sensitive indicators.

Past trophic changes, unrelated to human activity, usually involve the use of

Table 2 The pH classification system for diatoms, as set up by Hustedt (1937-39)

alkalibiontic	occurring at pH values above 7.0
alkaliphilous	occurring at pH values about 7, with widest distribution >7
indifferent	occurring equally on both sides of pH = 7
acidophilous	occurring at pH values about 7, with widest distribution <7
acidobiontic	occurring at pH values under 7, with optimum distribution at pH = 5.5 and under

cores spanning a lake's entire postglacial sedimentary history. Attempts at estimating paleoproductivity was a driving force for much of the early paleolimnological work (Wetzel, 1970; Adams and Duthie, 1976). Often, the apparent trophic shifts are related to a complex of interdependent factors. For example, in Little Round Lake (Ontario) diatom and chrysophyte microfossils (Smol and Boucherle, 1985) and fossil pigments (Brown *et al.*, 1977; Daley *et al.*, 1977) showed that the lake's trophic status changed over the past 11,000 years. These changes were attributed to cold temperatures during the lake's early history, watershed processes regulating erosion and nutrient release, and fluctuations in the lake's meromictic stability. One marked eutrophication event, which occurred around 4,800 years ago, was attributed to the effects of the decimation of hemlock trees in the lake's catchment (Boucherle *et al.*, 1986); a major change in forest composition that was most likely caused by a pathogen (Davis, 1981). The algal species shifts recorded at the "hemlock decline" were similar to those noted in Little Round Lake's recent sediments, with the latter coinciding with the arrival of European settlers and land-clearance in the late-1800s (Smol *et al.*, 1983).

Stockner (1978) reviewed most of the early Canadian work on eutrophication. Space limitations preclude a comprehensive update of the large number of studies that have focussed on cultural eutrophication, except to note that the limnological repercussions of activities such as road construction (Smol and Dickman, 1981), agriculture (Bradbury, 1975; Brugam, 1978), and sewage inputs (Stockner and Benson, 1967) have all been considered. Multidisciplinary approaches are the most instructive (Engstrom *et al.*, 1985).

Of particular interest to many North Americans is the quality of the water in the Great Lakes. Such large and deep bodies of water may provide special challenges to paleoecologists, but environmental reconstructions are still possible. Changes in diatom, chrysophyte, and green algal populations have all been used to document recent eutrophication changes (*e.g.*, Duthie and Sreenivasa, 1971; Frederick, 1981; Harris and Vollenweider, 1982; Stoermer *et al.*, 1985). An offshoot of these studies has been the quantification of biogenic silica in sedimentary profiles as an estimate of past diatom and chrysophyte populations (Schelske *et al.*, 1983, 1986). Such techniques have been applied to smaller lake basins (Newberry and Schelske, 1986).

In North America, cultural eutrophication is usually restricted to about the last 150 years; in Europe, cultural disturbances may date back thousands of years. For example, Cronberg (1982, 1986) recorded stratigraphic changes in green and blue-green algal microfossils from Lake Vaxjösjön (Sweden). Originally, the lake was oligotrophic, but it increased in primary production around A.D. 600 as a result of human activity in its drainage. Other investigators have traced the limnological impact of slash and burn agriculture and peatland drainage (Simola, 1983), hemp soaking (Battarbee *et al.*, 1980), fertilization by sea birds (Moss, 1978, 1979), and the effects of forest fires (Bradbury, 1986).

pH/acidification.

In recent years, paleolimnological studies on lake acidification have earned much credibility for paleoecological research in general. Presently, a large number of lakes in North America and northern Europe are acidic and devoid of higher life forms, such as fish. Acidic precipitation, related to cultural activities such as the smelting of ores or the burning of coal, has been suspected of acidifying these lakes. Without historical water chemistry data, causative factors are difficult to identify. Because a large number of diatoms and scaled chrysophytes are characteristic of restricted pH regimes, their stratigraphic distribution in affected lake sediments can be used to trace the timing, rates, and patterns of lake acidification. Much progress has been made in demonstrating the predictive information contained in fossil assemblages (Battarbee, 1984; Charles and Norton, 1986; papers in Smol *et al.*, 1986). In many ways, acidification studies provide a stellar example of the practical application of paleoecological techniques to environmental problems.

Hustedt (1937-1939) was the first to clearly recognize the pH specificity of diatom taxa. He identified diatoms from over 650 samples from Java, Bali, and Sumatra, concluding that the water's hydrogen ion concentration (pH) had the greatest influence on the composition of assemblages. The five categories he delineated in his pH classification system continue to be used commonly today (Table 2).

Nygaard (1956) expanded Hustedt's system by attempting to quantify past changes in lakewater pH. Nygaard proposed a series of indices that related the relative abundances of taxa in the various Hustedt categories to lakewater pH. The next major step towards quantification was by Meriläinen (1967), who used linear regression analysis to establish quantitative relationships between logarithms of Nygaard's indices and measured pH values. In recent years, many statistical techniques have been applied to diatom calibration sets (*e.g.*, Charles, 1985; Anderson *et al.*, 1986; Huttunen and Meriläinen, 1986; Taylor, 1986), and the predictive tools available to paleoecologists continue to be refined and improved (Davis and Smol, 1986; Stewart *et al.*, 1989; Dixit *et al.*, 1989b). Most of the literature again relates to diatoms (Battarbee, 1984), although chrysophyte scales increasingly are being used to enhance and supplement interpretations (Smol *et al.*, 1984; Smol, 1986; Steinberg and Hartmann, 1986; Hartmann and Steinberg, 1986).

Paleolimnological studies have now been completed or are in progress for numerous affected regions, and recent declines in lakewater pH have been inferred for many sites using diatoms and chrysophytes (Charles and Norton, 1986). Inferred pH changes in the recent sediments of the affected lakes are far greater than those recorded throughout the Holocene. For example, in Upper Wallface Pond (New York) the diatom inferred pH declined from above 7 to about 5 from *circa* 12,000 to 8,000 years ago. The diatom inferred pH then remained relatively stable throughout the Holocene (Whitehead *et al.*, 1986) until the recent acidification stage, which was of a larger magnitude than recorded previously (Charles *et al.*, 1986). Similar trends were inferred using chrysophyte scales (Christie and Smol, 1986). Such studies have identified naturally acidic sites (*e.g.*, Ford, 1984, 1986), and, more recently, lake recovery (Dixit *et al.*, 1989b).

Battarbee *et al.* (1985) used paleolimnological techniques to assess competing hypotheses for the onset of lake acidification in British lakes. Acid precipitation, heathland regeneration, afforestation, and natural acidification were all considered. The acid precipitation hypothesis was substantiated.

Although most pH reconstructions are concerned with the effects of acid deposition, a number of other types of studies have been completed. For example, Fritz and Carlson (1982) and Brugam and Lusk (1986) traced the recovery of acid strip-mine lakes in the mid-western states following the cessation of acid mine inputs; McKee *et al.* (1987) studied the effects of uranium mining and milling on Quirke Lake (Ontario); and Renberg (1986) showed that the present acid (pH = 3) state of Lake Blamissusjön (Sweden) was related to the oxidation of sulphides from nearby marine sediments, which were exposed as a result of isostatic uplift a few hundred years ago.

As with other aspects of paleophycological work, continued refinements of pH calibrations add additional autecological data for indicators. For example, Davis *et al.* (1985) showed that diatom assemblages may be quantified with respect to the water's total organic carbon (TOC), which can be used as an estimate of lakewater colour. With acidification, precipitation of coloured organic compounds with acid-mobilized metals often occurs, thus explaining the clarity of most acidified lakes. Metal concentrations, whose solubilities often are enhanced by acidic waters, have been implicated as factors responsible for the composition of diatom and chrysophyte assemblages (Charles, 1986; Dixit, 1986; Gibson *et al.*, 1987). Calibration sets that can quantify these relationships are now becoming available (*e.g.*, Dixit *et al.*, 1989a).

CONCLUSIONS

Paleophycological work has greatly increased our understanding of past aquatic environments. I have only gleaned a few studies from a myriad of publications dealing with sedimentary deposits. In addition, fossil algae have been used to reconstruct events that are totally unrelated to stratigraphic sequences. For example, archaeologists can trace the origins of pottery by the diatoms present in the potter's clay (Jansma, 1984), or forensic scientists used diatoms isolated from human tissues to determine if drowning was the cause of death of eight Greenlanders, who died in the fifteenth century (Foged, 1982). The applications are almost endless.

There are, of course, problems. The autecologies of many algal taxa are still poorly known, and this seriously hampers many investigations. Continued work with calibration sets, and the application of new and more powerful statistical approaches, will improve environmental reconstructions and predictive capabilities. This cannot be accomplished without continued improvements in taxonomic precision — often the biggest source of error in paleoecological studies. Other problems, such as poor preservation of key indicators (Round, 1964), bioturbation of the sedimentary sequence (Davis, 1974), or spatial variability in microfossil distribution (Dixit and Evans, 1986), can seriously limit interpretations. These problems, however, are not unique to paleophycological work.

Clearly, the best approach is a multidisciplinary one that uses the widest variety of indicators possible; therefore any redundancies in the data set will provide internal checks on interpretations, and a more holistic trajectory of the lake's ontogeny can be inferred. In North America, this was demonstrated decisively by the PIRLA (Paleoecological Investigation of Recent Lake Acidification) project, where a variety of botanical, zoological, chemical, and physical paleolimnological techniques, coupled with state-of-the-art multivariate statistical treatments, were applied to stratigraphic studies of lakes in acid sensitive areas (Charles and Whitehead, 1986). The results exemplify the power of integrated approaches (Charles *et al.*, 1986, 1987).

During the last decade, there has been a tremendous surge of interest in paleolimnology, with fossil algal work often taking a leading role. The transition from purely descriptive studies, to attempts at using "indicator species", to the calculation of indices, to the development of transfer functions using sophisticated statistical and computer techniques has occurred over a relatively short time. Similarly, with advances in the technological tools available to paleoecologists, such as electron microscopy or high pressure liquid chromatography (HPLC), the suite of potential indicators will continue to grow. As the number of qualified researchers increases, progress will accelerate at an even faster pace. As a result, we can expect to have a clearer understanding of ecological and environmental processes — both modern and in the distant past.

ACKNOWLEDGEMENTS

I would like to thank the following people for their constructive comments on the manuscript: J.C. Kingston, M. Douglas, J. Glew, C. Christie, K. Duff, W. Blake, Jr., R. Hall, S. Dixit, R. McNeely, D.F. Charles, H.C. Duthie, R. Brugam, M. Krawetz, K. Law, K. Marentette, B. Cumming, and B. Warner. I am grateful to Kate Duff for her assistance in compiling the bibliography and help with the photography. Ongoing research in my laboratory on paleolimnology is supported by grants from the Natural Sciences and Engineering Research Council of Canada.

REFERENCES

Aaby, B. and Digerfeldt, G., 1986, Sampling techniques for lakes and bogs, *in* Berglund, B.E., ed., Handbook of Holocene Palaeoecology and Palaeohydrology: John Wiley and Sons, Toronto, p. 181-194.

Adams, R.W. and Duthie, H.C., 1976, Relationships between sediment chemistry and postglacial production rates in a small Canadian lake: Internationale Revue gesamten Hydrobiologie, v. 61, p. 21-36.

Alhonen, P., 1971, The stages of the Black Sea as indicated by the diatom stratigraphy: Acta Botanica Fennica, v. 92, p. 1-18.

Anderson, D.S., Davis, R.B. and Berge, F., 1986, Relationships between diatom assemblages in lake surface-sediments and limnological characteristics in southern Norway, *in* Smol, J.P., Battarbee, R.W., Davis, R.B. and Meriläinen, J., eds., Diatoms and Lake Acidity: Dr. W. Junk Publishers, Dordrecht, p. 97-114.

Asmund, B. and Kristiansen, J., 1986, The genus *Mallomonas* (Chrysophyceae). A taxonomic survey based on the ultrastructure of silica scales and bristles: Opera Botanica, v. 85, p. 1-128.

Barber, H.G. and Haworth, E.Y., 1981, A Guide to the Morphology of the Diatom Frustule: Freshwater Biological Association, Scientific Publication No. 44, Ambleside, 112 p.

Battarbee, R.W., 1973, A new method for the estimation of absolute microfossil numbers, with reference especially to diatoms: Limnology and Oceanography, v. 18, p. 647-653.

Battarbee, R.W., 1981, Diatom and Chrysophyceae microstratigraphy of the annually laminated sediments of a small meromictic lake: Striae, v. 14, p. 105-109.

Battarbee, R.W., 1984, Diatom analysis and the acidification of lakes: Royal Society, Philosophical Transactions, v. 305, p. B451-B477.

Battarbee, R.W., 1986, Diatom analysis, *in* Berglund, B.E., ed., Handbook of Holocene Palaeoecology and Palaeohydrology: John Wiley and Sons, Toronto, p. 527-570.

Battarbee, R.W., Cronberg, G. and Lowry, S., 1980, Observations on the occurrence of scales and bristles of *Mallomonas* spp. (Chrysophyceae) in the micro-laminated sediments of a small lake in Finnish North Karelia: Hydrobiologia, v. 71, p. 225-232.

Battarbee, R.W., Flower, R.J., Stevenson, A.C. and Rippey, B., 1985, Lake acidification in Galloway: a palaeoecological test of competing hypotheses: Nature, v. 314, p. 350-352.

Battarbee, R.W. and Kneen, M.J., 1982, The use of electronically counted microspheres in absolute diatom analysis: Limnology and Oceanography, v. 27, p. 184-188.

Björck, S. and Keister, C.M., 1983, The Emerson Phase of Lake Agassiz, independently registered in northwestern Minnesota and northwestern Ontario: Canadian Journal of Earth Science, v. 20, p. 1536-1542.

Boucherle, M.M., Smol, J.P., Oliver, T.C., Brown, S.R. and McNeely, R., 1986, Limnologic consequences of the decline in hemlock 4800 years ago in three Southern Ontario lakes: Hydrobiologia, v. 143, p. 217-225.

Boyer, C.S., 1926, List of Quaternary and Tertiary Diatomaceae from deposits in Southern Canada: Canada Department of Mines, Museum Bulletin No. 45.

Bradbury, J.P., 1971, Paleolimnology of Lake Texcoco, Mexico. Evidence from diatoms: Limnology and Oceanography, v. 16, p. 180-200.

Bradbury, J.P., 1975, Diatom stratigraphy and human settlement in Minnesota: Geological Society of America, Special Paper 171, p. 1-74.

Bradbury, J.P., 1986, Effects of forest fire and other disturbances on wilderness lakes in northeastern Minnesota II — Paleolimnology: Archiv für Hydrobiologie, v. 106, p. 203-217.

Brown, S.R., 1962, Diagenesis of chlorophylls in lacustrine sediments, unpublished Ph.D. thesis, Yale University, New Haven, 52 p.

Brown, S.R., 1969, Paleolimnological evidence from fossil pigments: Mitteilungen Internationale Vereinigung für Limnologie, v. 17, p. 95-103.

Brown, S. and Colman, B., 1963, Oscillaxanthin in lake sediments: Limnology and Oceanography, v. 8, p. 352-353.

Brown, S.R., Daley, R.J. and McNeely, R.N., 1977, Composition and stratigraphy of the fossil phorbin derivatives of Little Round Lake, Ontario: Limnology and Oceanography, v. 22, p. 336-348.

Brugam, R.B., 1978, Human disturbance and the historical development of Linsley Pond: Ecology, v. 59, p. 19-36.

Brugam, R.B. and M. Lusk, 1986, Diatom evidence for neutralization in acid surface mine lakes, *in* Smol, J.P., Battarbee, R.W., Davis, R.B. and Meriläinen, J., eds., Diatoms and Lake Acidity: Dr. W. Junk Publishers, Dordrecht, p. 115-130.

Burne, R.V., Bauld, J. and De Deckker, P., 1980, Saline lake charophytes and their geological significance: Journal of Sedimentary Petrology, v. 50, p. 281-293.

Buzer, J.S., 1981, Diatom analyses of sediments from Lough Ine, Co. Cork, Southwest Ireland: New Phytologist, v. 89, p. 511-533.

Camburn, K.E., 1982, Subaerial diatom communities in Eastern Kentucky: American Microscopical Society, Transactions, v. 101, p. 375-387.

Carney, H.J. and Sandgren, C.D., 1983, Chrysophycean cysts: indicators of eutrophication in the recent sediments of Frains Lake, Michigan, U.S.A.: Hydrobiologia, v. 101, p. 195-202.

Charles, D.F., 1985, Relationships between surface sediment diatom assemblages and lakewater characteristics in Adirondack lakes: Ecology, v. 66, p. 994-1011.

Charles, D.F., 1986, A new diatom species, *Fragilaria acidiobiontica*, from acidic lakes in northeastern North America, *in* Smol, J.P., Battarbee, R.W., Davis, R.B. and Meriläinen, J., eds., Diatoms and Lake Acidity: Dr. W. Junk Publishers, Dordrecht, p. 35-44.

Charles, D.F. and Norton, S.A., 1986, Paleolimnological evidence for trends in atmospheric deposition of acids and metals, *in* National Research Council, Acid Deposition: Long-term trends: National Academy Press, Washington, p. 335-431.

Charles, D.F. and Whitehead, D.R., 1986, The PIRLA project: Paleoecological investigations of recent lake acidification: Hydrobiologia, v. 143, p. 13-20.

Charles, D.F., Whitehead, D.R., Anderson, D.S., Bienert, R., Camburn, K.E., Cook, R.B., Crisman, T.L., Davis, R.B., Ford, J., Fry, B.D., Hites, R.A., Kahl, J.S., Kingston, J.C., Kreis, R.G., Jr., Mitchell, M.J., Norton, S.A., Roll, L.A., Smol, J.P., Sweets, P.R., Uutala, A.J., White, J.R., Whiting, M.C. and Wise, R.J., 1986, The PIRLA project (Paleoecological Investigation of Recent Lake Acidification): Preliminary results for the Adirondacks, New England, N. Great Lakes states, and N. Florida: Water, Air and Soil Pollution, v. 30, p. 355-365.

Charles, D.F., Whitehead, D.R., Engstrom, D.R., Fry, B.D., Hites, R.A., Norton, S.A., Owen, J.S., Roll, L.A., Schindler, S.C., Smol, J.P., Uutala, A.J., White, J.R. and Wise, R.J., 1987, Paleolimnological evidence for recent acidification of Big Moose Lake, Adirondack Mountains, N.Y. (USA): Biogeochemistry, v. 3, p. 267-296.

Chivas, A.R., De Deckker, P., Nind, M., Thiriet, D. and Watson, G., 1986, The Pleistocene palaeoenvironmental record of Lake Buchanan: an atypical Australian playa: Palaeogeography, Palaeoclimatology, Palaeoecology, v. 54, p. 131-152.

Christie, C. and Smol, J.P., 1986, Recent and long-term acidification of Upper Wallface Pond (N.Y.) as indicated by mallomonadacean microfossils: Hydrobiologia, v. 143, p. 355-360.

Cox, E.R. and J. Hightower, 1972, Some corticolous algae of McMinn County, Tennessee, U.S.A.: Journal of Phycology, v. 8, p. 203-205.

Cronberg, G., 1982, *Pediastrum* and *Scenedesmus* (Chlorococcales) in sediments from Lake Vaxjösjön, Sweden: Archiv für Hydrobiologie (Supplement), v. 60, p. 500-507.

Cronberg, G., 1986a, Blue-green algae, green algae and chrysophyceae in sediments, *in* Berglund, B.E., ed., Handbook of Holocene Palaeoecology and Palaeohydrology: John Wiley and Sons, Toronto, p. 507-526.

Cronberg, G., 1986b, Chrysophycean cysts and scales in lake sediments: a review, *in* Kristiansen, J. and Anderson, R.A., eds., Chrysophytes: Aspects and Problems: Cambridge University Press, New York, p. 281-315.

Cronberg, G. and Sandgren, C.D., 1986, Proposal for the development of standardized nomenclature and terminology for chrysophycean statospores, *in* Kristiansen, J. and Anderson, R.A., eds., Chrysophytes: Aspects and Problems: Cambridge University Press, New York, p. 317-328.

Daley, R.J., Brown, S.R. and McNeely, R.N., 1977, Chromatographic and SCDP measurement of fossil phorbins and the postglacial history of Little Round Lake, Ontario: Limnology and Oceanography, v. 22, p. 349-360.

Davis, M.B., 1981, Outbreaks of forest pathogens in Quaternary history: Proceedings of the IV International Palynological Conference, Lucknow, (1976-77), v. 3, p. 216-227.

Davis, R.B., 1974, Stratigraphic effects of tubificids in profundal lake sediments: Limnology and Oceanography, v. 19, p. 466-488.

Davis, R.B., Anderson, D.S. and Berge, F., 1985, Palaeolimnological evidence that lake acidification is accompanied by loss of organic matter: Nature, v. 316, p. 436-438.

Davis, R.B. and Smol, J.P., 1986, The use of sedimentary remains of siliceous algae for inferring past chemistry of lake water — problems, potential and research needs, *in* Smol, J.P., Battarbee, R.W., Davis, R.B. and Meriläinen, J., eds., Diatoms and Lake Acidity: Dr. W. Junk Publishers, Dordrecht, p. 291-300.

Dawson, J.W., 1893, The Canadian Ice Age: W.V. Dawson, Montreal.

Dixit, S.S., 1986, Algal microfossils and geochemical reconstructions of Sudbury lakes: A test of the paleo-indicator potential of diatoms and chrysophytes, unpublished Ph.D. thesis, Queen's University, Kingston, 190 p.

Dixit, S.S., Dixit, A.S. and Smol, J.P., 1989a, Relationship between chrysophyte assemblages and environmental variables in seventy-two Sudbury lakes as examined by canonical correspondence analysis (CCA): Canadian Journal of Fisheries and Aquatic Sciences, v. 46, p. 1667-1676.

Dixit, S.S., Dixit, A.S. and Smol, J.P., 1989b, Lake acidification recovery can be monitored using chrysophyte microfossils: Canadian Journal of Fisheries and Aquatic Sciences, v. 46, p. 1309-1312.

Dixit, S.S. and Evans, R.D., 1986, Spatial variability in sedimentary algal microfossils and its bearing on diatom-inferred pH reconstructions: Canadian Journal of Fisheries and Aquatic Sciences, v. 43, p. 1836-1845.

Duthie, H.C. and Mannada Rani, R.G., 1967, Diatom assemblages from Pleistocene interglacial beds at Toronto, Ontario: Canadian Journal of Botany, v. 45, p. 2249-2261.

Duthie, H.C. and Sreenivasa, M.R., 1971, Evidence for the eutrophication of Lake Ontario from the sedimentary succession: Proceedings of the 14th Conference on Great Lakes Research, p. 1-13.

Ehrenberg, C.G., 1854, Microgeologie: Leopold Voss, Leipzig, 374 p.

Engstrom, D.R., Swain, E.B. and Kingston, J.C., 1985, A palaeolimnological record of human disturbance from Harvey's Lake, Vermont: geochemistry, pigments and diatoms: Freshwater Biology, v. 15, p. 261-288.

Florin, M.B. and Wright, H.E., 1969, Diatom evidence for the persistence of stagnant glacial ice in Minnesota: Geological Society of America, Bulletin, v. 80, p. 695-704.

Foged, N., 1982, Diatoms in human tissues: Nova Hedwigia, v. 36, p. 345-379.

Ford, M.S. (J.), 1984, the influence of lithology on ecosystem development in New England: a comparative paleoecological study, unpublished Ph.D. thesis, University of Minnesota, Minneapolis, 203 p.

Ford, J., 1986, The recent history of a naturally acidic lake (Cone Pond, N.H.), *in* Smol, J.P., Battarbee, R.W., Davis, R.B. and Meriläinen, J., eds., Diatoms and Lake Acidity: Dr. W. Junk Publishers, Dordrecht, p. 131-148.

Frederick, V.R., 1981, Preliminary investigation of the algal flora in the sediments of Lake Erie: Journal of Great Lakes Research, v. 7, p. 404-408.

Fritz, S.C. and Carlson, R.E., 1982, Stratigraphic diatom and chemical evidence for acid strip-mine recovery: Water, Air and Soil Pollution, v. 17, p. 151-163.

Gasse, F., 1986, East african diatoms and water pH, *in* Smol, J.P., Battarbee, R.W., Davis, R.B. and Meriläinen, J., eds., Diatoms and Lake Acidity: Dr. W. Junk Publishers, Dordrecht, p. 149-168.

Gibson, K.G., Smol, J.P. and Ford, J., 1987, Chrysophycean microfossils provide new insight into the recent history of a naturally acidic lake (Cone Pond, N.H.): Canadian Journal of Fisheries and Aquatic Sciences, v. 44, p. 1584-1588.

Harris, G.P. and Vollenweider, R.A., 1982, Paleolimnological evidence of early eutrophication in Lake Erie: Canadian Journal of Fisheries and Aquatic Sciences, v. 39, p. 618-626.

Hartmann, H. and Steinberg, C., 1986, Mallomonadacean (Chrysophyceae) scales: Early biotic paleoindicators of lake acidification: Hydrobiologia, v. 143, p. 87-91.

Haworth, E.Y., 1976, Two late-glacial (late Devensian) diatom assemblage profiles from northern Scotland: New Phytologist, v. 77, p. 227-256.

Hickman, M. and Klarer, D.M., 1981, Paleolimnology of Lake Isle, Alberta, Canada (Including sediment chemistry, pigments and diatom stratigraphy): Archiv für Hydrobiologie, v. 91, p. 490-508.

Hinde, G.J., 1877, Glacial and interglacial strata of Scarborough heights: Journal of the Canadian Institute (Royal Canadian Institute), v. 15, p. 388-413.

Hustedt, F., 1937-1939, Systematische und ökologische Untersuchungen über den Diatomeenflora von Java, Bali, Sumatra: Archiv für Hydrobiologie (Supplement), v. 15 and 16.

Hustedt, F., 1957, Die Diatomeenflora des Fluss-systems der Weser im Gebiet der Hansestadt Bremen: Abhandlungen herausgegeben vom Naturwissenschaftlichen Verein zu Bremen, v. 34, p. 181-440.

Huttunen, P. and Meriläinen, J., 1986, Applications of multivariate techniques to infer limnological conditions from diatom assemblages, *in* Smol, J.P., Battarbee, R.W., Davis, R.B. and Meriläinen, J., eds., Diatoms and Lake Acidity: Dr. W. Junk Publishers, Dordrecht, p. 201-212.

Jansma, M.J., 1984, Diatom analysis of prehistoric pottery, *in* Mann, D.G., ed., Proceedings of the Seventh International Diatom Symposium: Otto Koeltz, Koenigstein, p. 529-536.

Johansen, J.R., Rushforth, S.R. and J.D. Brotherson, 1983, The algal flora of Navajo National Monument, Arizona, U.S.A.: Nova Hedwigia, v. 38, p. 501-553.

Jones, R., Benson-Evans, K., Chambers, F.M., Seddon, B.A. and Tai, Y.C., 1978, Biological and chemical studies of sediments from Llangorse Lake, Wales: Internationale Vereinigung für Theoretische und Angewandte Limnologie, v. 20, p. 642-648.

Kilham, P., Kilham, S.S. and Hecky, R.E., 1986, Hypothesized resource relationships among African planktonic diatoms: Limnology and Oceanography, v. 31, p. 1169-1181.

Kjemperud, A., 1981, Diatom changes in sediments of basins possessing marine/lacustrine transitions in Frosta, Nord-Trondelag, Norway: Boreas, v. 10, p. 27-38.

Kolbe, R.W., 1927, Sur Ökologie, Morphologie und Systematik der Brackwasser-Diatomeen: Pflanzenforschung, v. 7, p. 1-146.

Lasalle, P., 1966, Late Quaternary vegetation and glacial history in the St. Lawrence Lowlands, Canada: Leidse Geologische Mededelingen, v. 38, p. 91-128.

Lee, R.E., 1980, Phycology: Cambridge University Press, London, 478 p.

Lichti-Federovich, S., 1980, Diatom flora of red snow from Isbjorneo, Carey Oer, Greenland: Nova Hedwigia, v. 33, p. 395-431.

Livingstone, D., 1984, The preservation of algal remains in recent lake sediments, *in* Haworth, E.Y. and Lund, J.W.G., eds., Lake Sediments and Environmental History: University of Minnesota Press, Minneapolis, p. 191-202.

Manny, B.A., Wetzel, R.G. and Bailey, R.E., 1978, Paleolimnological sedimentation of organic carbon, nitrogen, phosphorus, fossil pigments, pollen, and diatoms in a hypereutrophic lake: a case history of eutrophication: Polskie Archiwum Hydrobiologii, v. 25, p. 243-267.

Meriläinen, J., 1967, The diatom flora and the hydrogen-ion concentration of water: Annales Botanica Fennica, v. 4, p. 51-58.

McKee, P.M., Snodgrass, W.J., Hart, D.R., Duthie, H.C., McAndrews, J.H. and Keller, W., 1987, Sedimentation rates and sediment core profiles of ^{238}U and ^{232}Th decay chain radionuclides in a lake affected by uranium mining and milling: Canadian Journal of Fisheries and Aquatic Sciences, v. 44, p. 390-398.

Moss, B., 1978, The ecological history of a medieval man-made lake, Hickling Broad, Norfolk, United Kingdom: Hydrobiologia, v. 60, p. 23-32.

Moss, B., 1979, Algal and other fossil evidence for major changes in Strumpshaw Broad, Norfolk, England in the last two centuries: British Phycological Journal, v. 14, p. 263-283.

Munch, C.S., 1980, Fossil diatoms and scales of Chrysophyceae in the recent history of Hall Lake, Washington: Freshwater Biology, v. 10, p. 61-66.

Newberry, T.L. and Schelske, C.L., 1986, Biogenic silica record in the sediments of Little Round Lake, Ontario: Hydrobiologia, v. 143, p. 293-300.

Norris, G. and McAndrews, J.H., 1970, Dinoflagellate cysts from post-glacial lake muds, Minnesota (U.S.A.): Review of Palaeobotany and Palynology, v. 10, p. 131-156.

Nygaard, G., 1956, Ancient and recent flora of diatoms and chrysophyceae in Lake Gribso. Studies on the humic acid Lake Gribso: Folia Limnologica Scandinavica, v. 8, p. 32-94.

Osborne, P.L. and Moss, B., 1977, Paleolimnology and trends in the phosphorus and iron budgets of an old man-made lake, Barton Broad, Norfolk: Freshwater Biology, v. 7, p. 213-233.

Patrick, R., 1984, The history of the science of diatoms in the United States of America, *in* Mann, D.G., ed., Proceedings of the Seventh International Diatom Symposium: Otto Koeltz, Koenigstein, p. 11-20.

Patrick, R. and Reimer, C.W., 1966, The Diatoms of the United States, v. 1: The Academy of Natural Sciences of Philadelphia, Philadelphia, 688 p.

Peglar, S.M., Fritz, S.C., Alapieti, T., Saarnisto, M. and Birks, H.J.B., 1984, Composition and formation of laminated sediments in Diss Mere, Norfolk, England: Boreas, v. 13, p. 13-28.

Renberg, I., 1986, A sedimentary diatom record of severe acidification in Lake Blamissusjön, N. Sweden, through natural soil processes, *in* Smol, J.P., Battarbee, R.W., Davis, R.B. and Meriläinen, J., eds., Diatoms and Lake Acidity: Dr. W. Junk Publishers, Dordrecht, p. 213-220.

Richardson, J.L. and Richardson, A.E., 1972, History of an African rift lake and its climatic implications: Ecological Monographs, v. 42, p. 499-534.

Ritchie, J.C. and Koivo, L.K., 1975, Postglacial diatom stratigraphy in relation to the recession of glacial Lake Agassiz: Quaternary Research, v. 5, p. 529-540.

Round, F.E., 1964, The diatom sequence in lake deposits: some problems of interpretation: Internationale Vereinigung für Theoretische und Angewandte Limnologie, v. 15, p. 1012-1020.

Round, F.E., 1973, The Biology of the Algae, Second Edition: Edward Arnold (Publishers) Ltd., London, 278 p.

Round, F.E., 1981, The Ecology of Algae: Cambridge University Press, New York, 653 p.

Rybak, M., 1986, The chrysophycean paleocyst flora of the bottom sediments of Kortowskie Lake (Poland) and its ecological significance: Hydrobiologia, v. 140, p. 67-84.

Schelske, C.L., Conley, D.J., Stoermer, E.F., Newberry, T.L. and Campbell, C.D., 1986, Biogenic silica and phosphorus accumulation in sediments as indices of eutrophication in the Laurentian Great Lakes: Hydrobiologia, v. 143, p. 79-86.

Schelske, C.L., Stoermer, E.F., Conley, D.J., Robbins, J.A. and Glover, R.M., 1983, Early eutrophication in the Lower Great Lakes: New evidence from biogenic silica in sediments: Science, v. 222, p. 320-322.

Schopf, J.W., 1975, The age of microscopic life: Endeavour, v. 34, p. 51-58.

Simola, H., 1977, Diatom succession in the formation of annually laminated sediment in Lovojarvi, a small eutrophic lake: Annales Botanica Fennica, v. 14, p. 143-148.

Simola, H., 1983, Limnological effects of peatland drainage and fertilization as reflected in the varved sediment of a deep lake: Hydrobiologia, v. 106, p. 43-57.

Smol, J.P., 1980, Fossil synuracean (Chrysophyceae) scales in lake sediments: a new group of paleoindicators: Canadian Journal of Botany, v. 58, p. 458-465.

Smol, J.P., 1981, Problems associated with the use of "species diversity" in paleolimnological studies: Quaternary Research, v. 15, p. 209-212.

Smol, J.P., 1983, Paleophycology of a high arctic lake near Cape Herschel, Ellesmere Island: Canadian Journal of Botany, v. 61, p. 2195-2204.

Smol, J.P., 1985, The ratio of diatom frustules to chrysophycean statospores: A useful paleolimnological index: Hydrobiologia, v. 123, p. 199-208.

Smol, J.P., 1986, Chrysophycean microfossils as indicators of lakewater pH, *in* Smol, J.P., Battarbee, R.W., Davis, R.B. and Meriläinen, J., eds., Diatoms and Lake Acidity: Dr. W. Junk Publishers, Dordrecht, p. 275-290.

Smol, J.P., 1988a, Chrysophycean microfossils in paleolimnological studies: Palaeogeography, Palaeoclimatology, Palaeoecology, v. 62, p. 287-297.

Smol, J.P., 1988b, Paleoclimate proxy data from freshwater arctic diatoms: Internationale Vereinigung für Theoretische und Angewandte Limnologie, v. 23, p. 837-844.

Smol, J.P., Battarbee, R.W., Davis, R.B. and Meriläinen, J., 1986, eds., Diatoms and Lake Acidity: Dr. W. Junk Publishers, Dordrecht, 307 p.

Smol, J.P. and Boucherle, M.M., 1985, Postglacial changes in algal and cladoceran assemblages in Little Round Lake, Ontario: Archiv für Hydrobiologie, v. 103, p. 25-49.

Smol, J.P., Brown, S.R. and McNeely, R.N., 1983, Cultural disturbances and trophic history of a small meromictic lake from central Canada: Hydrobiologia, v. 103, p. 125-130.

Smol, J.P., Charles, D.F. and Whitehead, D.R., 1984, Mallomonadacean microfossils provide evidence of recent lake acidification: Nature, v. 307, p. 628-630.

Smol, J.P. and Dickman, M.D., 1981, The recent histories of three Canadian Shield lakes: a paleolimnological experiment: Archiv für Hydrobiologie, v. 93, p. 83-108.

Sreenivasa, M.R. and Duthie, H.C., 1973, The postglacial diatom history of Sunfish Lake, southwestern Ontario: Canadian Journal of Botany, v. 51, p. 1599-1609.

Stager, J.C., 1984, The diatom record of Lake Victoria (East Africa): the last 17,000 years, *in* Mann, D.G., ed., Proceedings of the Seventh International Diatom Symposium: Otto Koeltz, Koenigstein, p. 455-476.

Steinberg, C. and Hartmann, H., 1986, A biological paleoindicator for early lake acidification: Mallomonadacean (Chrysophyceae) scale abundance in sediments: Naturwissenschaften, v. 73, p. 37-38.

Stevenson, A.C., Birks, H.J.B., Flower, R.J. and Battarbee, R.W., 1989, Diatom-based pH reconstruction of lake acidification using canonical correspondence analysis: AMBIO, v. 18, p. 228-233.

Stockner, J.G., 1971, Preliminary characterization of lakes in the Experimental Lakes Area, northwestern Ontario, using diatom occurrences in sediments: Journal of the Fisheries Research Board, Canada, v. 28, p. 265-275.

Stockner, J.G., 1978, Man's impact on lakes and watersheds in Canada as assessed from paleolimnological studies: Polskie Archiwum Hydrobiologii, v. 25, p. 393-402.

Stockner, J.G. and Benson, W.W., 1967, The succession of diatom assemblages in the recent sediments of Lake Washington: Limnology and Oceanography, v. 12, p. 513-532.

Stoermer, E.F., Taylor, S.M. and Callender, E., 1971, Paleoecological interpretation of the holocene diatom succession in Devil's Lake, North Dakota: American Microscopical Society, Transactions, v. 90, p. 195-206.

Stoermer, E.F., Wolin, J.A., Schelske, C.L. and Conley, D.J., 1985, An assessment of ecological changes during the recent history of Lake Ontario based on siliceous algal microfossils preserved in the sediments: Journal of Phycology, v. 21, p. 257-276.

Stoffers, P. and Hecky, R.E., 1978, Late Pleistocene-Holocene evolution of the Kivu-Tanganyika Basin: International Association of Sedimentologists, Special Publications, v. 2, p. 43-55.

Taylor, M.C., 1986, Surficial sedimentary diatoms — limnological relationships in Canadian Shield lakes, M.Sc. thesis, University of Waterloo, Waterloo, 159 p.

Terasmae, J., 1960, A palynological study of the Pleistocene interglacial beds at Toronto, Ontario: Geological Survey of Canada, Bulletin 56, p. 23-41.

Tippett, R., 1964, An investigation into the nature of the layering of deep water sediments in two eastern Ontario lakes: Canadian Journal of Botany, v. 42, p. 1693-1709.

Trainor, F.R., 1978, Introductory Phycology: J. Wiley and Sons, Inc., Toronto, 525 p.

Tuchman, M.L., Stoermer, E.F. and Carney, H.J., 1984, Effects of increased salinity on the diatom assemblage in Fonda Lake, Michigan: Hydrobiologia, v. 109, p. 179-188.

Vallentyne, J.R., 1955, Sedimentary chlorophyll determination as a paleobotanical method: Canadian Journal of Botany, v. 33, p. 304-313.

Vallentyne, J.R., 1956, Epiphasic carotenoids in post-glacial lake sediments: Limnology and Oceanography, v. 1, p. 252-262.

Van Geel, B., 1986, Application of fungal and algal remains and other microfossils in palynological analyses, *in* Berglund, B.E., ed., Handbook of Palaeoecology and Palaeohydrology: John Wiley and Sons, Toronto, p. 497-506.

Van Geel, B., Bohncke, S.J.P. and Dee, H., 1981, A palaeolimnological study of an upper Late Glacial and Holocene sequence from "De Borchert", The Netherlands: Review of Palaeobotany and Palynology, v. 31, p. 367-448.

Wetzel, R.G., 1970, Recent and postglacial production rates of a marl lake: Limnology and Oceanography, v. 15, p. 491-503.

Whitehead, D.R., Charles, D.F., Jackson, S.J., Reed, S.E. and Sheehan, M.C., 1986, Late-glacial and Holocene acidity changes in Adirondack (N.Y.) lakes, *in* Smol, J.P., Battarbee, R.W., Davis, R.B. and Meriläinen, J., eds., Diatoms and Lake Acidity: Dr. W. Junk Publishers, Dordrecht, p. 251-274.

Whitford, L.A., 1968, Notes on the history of freshwater phycology: Journal of Phycology, v. 4, p. 169-173.

Originally published in
Geoscience Canada Volume 14 Number 4
(December 1987)

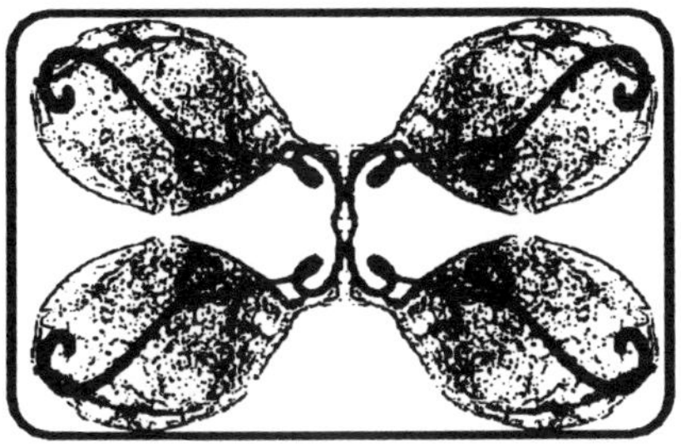

Fungi

K.A. Pirozynski
Paleobiology Division
National Museum of Natural Sciences
P.O. Box 3443, Station "D"
Ottawa, Ontario K1P 6P4

SUMMARY

Fungi are ubiquitous in Quaternary fossiliferous deposits. Most are incidentally encountered during paleobotanical, including palynological, and paleozoological procedures and are rarely studied. Yet, explicitly paleomycological investigations reveal that fungi have considerable potential to provide data sharpening the focus of biotic and environmental reconstructions hitherto based solely on the more traditional paleontological evidence.

INTRODUCTION

Fungi are everywhere. Hyphae and diverse "fruiting" structures bearing spores of many kinds grow in or on living plants and animals or their remains on land and in the water. The spores are in the air, water and soil. Dispersed spores and other components, sometimes still associated with specific substrates, occur in, or find their way to, sites where conditions favour preservation. The special circumstances are anoxic sedimentation or, more rarely, burial in volcanic ash, freezing and desiccation. Vast amounts of fungal material, chiefly dispersed spores, accumulate in aquatic sediments (Muller, 1959; Saad and Sani, 1967; Jarzen and Elsik, 1986). Cross *et al.* (1966) calculated that one gram of dried sediment from the Gulf of California contained up to 22,500 fungal spores, exceeding even the superabundant pine pollen. The spores have already "survived" physical and biological degradation as well as processing in the laboratory. The occurrence of fungi in older sediments, therefore, was to be expected. Indeed, Wolf (1967b) estimated that 1 mL of Holocene sediment cores from East African lakes contained up to 60,000 fungal spores. In Holocene peatlands of northern Europe and southeastern United States, fungi outnumber algal and animal remains (Van Geel, 1972, 1978; TeStrake, 1979). In a late Wisconsinan Oklahoma blue clay swamp containing a mammoth skeleton, "fungal spores far outnumber ... all other palynological fossils" (Wilson, 1966). Yet, in that study, fungi were not considered further; in most other palynological studies they are not even mentioned. Fossil fungi are deemed too difficult to interpret and identify or even distinguish from living inhabitants of a given site (Traverse and Ginsburg, 1966). The identification and interpretation of Quaternary fungi, like other fossils, pivots on the familiarity with the living. Few mycologists, however, showed interest in fossil fungi.

QUATERNARY FUNGI: WHAT AND WHERE?

Fungal spores (Figure 1a-d) are "seeds" adapted for dispersal by air, water or animals, and for withstanding physical and biological adversities during the sometimes extended periods of conditions inimical to growth. Their more striking analogy, however, is to pollen which they resemble in size, structure and composition, and in being encountered by palynologists. The analogy to pollen extends further: a sample may be skewed toward species adapted for abiotic dispersal, especially by air; air-dispersed spores may be more robust, for example to resist dessication, and consequently could be selectively transported, preserved and re-worked to further increase sample bias. The details of the site may be obscured because airborne spores can traverse vast distances, and the direction of travel may not be polarized by biological and ecological traits of parent organisms. In addition, the information content of an individual sample may be grossly distorted if it originally included spore-bearing organs containing, as they sometimes do, hundreds of thousands of individual propagules. The preservation itself depends on rapid burial in an anoxic, usually aquatic milieu, most frequently washed-in sediment, which necessarily involves allochthonous elements from a potentially wide range of habitats, including airborne components (Jacobson and Bradshaw, 1981). Clearly the spectre of a large margin of error weakens interpretations of isolated samples, but is reduced if studies extend to assemblages from different horizons and sites. Palynological methods designed to minimize interpretative errors largely apply to fungal spores, if for no other reason than that there is little else available to students of dispersed paleofungi. Some peculiarities of fungi, however, are already apparent and need to be accommodated. For example, Muller (1959) claimed that estuarine circulation tends to precipitate the relatively smaller fungal spores in bottom sediments closer to shore than pollen, much of which is carried seaward. Hydrodynamic properties of spores of terrestrial fungi have not been investigated in relation to sedimentation. Unlike most pollen, a significant portion of an assemblage may be derived from autochthonous populations of fungi both in water and in the soil. Some will be anachronistic contaminants derived from sediment- or soil-dwelling fungi or from intrusive mycorrhizae (Traverse and Ginsburg, 1966; Elsik, 1986; Pirozynski and Dalpé, 1989). Non-melanized spores or cells are less likely to become fossilized, or to withstand laboratory extraction, than their melanized counterparts. However, the near absence of unpigmented spores in palynologically processed samples may be also partly due to the darkening of walls during acetolysis (Wolf, 1966). Finally, it should be noted that although mycological studies of East African lake-sediments (Wolf, 1967a, 1969) revealed some correspondence between spore content and vegetation, inexplicably some horizons are mycologically sterile. Furthermore, the spore content can decline with depth of sediment for no apparent reason, or fungi mysteriously disappear from refrigerated samples (Wolf, 1969).

There will, of course, always be the risk of misdetermination of isolated spores or their confusion with similar, but unrelated, objects. Particular care is called for when interpreting one-celled spores, even when they are judged to be distinctive, because, this economical format has been arrived at independently by very different organisms (Pirozynski, 1976). Hence ascospores of Xylariaceae resemble some plant microspores (*e.g.*, those of *Isoetes*), and some conidia and sordariaceous as-

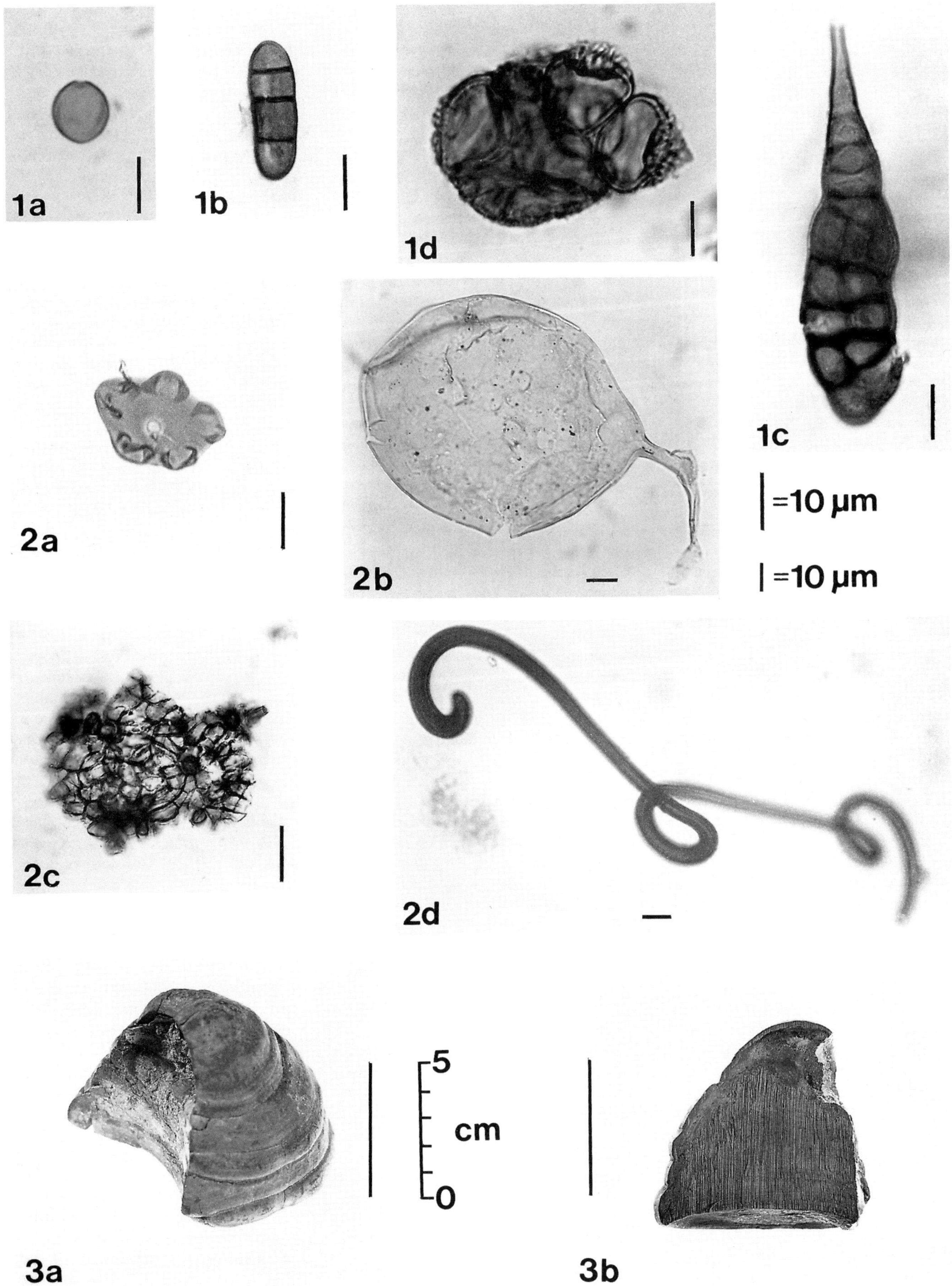
1a
1b
1d
1c
2a
2b
=10 µm
=10 µm
2c
2d
5
cm
0
3a
3b

cospores can be confused with cysts of the Chrysophyceae (Duff and Smol, 1988) or cyanobacteria (Van Geel, 1986).

The same juxtaposition of circumstances that leads to concentration and preservation of dispersed spores also applies to other microscopic parts of fungi, although predictably most vegetative or reproductive organs are more prone to decay and, consequently, are less likely to become fossilized (Van Geel, 1978). Those that do, comprise some distinctive hyphae and their modifications as hyphopodia, appressoria, terminal or intercalary swelling or "chlamydospores", conidiophores, bristle-like setae, bulbil-like sclerotia, and many kinds of conceptacular fructifications or their more unusual components (Pirozynski *et al.*, 1984) (Figure 2a-d). Pitfalls of convergent evolution, too, await their identification and interpretation. Hence some shield-shaped fruit-bodies (thyriothecia) of leaf-dwelling microthyriaceous and micropeltaceous fungi (Pirozynski, 1978) not only resemble thalli of co-inhabiting green algae (*e.g., Phycopeltis*), but can also be confused with those of aquatic encrusting *Ulvella*. Some "thyriothecia" identified in the Arabian Sea sediments up to 45 km offshore India (Ratan and Chandra, 1982) may be algae, and some shield-shaped thalli attributed to the algae (Hansen, 1980) may be fungi.

Figure 1 **(A)** *ascospore of* Chaetomium simile*;*
(B) *indetermined ascospore;*
(C) *conidium of* Alternaria *sp.;*
(D) *Spore ball of* Thecaphora deformans.
(A) (C) and (D) from dung of Arctic ground squirrel (Pirozynski et al.*, 1984), (B) from clay balls accompanying the "Hillsborough mastodon" (Pirozynski* et al.*, 1988). (All* 1000 ×).

Figure 2 **(A)** *appressorium* (1000 ×);
(B) *chlamydospore of* Glomus *sp.* (500 ×);
(C) *portion of fructification (ascoma) wall* (1000 ×);
(D) *ascomatal appendage of* Chaetomium simile (500 ×).
(A) and (B) from clay balls accompanying the "Hillsborough mastodon" (Pirozynski et al.*, 1988), (C) and (D) from dung of Arctic ground squirrel (Pirozynski* et al.*, 1984).*

Figure 3 Fomes fomentarius **(A)** *complete and* **(B)** *cross-section of fruit-body (from DAOM 199110) (one-half natural size).*

A superior source of the more delicate vegetative and reproductive expressions of fungi are plant macrofossils. In addition, fossilized macrofungi are occasionally collected by students of plant or animal macrofossils. Together with fungi associated with animal remains, including dung, they represent an important, but largely untapped, paleoecological resource.

METHODS OF SAMPLING AND LABORATORY EXTRACTION

Most Quaternary fungi are preserved as empty cells without permineralization or petrifaction (Wolf, 1966, 1967a,b, 1969; Vishnu-Mittre, 1975). As noted earlier, most are encountered in paleontological samples processed for a different purpose. Dispersed spores in sediment cores or sections of peat are disclosed by palynological acetolysis following the removal of mineral matrices with appropriate acids (HCl, HF and rarely, HNO_3) (Wolf, 1967b; Pals *et al.*, 1980; MacDonald, 1988). A refinement of this procedure in processing gyttja, substitutes HF with bromoform-alcohol mixture (Van Geel *et al.*, 1983). A simple aqueous suspension of sediment is not satisfactory (Wolf, 1966), but repeated decanting to remove sand eventually concentrates the spores sufficiently to allow direct mounting of the residue in Canada balsam on microscope slides (Wolf, 1969). A similar problem of low concentration and, consequently, insufficient sample size from any one horizon in a depositional sequence, may apply to airborne palynomorphs trapped in perennial snow or ice. Cores are removed by thermal drills, melted, and the meltwater laboriously filtered through micropore acetate filters under rigorous contamination-proof conditions prior to acetolysis or direct mounting of the residue (Lichti-Federovich, 1975a,b). Although the resulting pollen spectra are of some value in interpreting broad climatic changes, fungal spores are too infrequent and the types recovered too widespread to justify the effort.

Some fungal fructifications, chiefly shield-shaped thyriothecia, are sometimes left intact or almost intact in palynologically processed concentrates (Eriksson, 1978), but they invariably lack spores. Other kinds of fruit-bodies occasionally with spores, and sclerotia accompany macrofossil samples of fruits, seeds or insects, obtained by screening or boiling of sieved-out sediment or peat in 5% KOH for 5 to 10 minutes (Van Geel *et al.*, 1981).

Fossil leaves and, to a lesser extent, wood are the best sources of complete fructifications of fungi. Those embedded in leaf tissues are prepared by peeling off leaves from the matrix and immersing them for two to five days in 5-10% KOH. Cleared leaves are washed in distilled water and can be mounted directly on slides in standard mounting media (Doi and Amano, 1982). Foliicolous fungi growing and fruiting on leaf cuticles are revealed by clearing the leaves in 5% KOH, washing, dehydrating in a series of alcohols, fixing in xylene and mounting in synthetic resin. This procedure was successfully applied to Eocene fungi (Dilcher, 1965). However, for at least some Quaternary foliicolous fungi, the acetate peel technique, which is the standard method of preparing living material (Ellis, 1950), could be employed to advantage. A novel application of the peel technique was to lift fungal spores from archeological potsherds (Stewart and Robertson, 1968). The preparation of Quaternary fungi contained in leaves, or in wood that is not permineralized, can be done by normal mycological procedures of microtome sectioning, or squashing or teasing in standard laboratory reagents (Romans *et al.*, 1966; De Vries, 1966). This certainly applies to fruit bodies of larger basidiomycetes, usually bracket fungi and puff-balls that sometimes still contain spores (Chaney and Mason, 1936).

INTERPRETATION

Fungi and Physical Environment.

Climate. A late glacial stony clay-gyttja from Saskatchewan yielded fragments of *Carex* bearing telia of the rust fungus *Puccinia caricina* (De Vries, 1966). The fungus is heteroecious, with *Ribes* as the alternate host. The inferred requirement for *Ribes*, coupled with the knowledge that *P. caricina* has a narrower tolerance of harsh climate than the potential host, indicated to De Vries (1966) climatic amelioration of the site at the time of deposition of the specimens. Similarly, Pirozynski *et al.* (1984) interpreted the presence of *Thecaphora deformans* spores (Figure 1d) in dung from a ground squirrel's nest in Yukon, some 12,000 years B.P., as indicating a climate more temperate than at present, because modern distribution of this smut of *Astragalus* is farther south. A more extreme climatic shift was implied by Prasad and Ramesh (1984) who found a distinctive conidium of *Circinoconis* in a Holocene peat bog in Tripura, NW India. The fungus is no longer extant on the subcontinent: the sole representative, *C. paradoxa*, occurs in Malaya and Indonesia.

Fungi may also offer clues to more specific climatic factors. For example, Quaternary occurrences of *Sporormiella*, like the present-day saprotrophic species, appear to correlate with aridity. Certain peat bog fungi are in strongly humified horizons that are presumed to have been formed during drier phases. The occurrence of others correlates more strongly with wetter phases (Van Geel, 1972, 1986; Van der Wiel, 1982). Van Geel *et al.* (1981) also sought correlation between the occurrence of specific kinds of fungal spores in peat and temperature regimes and fluctuations, as did Vishnu-Mittre (1975) in East Anglian fens. The latter found evidence of fungi that no longer live in England, but occur farther south in Europe. The disappearance of some may be linked with the extinction of specific host plants, but the discontinuity of others is more satisfactorily explained by climatic cooling.

Hydrology. Freshwater, brackish and marine habitats are occupied by different fungi that can leave a tell-tale record of salinity (Pirozynski *et al.*, 1988). Thus, mycological data contributed to the tracing of stages in the evolution of a tidal flat (Bakker and Van Smeerdijk, 1982) or a brackish lake (Pals *et al.*, 1980) into a swamp and, eventually a bog; of a rising water table and replacement of carr by open water (Van der Wiel, 1982); and of periodic freshwater inundation of fens (Vishnu-Mittre, 1975). Fungal remains may have considerable potential in paleolimnological reconstructions concerned with currents, rate of flow, pH and seasonal drying out (Sherwood-Pike, 1988).

Soils. The role of saprophytic organisms, including fungi, in ontogeny and characterization of soils and peats has been recognized (Anderson, 1984). However, attempts to implicate fungi in Quaternary pedology failed (Vishnu-Mittre, 1975), though there remains an indirect link between some soil-borne fungi and certain types of vegetation (Van Geel, 1978). The mycorrhizal fungi, especially, can play an important role in characterization of soils.

Fire. Several kinds of fungi are known pioneer colonizers of burnt sites, either as saprotrophs or as mycorrhizal symbionts. Distinctive ascospores of *Gelasinospora* and *Neurospora* are often associated with charcoal or burnt *Molinia*, and their stratigraphy has been widely used in tracing periodic burning of peat bogs (Van Geel, 1978, 1986; Bakker and Van Smeerdijk, 1982; Garneau, 1987).

Fungi and Plants.

Microfungi. Unlike Wolf (1966, 1967a,b, 1969), who found little evidence that fungi in East African lake-sediments are derived from and, consequently, represent local vegetation, Van Geel and collaborators (Van Geel, 1978, 1986; Van Geel *et al.*, 1981, 1983, 1986) were more successful in documenting such links. In fairness, Van Geel assessed the significance of fungal spores in the broader context of assemblages that included plant and animal macrofossils, pollen and algae, whereas Wolf tried to rely on fungal spores alone. He did, however, note that "it seems reasonable to assume that eventually spore analyses and pollen analyses may both be used in interpreting floristic changes" (Wolf, 1967a). Today, there is evidence that some fungal spores better differentiate local vegetation than pollen. Some of this evidence comes from studies of assemblages of fungal spores in modern soils, that more effectively discriminate ground-cover communities than the co-occurring assemblages of pollen (Kurmann, 1985). This may be also true of fossils. Elsik (1986) found abundant spores with a skeleton of a giant ground sloth in Texas. The mycological data, combined with pollen data, permitted reconstruction suggestive of a low-lying coastal grassland interspersed with open woodland and traversed by bayous. The evidence for the grassland and the bayous came mainly from the fungi. Elsewhere, a comparative study of *Ustulina deusta* in Quaternary deposits indicated that the ascospores of this weak parasite of broad-leaved forest trees are far less easily transported than the pollen of the hosts (Van Geel and Andersen, 1988).

Vishnu-Mittre (1975) and Van Geel (1978) claimed that most spores in peat originated either from fungi growing on plants on or near bogs or from saprotrophic fungi growing in peat. In subsequent contributions (Van Geel, 1986; Van Geel *et al.*, 1981, 1983, 1986; Pals *et al.*, 1980; Bakker and Van Smeerdijk, 1982; Kuhry, 1985), specific fungi or fungal spores were linked with specific plants or plant communities. Additional examples are *Thecaphora deformans* whose spores (Figure 1d), found in an arctic ground squirrel's nest, identified *Astragalus* in the diet and, consequently, in the foraging territory of the animal (Pirozynski *et al.*, 1984), and *Puccinia caricina* which, even if not found on identifiable leaves of *Carex* would have indicated the presence of this plant — as well as that of the alternate host, *Ribes* — in the ecosystem (De Vries, 1966).

Appressoria or hyphopodia of *Gaeumannomyces, Buergenerula* and *Clasterosporium* (Figure 2a), the last frequently accompanied by conidia, commonly found in coastal sediments, identify palustrine monocots (Pirozynski *et al.*, 1988) and, consequently, a palustrine stage in terrestrial succession (Cowley and Colquhoun, 1966).

Macrofungi. Of considerable interest, and as yet untapped potential in identifying woody plants and their condition at the time of burial, are bracket fungi which are often remarkably well preserved, sometimes even with spores (Chaney and Mason, 1936). The purportedly Quaternary finds of *Fomes* (Andrews and Lenz, 1947; Galbreath, 1947; Rosendahl, 1948) and *Ganoderma* (Bondarcev, 1960) are assignable to modern *F. applanatus, F. fomentarius* and *F. pinicola* (Mason, 1934; Chaney and Mason, 1936; Buchwald, 1970, 1971) and *G. lucidum* (Chaney and Mason, 1936). Their age has been inferred from stratigraphic correlations, except when attached to trees in a buried forest.

Here I record another detached specimen of *Fomes fomentarius* (Figure 3a) found by C.R. Harington and R. Johnson on 13 August 1979, washed out from frozen "muck" (dark organic silt) at Gay Gulch, Eldorado Creek, Dawson (Loc. 45), Yukon Territory, Canada (DAOM 199110), and having a radiocarbon age of 5030 ± 70 years B.P. (Beta 28764). The usual boreal host of *F. fomentarius* is *Betula*. This can be verified because even detached basidiomata often incorporate fragments of host bark. The date marks the onset of cooling that ended the Hypsithermal, but the transition is not inferred from this particular find: both *F. fomentarius* and *Betula* are still extant in the area. The curvature of the adnate side of the fructification indicates that the supporting trunk or limb was about 8 cm in diameter (Figure 3a); darker bands seen in cross-section (Figure 3b), like annual growth rings, reveal a seven-year association with a particular tree (*cf.* Andrews, 1948).

Mycorrhiza.

Different kinds of fungi involved in vesicular-arbuscular, ericoid and ectomycorrhizae identify different communities of plants and should record their succession.

Mycorrhizal fungi should not be expected in meso- and oligotrophic peat because wet conditions select for non-mycorrhizal plants, *Sphagnum* and aquatic or semi-aquatic angiosperms, including sedges and rushes. Sclerotia of *Cenococcum* were once claimed to be abundant and ubiquitous in all kinds of peat (*cf.* Vishnu-Mittre, 1975, and refs. therein), but this is an oversimplification. Vishnu-Mittre (1975) noted that *Cenococcum* was more abundant in wood peat than in raised peat bogs, and Van Geel (1978) did not find sclerotia in wet peats except when accompanying intrusive roots from surface vegetation. Contamination from living mycorrhizae may have contributed to the former generalization. Another contributing factor is that sclerotia are hollow, they float and are apparently dispersed by seasonal inundations (Malloch *et al.*, 1987). This complication notwithstanding, even in raised peat bogs it has been possible to correlate the occurrence of *Cenococcum* with drier conditions (Van Geel, 1978) during which bogs may be invaded by mycorrhizal plants. Indeed, abundant sclerotia in *Sphagnum* bogs mark a transition to *Alnus* carr (Van Geel *et al.*, 1981), and they become characteristic of alder peat (Vishnu-Mittre, 1975). *Cenococcum graniforme* is frequently mycorrhizal with alders pioneering bog colonization.

Intrusive roots or root fragments of the Ericaceae growing on peat are characteristically packed with hyphae of mycorrhizal fungi. However, the sometimes implied or claimed (Garneau, 1987) association of the Ericaceae with *Cenococum* is enigmatic inasmuch as this fungus is not known to form "ericoid" mycorrhiza (such as characterizes peat bog Ericaceae) though it may form "arbutoid" mycorrhiza with, for example, *Arctostaphylos*. The hyphae and chlamydospores accompanying *Calluna vulgaris* (Van Geel, 1978; "type 10"; Kuhry, 1985) cannot be considered in the light of present knowledge as belonging to *Cenococcum* or even as being mycorrhizal.

The establishment of ectomycorrhizal *Alnus, Betula, Picea, Pinus, etc.* cannot be determined from the more usual and often specific fungal symbionts, because their fruit-bodies (mushrooms) and spores are delicate and ephemeral. However, symbiosis alters, to some extent, the anatomy and morphology of roots, and there are grounds for predicting that fragments of mycorrhizae will eventually provide clues to the identity of both plants and the fungi involved. It has already been possible to identify "short roots" of *Pinus sylvestris* in Quaternary peat (Van Geel *et al.*, 1981).

Vesicular-arbuscular mycorrhizal (VAM) fungi are rarely found in peat, presumably because VAM plants are rare in *Sphagnum* bogs. But although VAM fungi have not been recorded in studies of northern European peats (Van Geel, 1978, 1986; Van Geel *et al.*, 1981, 1983, 1986; Vishnu-Mittre, 1975; Pals *et al.*, 1980; Bakker and Van Smeerdijk, 1982; Kuhry, 1985), it would not be prudent to extrapolate because the probability of contamination from intrusive mycorrhizae, animals or sediments is very real: the vast majority of plants form VAM, especially with the ubiquitous *Glomus* (Figure 2b). Thus *Glomus* (= *Rhizophagites*) was found in a glacial bog in Alberta (Butler, 1939); in a well core section indicative of a spruce-tamarack bog in Minnesota (Rosendahl, 1943, 1948); in post-glacial marine silt and silty clay in Prince Edward Island (Terasmae, 1958); in forest soil horizon in Oklahoma (Wilson, 1965); in lacustrine marl sediment in North Carolina (Wolf, 1969); in glacial lake sediment in New York (Miller, 1973); in lake-sediment cores identifying post-glacial pioneer tundra vegetation in Maine (Anderson *et al.*, 1948: *Glomus fasciculatum* complex); in gyttja and silty clay in Ontario (Berch and Warner, 1985: *G. fasciculatum*); and in *Scirpus – Cladium* peat and silty clay also in Ontario (Berch and Warner, 1985: *G. deserticola*). In deposits representing swamps, non-mycorrhizal *Scirpus – Cladium* or ectomycorrhizal spruce-tamarack, VAM fungi are likely to be allochthonous. The effectiveness of the physical transport of *Glomus* is apparent in Terasmae's (1958) sample in which chlamydospores occur with marine diatoms. Pirozynski *et al.* (1988) found *Glomus* and the closely related *Sclerocystis* among spores of fungi that may have been ingested with herbage by a late Pleistocene American mastodon. Dispersal of VAM spores "by gut" appears to be widespread in nature and is both "horizontal", by widely foraging mammals and birds, and "vertical", by earthworms and burrowing mammals (Pirozynski and Malloch, 1988). All these observations cumulatively call for caution in extrapolating even broad vegetational changes from sporadic occurrences of VAM fungi.

Plant Decline and Extinction.

The dramatic decline of the American chestnut and elm caused by parasitic fungi during this century has justifiably led to suggestions that similar declines recorded in paleopalynological profiles could have also been caused by epidemics of pathogenic fungi. Particular attention has been focussed on the decline of elm in Europe (Moore, 1985) and of hemlock in North America (Davis, 1981; Allison *et al.*, 1986) about 5000 years B.P. Both claims rely on indirect evidence from (a) comparative analysis of pollen profiles accompanying the recent decline of chestnut and elm and of the ensuing compensatory succession, and (b) elimination of other potential physical and biological causes, notably climatic shift and human impact. Judging from the morphology and ecology of the fungi responsible for the recent extinctions, the probability of finding direct paleomycological evidence for the Holocene outbreaks seems remote. Even when a known parasite has left a record of distinctive spores, *e.g.*, *Ustulina deusta*, the relative abundance of which correlates with the occupation of the site by more or less susceptible trees, the floristic changes are more likely to have resulted from climatic shifts than from disease, however seductive the biological explanation may be (Van Geel *et al.*, 1986; Van Geel and Andersen, 1988).

Fungi and Animals.

Fungi parasitic on arthropods are known from Tertiary ambers. Quaternary cases may be equally common, but remain to be diagnosed. Fungi pathogenic to vertebrates are less likely to outlast nonskeletal remains, but, theoretically, should persist on infected mummified animals. Not to be confused are anachronistic biodegraders such as found on the mummy of Rameses II (Mouchacca, 1978). The best documented fossil associations are non-pathogenic and involve fungi of stored food or faeces inside burrows of sedentary animals or in digested food within carcasses, skeletons or in dung (Figures 1 and 2).

Stomachs and skeletons. If "palaeofaeces are pearls beyond price" (Hillman, 1986), it is because the stomachs and guts of mummified animals, including man, or their presumed contents preserved within skeletal rib cages, have provided direct evidence of the dietary habits and, consequently, of the immediate physical and biotic environment of the consumer. Microscopic fungi are rarely included in the analyses (perhaps partly because modern *Homo sapiens "supermarketalis"* is an atypical, barren control) despite a precedent set by Weber (1915) who listed sclerotia of *Cenococcum graniforme* and spores of *Uredo* and *Uromyces* among 70 plants identified in the diet of the "Borna mammoth". A largely mycological investigation of clay balls associated with the skeleton of the "Hillsborough mastodon" from New Brunswick attempted to redress the balance (Pirozynski *et al.*, 1988). The sample yielded 137 kinds of fungi which cumulatively suggested an enterogastric environment despite a large volume of the accompanying clay. Many were identified as coprophilous; others represented brackish or even marine habitats, exotic to the inferred site of deposition. A suggestion that tidal clay may have been ingested to provide salt is underscored by an independent claim of the importance of salt in a mastodon's diet and the effect of its availability on the distribution and, ultimately, the demise of this animal in North America (Holman *et al.*, 1988).

Dung. Dung has long been recognized as an important source of data on food, feeding and, consequently, on the immediate environment of the consumer. Recent literature pertaining to Quaternary North America include Spaulding and Martin (1979), Davis *et al.* (1984) and Mead *et al.* (1984, 1986a,b). An explicitly mycological investigation of fecal pellets found inside a ground squirrel's nest in the Canadian arctic (Pirozynski *et al.*, 1984) yielded accidental contaminants from air or arthropods, fungi incidentally ingested with plants and often betraying their identity, and specialists which are either adapted for dispersal by ingestion (enterophiles) or are both dispersed "by gut" and utilize dung as the normal growth substrate (coprophiles). The coprophiles frequently offer clues to the identity of the consumers as well as their diets, and are thus important in the interpretation of "dispersed" dung. The pioneering attempts by Davis (1976) and Davis *et al.* (1977, 1984) in western North America and by Pals *et al.* (1980), Van Geel *et al.* (1981, 1983) and Witte and Van Geel (1985) in northern Europe attributed the increased abundance of coprophilous fungi in lake sediments and peat to the invasion of, or introduction to, the sites of large grazing herbivores. The concomitant disturbance is independently verifiable by the elevated counts of pollen of weeds and, in historic times, by the appearance of pathogenic fungi characteristic of agroecosystems. Davis (1976, 1987) singled out *Sporormiella* as quantifying grazing pressure both before the megafaunal extinction in the Pleistocene and after the introduction of grazing livestock in the Holocene. He noted that, although not all heavily grazed sites studied in the western USA yielded *Sporormiella*, all *Sporormiella* sites show signs of abundant grazers. In contrast, Pirozynski *et al.* (1988) did not find *Sporormiella* among 175 spores associated with the skeleton of an American mastodon in Atlantic Canada, though admittedly the fungus may be more characteristic of grasslands and grazers. Nor has *Sporormiella* been found in northern European peats, even in horizons indicating relatively dry conditions that opened these sites to large herbivores (Van Geel *et al.*, 1981). Saprotrophic, non-coprophilous species of *Sporormiella* are particularly abundant in arid regions of North America (D.W. Malloch, pers. comm.) suggesting that the distribution of the coprophilous elements too may correlate more strongly with dry environments.

Fungi and Man.

Since time immemorial, fungi have served as man's food and medicine, contributed to rituals, provided escape from reality, competed for crops, and became "domesticated" by adapting to man's habits as a sedentary social hoarder of food. But although fungi are sporadically mentioned in archeological literatures (*e.g.*, Sharma, 1976) only very rarely did they invite a more detailed study. Among exceptions are fungi, including sclerotia of psychoactive ergot, in paleofaeces of "bog people" (Hillman, 1986; Scaife, 1986 and refs. therein) and of a Peruvian mummy (Vreeland and Korf, 1978). Potsherds excavated in Iraq yielded spores of fungi characteristic of grasses which, as Stewart and Robertson (1968) inferred, came from chaff and manure used for tamper in making clay pots. But, in addition, spores of fungi pathogenic to cultivated grain crops provided indirect evidence of the kind of cereals planted, and of the problems encountered in their cultivation. Dried puff-balls in pre-Hadrianic and Roman sites in Great Britain were thought (Watling and Seaward, 1976) to have served as haemostatic agents or as tinder.

CONCLUSION

The few examples of paleomycological approaches to Quaternary studies reviewed above, indicate that fungi can substantially contribute to the accuracy and completeness of Quaternary paleoenvironmental reconstructions, but at present this potential of fungi is underutilized and hampered by often real, but mostly imagined, difficulties in identification of the fossils.

ACKNOWLEDGEMENTS

I thank Dr. C.R. Harington for making the specimen of *Fomes fomentarius* from Yukon Territory available for study and for helpful comments on the manuscript, and R.G. Day and K. Shepherd for the photographs.

REFERENCES

Allison, T.D., Moeller, R.E. and Davis, M.B., 1986, Pollen in laminated sediments provides evidence for a mid-Holocene forest pathogen outbreak: Ecology, v. 67, p. 1101-1105.

Andersen, S.T., 1984, Stages in soil development reconstructed by evidence from hypha fragments, pollen and humus contents in soil profiles, *in* Haworth, E.Y. and Lund, J.W.G., eds., Lake Sediments and Environmental History: University of Minnesota Press, Minneapolis, p. 295-316.

Anderson, R.S., Homola, R.L., Davis, R.G. and Jacobson, G.L., 1984, Fossil remains of the mycorrhizal fungi *Glomus fasciculatum* complex in postglacial lake sediments from Maine: Canadian Journal of Botany, v. 62, p. 2325-2328.

Andrews, H.N., 1948, A note on *Fomes idahoensis* Brown: Annals of the Missouri Botanical Garden, v. 35, p. 207.

Andrews, H.N. and Lenz, L.W., 1947, Fossil polypores from Idaho: Annals of the Missouri Botanical Garden, v. 34, p. 113-114.

Bakker, M. and Van Smeerdijk, D.G., 1982, A palaeoecological study of a Late Holocene section from "Het Ilperveld", Western Netherlands: Review of Palaeobotany and Palynology, v. 36, p. 95-163.

Berch, S.M. and Warner, B.G., 1985, Fossil vesicular-arbuscular mycorrhizal fungi: two *Glomus* species (Endogonaceae, Zygomycetes) from Late Quaternary deposits in Ontario, Canada: Review of Palaeobotany and Palynology, v. 45, p. 229-237.

Bondarcev, A.S., 1960, A fossil fungus of the genus *Ganoderma*: Botanicki Zhurnal S.S.S.R., v. 45, p. 1504-1506.

Buchwald, N.F., 1970, *Fomes idahoensis* Brown a fossil polypore fungus from the late Tertiary of Idaho, U.S.A.: Friesia, v. 9, p. 339-340.

Buchwald, N.F., 1971, Ein fossiler Röhrenpilz: Schweizerische Zeitschrift für Pilzkunde, v. 49, p. 58-59.

Butler, E.J., 1939, The occurrences and systematic position of the vesicular-arbuscular type of mycorrhizal fungi: British Mycological Society, Transactions, v. 22, p. 274-301.

Chaney, R.W. and Mason, H.L., 1936, A Pleistocene flora from Fairbanks, Alaska: American Museum of Natural History, Novitates, v. 887, p. 1-17.

Cowley, G.T. and Colquhoun, D.J., 1966, A Pleistocene fungus from South Carolina: Mycologia, v. 58, p. 483-486.

Cross, A.T., Thompson, G.C. and Zaitzeff, J.B., 1966, Source and distribution of palynomorphs in bottom sediments, southern part of Gulf of California: Marine Geology, v. 4, p. 467-524.

Davis, M.B., 1981, Outbreaks of forest pathogens in forest history: IV International Palynological Conference, Lucknow, Proceedings, v. 3, p. 216-227.

Davis, O.K., 1976, Pollen analysis of Wildcat Lake, Whitman County, Washington — the fungal spores: Geoscience and Man, v. 15, p. 143-144.

Davis, O.K., 1987, Spores of the dung fungus *Sporormiella*: increased abundance in historic sediments and before Pleistocene megafaunal extinction: Quaternary Research, v. 28, p. 290-294.

Davis, O..K., Agenboard, L., Martin, P.S. and Mead, J.I., 1984, The Pleistocene dung blanket of Bechan Cave, Utah, *in* Genoways, H.H. and Dawson, M.R., eds., Contribution in Quaternary Vertebrate Paleontology: Carnegie Museum of Natural History, Special Publications, v. 8, p. 167-282.

Davis, O.K., Kolva, D.A. and Mehringer, P.J., Jr., 1977, Pollen analysis of Wildcat Lake, Whitman County, Washington: The last 1000 years: Northwest Science, v. 51, p. 13-29.

De Vries, B., 1966, *Puccinia caricina* DC., a rust from a Late-glacial deposit from the Missouri Coteau, Saskatchewan: Canadian Journal of Botany, v. 44, p. 1228-1230.

Dilcher, D.L., 1965, Epiphyllous fungi from Eocene deposits in western Tennessee, U.S.A.: Palaeontographica, v. 116, p. 54.

Doi, D. and Amano, N., 1982, Fossil fungi from Pleistocene plant deposits, Shiobara, Tochigi Pref., Japan, with some palaeomycological notes: Acta Phytotaxonomia Geobotanica, v. 33, p. 55-72. (in Japanese).

Duff, K.E. and Smol, J.P., 1988, Chrysophycean stomatocysts from the postglacial sediments of a High Arctic lake: Canadian Journal of Botany, v. 66, p. 117-128.

Ellis, M.B., 1950, A modification of the "Necol" technique for mounting microfungi: British Mycological Society, Transactions, v. 33, p. 22.

Elsik, W.C., 1986, Palynology of Late Pleistocene Giant Ground Sloth locality, southwest Harris County, Texas: Pollen et Spores, v. 28, p. 77-82.

Eriksson, B., 1978, Fossil microthyriaceous fungi from Tervola, northern Finland: Annales Botanici Fennici, v. 15, p. 112-127.

Galbreath, E.C., 1947, Additions to the flora of the late Pleistocene deposits at Ashmore, Illinois: Kansas Academy of Science, Transactions, v. 50, p. 60-61.

Garneau, M., 1987, Reconstitution paléoécologique d'une tourbière littorale de l'estuaire du Saint-Laurent: analyse macrofossile et sporopollinique: Géographie physique et Quaternaire, v. 41, p. 109-125.

Hansen, J.M., 1980, Morphological characterization of encrusting, palynomorph green algae from Cretaceous-Tertiary of central West Greenland and Denmark: Grana, v. 19, p. 67-77.

Hillman, G., 1986, Plant food in ancient diet: the archeological role of palaeofaeces in general and Lindow man's gut contents in particular, *in* Stead, I.M., Bourke, J.B. and Brothwell, D., eds., Lindow Man. The Body in the Bog: British Museum, London, p. 99-115.

Holman, J.A., Abraczinska, L.M. and Westjohn, D.B., 1988, Pleistocene proboscidians and Michigan salt deposits: National Geographic Research, Winter 1988, p. 4-5.

Jacobson, G.L. and Bradshaw, R.H.W., 1981, The selection of sites for paleovegetational studies: Quaternary Research, v. 16, p. 80-96.

Jarzen, D. and Elsik, W.C., 1986, Fungal palynomorphs recovered from recent river deposits, Luangwa Valley, Zambia: Palynology, v. 10, p. 35-60.

Kuhry, P., 1985, Transgression of a raised bog across a coversand ridge originally covered with an oak-lime forest: Review of Palaeobotany and Palynology, v. 44, p. 303-353.

Kurmann, M.H., 1985, An opal phytolith and palynomorph study of extant and fossil soils in Kansas (U.S.A.): Palaeogeography, Palaeoclimatology, Palaeoecology, v. 49, p. 217-235.

Lichti-Federovich, S., 1975a, Pollen analysis of ice core samples from the Devon Island ice cap: Geological Survey of Canada, Paper 75-1A, p. 441-444.

Lichti-Federovich, S., 1975b, Pollen analysis of surface snow from five Canadian arctic ice caps: Geological Survey of Canada, Paper 75-1B, p. 135-137.

MacDonald, G.M., 1988, Palynology: Geoscience Canada, v. 15, p. 29-43.

Malloch, D.W., Grenville, D. and Hubart, J.-M, 1987, An unusual subterranean occurrence of fossil fungal sclerotia: Canadian Journal of Botany, v. 65, p. 1281-1283.

Mason, H.L., 1934, Pleistocene flora of the Tomales Formation: Carnegie Institution of Washington, Publications, v. 415, no. 4, p. 143.

Mead, J.I., Agenbroad, L.D., Davis, O.K. and Martin, P.S., 1986a, Dung of *Mammuthus* in the arid southwest, North America: Quaternary Research, v. 25, p. 121-127.

Mead, J.I., Agenbroad, L.D., Martin, P.S. and Davis, O.K., 1984, The mammoth and sloth dung from Bechan Cave in southeastern Utah: Current Research, v. 1, p. 79-80.

Mead, J.I., O'Rourke, M.K. and Foppe, T.M., 1986b, Dung and diet of the extinct Harrington's Mountain Goat (*Oreamnos harringtoni*): Journal of Mammalogy, v. 67, p. 284-293.

Miller, N.G., 1973, Late glacial plants and plant communities in northwestern New York state: Journal of the Arnold Arboretum, Harvard University, v. 54, p. 123-159.

Moore, P., 1985, The death of the elm: New Scientist, no. 1466, p. 32-34.

Mouchacca, J., 1978, Elements biodeteriogenes de la momie de Ramesès II: Bulletin de la Société de Mycologie medicale, v. 7, p. 5-10.

Muller, J., 1959, Palynology of Recent Orinoco delta and shelf sediments: Reports of the Orinoco Shelf Expedition: Micropaleontology, v. 5, no. 1, p. 1-32.
Pals, J.P., Van Geel, B. and Delfos, A., 1980, Paleoecological studies in the Klokkeweel Bog near Hoogkarspel (Prov. of Noord-Holland): Review of Palaeobotany and Palynology, v. 30, p. 371-418.
Pirozynski, K.A., 1976, Fungal spores in fossil record: Biological Memoirs, v. 1, p. 104-120.
Pirozynski, K.A., 1978, Fungal spores through the ages — A mycologist's view: IV International Palynological Conference, Lucknow, Proceedings, v. 1, p. 327-330.
Pirozynski, K.A., Carter, A. and Day, R.G., 1984, Fungal remains in Pleistocene ground squirrel dung from Yukon Territory, Canada: Quaternary Research, v. 22, p. 375-382.
Pirozynski, K.A. and Dalpé, Y., 1989, Geological history of the Glomaceae with particular reference to mycorrhizal symbiosis: Symbiosis, v. 7, p. 1-36.
Pirozynski, K.A., Jarzen, D.M., Carter, A. and Day, R.G., 1988, Palynology and mycology of organic clay balls accompanying mastodon bones - New Brunswick, Canada: Grana, v. 27, p. 123-139.
Pirozynski, K.A. and Malloch, D.W., 1988, Seeds, spores and stomachs: coevolution in seed dispersal mutualisms, *in* Pirozynski, K.A. and Hawksworth, D.L., eds., Coevolution of Fungi with Plants and Animals: Academic Press, London, p. 227-246.
Prasad, M.N.V. and Ramesh, N.R., 1984, First record of *Circinoconis* conidia from the Holocene formation of Tripura, India: Current Science, v. 53, p. 38-39.
Ratan, R. and Chandra, A., 1982, Isolated thyriothecia from surface (bottom) sediments of the Arabian Sea: Geophytology, v. 12, p. 260-263.
Romans, J.C.C., Stevens, J.H., Robertson, L. and Jones, D., 1966, Semi-fossil lichen fungi in Scottish hills soils: Nature, v. 209, p. 96.
Rosendahl, C.O., 1943, Some fossil fungi from Minnesota: Bulletin of the Torrey Botanical Club, v. 70, p. 126-138.
Rosendahl, C.O., 1948, A contribution to the knowledge of the Pleistocene flora of Minnesota: Ecology, v. 29, p. 284-319.
Saad, S.I. and Sani, S., 1967, Studies of pollen and spores content of Nile delta deposits (Berenbal region): Pollen et Spores, v. 9, p. 467-503.
Scaife, R.G., 1986, Pollen in human palaeofaeces; and a preliminary investigation of the stomach and gut contents of Lindow man, *in* Stead, I.M., Bourke, J.B. and Brothwell, D. eds., Lindow Man. The Body in the Bog: British Museum, London, p. 126-135.
Sharma, Ch., 1976, Some fungal spores from Quaternary deposits of Malvan, Gujarat: The Palaeobotanist, v. 23, p. 79-81 (1974).
Sherwood-Pike, M., 1988, Freshwater fungi: fossil record and paleoecological potential: Palaeogeography, Palaeoclimatology, Palaeoecology, v. 62, p. 271-285.
Spaulding, W.G. and Martin, P.S., 1979, Ground Sloth dung of the Guadalupe Mountains, *in* Genoways, H.H. and Baker, R.J., eds., Biological Investigations in the Guadalupe Mountains, National Park, Texas: National Park Service, Proceedings and Transactions Series, no. 4, p. 259-269.
Stewart, R.B. and Robertson, W., 1968, Fungus spores from prehistoric potsherds: Mycologia, v. 60, p. 701-704.
Terasmae, J., 1958, "Microforaminifera" from Pleistocene deposits, Prince Edward Island, Canada: Micropaleontology, v. 4, p. 429-430.
TeStrake, D., 1979, Dispersed fungal spores associated with a Central Florida Pleistocene peat bog: Nova Hedwigia, v. 31, p. 667-675.
Traverse, A. and Ginsburg, R.N., 1966, Palynology of the surface sediments of Great Bahama Bank, as related to water movement and sedimentation: Marine Geology, v. 4, p. 417-459.
Van der Wiel, A.M., 1982, A palaeoecological study of a section from the foot of Hazendonk (Zuid-Holland, the Netherlands), based on the analysis of pollen, spores and macroscopic plant remains: Review of Palaeobotany and Palynology, v. 38, p. 35-90.
Van Geel, B., 1972, Palynology of a section from the raised peat bog "Wietmarscher Moor", with a special reference to fungal remains: Acta Botanica Neerlandica, v. 21, p. 261-284.
Van Geel, B., 1978, A palaeoecological study of Holocene peat bog sections in Germany and the Netherlands: Review of Palaeobotany and Palynology, v. 25, p. 1-120.
Van Geel, B., 1986, Application of fungal and algal remains and other microfossils in palynological analyses, *in* Berglund, B.E., ed., Handbook of Holocene Palaeoecology and Palaeohydrology: J. Wiley and Sons, p. 497-505.
Van Geel, B. and Andersen, S.T., 1988, Fossil ascospores of the parasitic fungus *Ustulina deusta* in Eemian deposits in Denmark: Review of Palaeobotany and Palynology, v. 56, p. 89-93.
Van Geel, B., Bohncke, S.J.P. and Dee, H., 1981, A palaeoecological study of an Upper Late Glacial and Holocene sequence from "De Borchert", The Netherlands: Review of Palaeobotany and Palynology, v. 31, p. 367-448.
Van Geel, B., Hallewas, D.P. and Pals, J.P., 1983, A Late Holocene deposit under the Westfriese Zeedijk near Enkhuizen (Prov. of Noord-Holland, The Netherlands): palaeoecological and archaeological aspects: Review of Palaeobotany and Palynology, v. 38, p. 269-335.
Van Geel, B., Klink, A.G., Pals, J.P. and Wigers, J., 1986, An Upper Eemian lake deposit from Twente, eastern Netherlands: Review of Palaeobotany and Palynology, v. 47, p. 31-61.
Vishnu-Mittre, 1975, Studies on fungal remains from the Flandrian deposits in the Whittlesey Mere region, Hunts, England: The Palaeobotanist, v. 22, p. 89-110.
Vreeland, J.M. and Korf, R.P., 1978, Fungal spores from a Peruvian mummy *ca.* 800 A.D.: Paleopathology Newsletter, v. 19, p. 11-12.
Warner, B.G., 1988, Plant macrofossils: Geoscience Canada, v. 15, p. 121-129.
Watling, R. and Seaward, M.R.D., 1976, Some observations on puff-balls from British archeological sites: Journal of Archeological Science, v. 3, p. 165-172.
Weber, C.A., 1915, Die Mammutfora von Borna: Abhandlungen Naturwissenschaftlichen Verein zu Bremen, v. 23, p. 1-69.
Wilson, L.R., 1965, *Rhizophagites*, a fossil fungus from the Pleistocene of Oklahoma: Oklahoma Geological Notes, v. 25, p. 257-260.
Wilson, L.R., 1966, Palynology of the Domebo site: Contributions of the Museum of Great Plains, v. 1, p. 1-7.
Witte, H.J.L. and Van Geel, B., 1985, Vegetational and environmental succession and net organic production between 4500 and 800 B.P. reconstructed from a peat deposit in the western Dutch coastal area (Assendelver Polder): Review of Palaeobotany and Palynology, v. 45, p. 239-300.
Wolf, F.A., 1966, Fungus spores in East African lake sediments. II: Journal of the Elisha Mitchell Scientific Society, v. 82, p. 57-61.
Wolf, F.A., 1967a, Fungus spores in East African lake sediments. IV: Bulletin of the Torrey Botanical Club, v. 94, p. 31-34.
Wolf, F.A., 1967b, Fungus spores in East African lake sediments. V: Mycologia, v. 59, p. 397-404.
Wolf, F.A., 1969, Nonpetrified fungi in late Pleistocene sediment from eastern North Carolina: Journal of the Elisha Mitchell Scientific Society, v. 85, p. 41-44.

Originally published in
Geoscience Canada Volume 16 Number 3
(September 1989)

Bryophytes

Jan A. Janssens
Department of Ecology, Evolution, and Behavior
University of Minnesota
318 Church Street S.E.
Minneapolis, Minnesota 55455 USA

SUMMARY
Mosses are common fossils in Quaternary sediments. The fragments can be readily identified to species, and no fossils found in Quaternary sediments represent extinct species. Because of the detailed ecological information available from living populations, the Quaternary record of fossil mosses is superbly suited for paleoenvironmental reconstruction of peatland development and of other moss-rich habitats, and particular information is here presented concerning bryophyte fossils in Canadian deposits. New numerical techniques are being developed to establish quantitative bryophyte paleoecology in its own right among Quaternary disciplines.

INTRODUCTION

Bryophytes are an artificial group of primitive embryophytes, comprising the mosses, liverworts, and hornworts (Schofield, 1985; Crandall-Stotler, 1984; Mishler and Churchill, 1984, 1985). They are mostly terrestrial, autotrophic (green) plants with two morphologically distinct generations during their life cycle: (1) an autotrophic gametophyte or sexual generation and (2) a partially parasitic sporophyte or asexual generation, **attached to the gametophore** (Figure 1). Bryophyte diversity, approximately 14,500 species worldwide, is higher than in any other group of terrestrial plants except angiosperms (Scagel *et al.*, 1982). In addition, the expansion of bryophyte, and particularly *Sphagnum*, dominated landscapes (peatlands) during the Quaternary Period and the preservation of *Sphagnum* remains in peat have resulted in the speculation that there is more carbon stored in *Sphagnum*, alive and dead, than in any other plant (Clymo and Hayward, 1982).

The fossil record of bryophytes begins in the Paleozoic (Schofield, 1985; Krassilov and Schuster, 1984; Oostendorp, 1987). During the later Tertiary (Neogene) and Quaternary Periods the record, owing to the presence of numerous fossils referable to extant species, conveys abundant paleoenvironmental information (Miller, 1984). Phylogenetic considerations are based, however, exclusively on the study of extant groups, with only secondary attention given to the fossil record because of poor preservation of relevant systematic characters. However, many pre-Tertiary fossils actually can be interpreted in light of cladistic study based on extant mosses and liverworts (Mishler and Churchill, 1984).

Quaternary moss remains are easily recovered from Quaternary deposits and most can readily be identified to species. Fossil liverworts and hornworts, however, are exceedingly rare (Steere, 1942; Miller, 1984). Their haphazard occurrence, mainly in Holocene or interstadial/glacial peats and organic-rich deposits, is evident from the catalogs published for North America by Miller (1980a) and Janssens (1984). The poor Quaternary record of the liverworts and hornworts is in sharp contrast to detailed descriptions of Paleozoic, Mesozoic, and Tertiary species (Harris, 1939; Hueber, 1961; Grolle, 1985; Oostendorp, 1987; Schuster and Janssens, 1989).

Moss fragments are commonly extracted from peats (Janssens, 1988), lake sediments (Birks, 1976), and detrital (allochthonous) deposits (Miller, 1973). Leafy stems are most commonly recovered, but *Sphagnum* remains are usually detached stem and branch leaves (Janssens, 1983a, 1988). Rarely

Figure 1 *A short turf of* Bartramia pomiformis *with exserted, mature sporophytes. The gametophores are clustered in a clonal population. (Photo provided by Dr. R.M. O'Clair, Juneau, Alaska.)*

sporophytes (Figure 2) or calyptrae are found (Conolly and Dickson, 1969; Brassard and Blake, 1978; Janssens and Baker, 1984; Miller, 1987). Except for those of *Sphagnum* and hornworts, spores are only occasionally differentiated into genera or species by palynologists (Dickson, 1973, 1986). Considerable potential exists, however, for the use of sculpture patterns in identifying bryophyte spores (Horton, 1982; Boros and Jarai-Komlodi, 1975).

In this paper, I discuss the development of Quaternary paleobryology by focussing on the more important sites in Canada and North America. Some field and laboratory procedures are outlined, with notes on taphonomy and paleoenvironmental reconstruction. Peatland development, paleobryogeography, evolution and putative extinct species, and future directions for research are discussed. Recent catalogs of fossil species and reviews with exhaustive bibliographies on Quaternary fossil bryophytes are found in Dickson (1973), H.A. Miller (1982), N.G. Miller (1976, 1980a,b,c, 1984), Odgaard (1981), Birks (1982), Janssens (1984), and Dickson (1986). Quantification of data has been attempted (Janssens, 1983a; Wasylikowa, 1986), and future developments will include refinements of those techniques (Birks, 1985; Janssens, 1988).

HISTORICAL DEVELOPMENT IN CANADA

Quaternary fossil bryophytes have been studied in North America for over a hundred years (H.A. Miller, 1982; N.G. Miller, 1976, 1980a, 1984; Birks, 1982). Only one published study from the United States predates those from Canada (Pratt, 1876). The first Canadian records were published by Hinde in 1877 (Miller, 1976). Fossil moss remains were extracted by G.J. Hinde and W.A. Johnson from exposures along the Scarborough Bluffs near Toronto and were identified by John Macoun and T.P. James. Several other studies published during the latter part of the nineteenth century reported moss fossils from the Toronto and Ottawa areas and from along the Missinaibi River (Miller, 1980a). Few records were published between 1900 and 1958, and detailed information started to accumulate only after the latter date. Published reports of North American bryophyte fossils were compiled by Miller (1980a) and Janssens (1984). A new catalog is in preparation by N.G. Miller and J.A. Janssens. Most of the 50 Canadian sites at which fossil bryophytes have been studied or are under study (51 papers) are located in the Arctic Islands (17 sites), Ontario (9), Yukon (6), Quebec (7), British Columbia (4), Alberta (2), Northwest Territories (2), Saskatchewan, Manitoba, and Labrador. Most of these sites (23) are Holocene. Many others represent Wisconsinan interstadial deposits or are interglacial, usually with poor direct-dating control (20). Another ten are late-glacial deposits. The Worth Point deposit described originally by Kuc (1974a) is probably of Early Quaternary age (Matthews, 1989). The Beaufort Formation localities on the Arctic Islands, once thought to be of Quaternary age, are not discussed here in detail because they are now known to be Neogene (Matthews, 1987, 1989; Matthews *et al.*, in press). These floras are clearly boreal, and all of the fossil mosses belong to extant species. The remains are preserved without diagenesis (Kuc, 1973, 1974b; Kuc and Hills, 1971).

The most significant lists of species for single locations with well-documented stratigraphy, radiometric dates, or both were published by de Vries and Bird (1965), Mathewes (1973), LaSalle *et al.* (1979), Hughes *et al.* (1981), and Stewart and England (1983). All are discussed below.

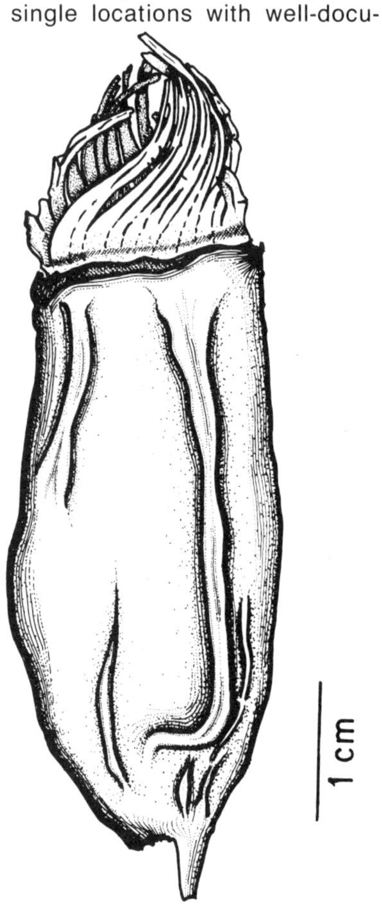

Figure 2 *Sporangium of* Barbula fallax *from a full-glacial assemblage, southeastern Iowa. (Modified from Janssens and Baker, 1984).*

Southern Saskatchewan.

A late-glacial moss assemblage of 16 species and 1 variety was identified by C.D. Bird from a kettle-hole deposit near Moose Jaw, Saskatchewan (de Vries and Bird, 1965). The mixed nature of the fossil assemblage is indicated by the presence of abundant well-preserved fragments of *Drepanocladus aduncus,* a semi-terrestrial wetland moss, and of upland species such as *Thuidium abietinum, Bryoerythrophyllum recurvirostrum,* and *Tortula mucronifolia*.

British Columbia.

W.B. Schofield identified 22 moss species, from a late-glacial and early-Holocene lake-sediment sequence in Marion Lake, southwestern British Columbia (Mathewes, 1973). All of the species presently occur in the area and represent either terrestrial or epiphytic habitats. The presence of epiphytes, as *Antitrichia curtipendula* and *Isothecium stoloniferum,* is exceptional. Furthermore, eight of the species represented are endemic to western North America.

Southern Quebec.

LaSalle *et al.* (1979) described an interstadial Mid-Wisconsinan arctic-subarctic deposit from Vallée-Jonction, Quebec, from which W.C. Steere identified 12 terrestrial and wetland species. Mosses that do not belong to the present regional flora include the arctic species *Aulacomnium turgidum* and *Didymodon asperifolius.* All are characteristic of aquatic, semi-terrestrial, or stream-side habitats.

Ellesmere Island.

An early-Holocene detrital assemblage of fossil bryophytes was recovered from emergent marine delta deposits at Clements Markham Inlet, northern Ellesmere Island, by Stewart (1981). I identified 25 taxa indicating rich-fen, dry-tundra, and fluvial habitats. All of these mosses occur in the region, and the fossil flora represent approximately one-fourth of the total extant bryoflora of Ellesmere Island (Brassard, 1976).

Northern Yukon.

An interstadial deposit from the Hungry Creek section in the Bonnet Plume Basin, northern Yukon Territory, contained a large fossil bryophyte flora

(Hughes *et al.*, 1981). The allochthonous assemblage is situated on bedding planes of superimposed sand and silt layers, which were overridden by the late Wisconsinan Hungry Creek glacial advance. Preservation of the abundant fossils is excellent, which suggests minimal reworking. I identified 39 species plus seven additional taxa to section and genus. The family Amblystegiaceae, indicating a highly minerotrophic peatland, was best represented. Other species represent communities ranging from lake-edge to rock-outcrop and dry upland habitat. Some arctic-alpine species are also present. *Hylocomium splendens,* represented by its luxuriant, sympodially branched wefts growthform, suggests the presence of spruce forests growing nearby during deposition of the alluvial unit; the site is presently subarctic–subalpine.

Summary.

Table 1 lists the 20 moss species that have 10 or more Quaternary records for Canada. It is compiled from published sources (Miller, 1980a; Janssens, 1983b, 1984) and from unpublished reports (Miller and Janssens, in preparation). Twelve of the 20 species belong to the family Amblystegiaceae (brown mosses). Nine of the 20 are rich-fen indicators and two *(Drepanocladus lycopodioides* var. *brevifolius* and *Aulacomnium turgidum)* are mainly subarctic or arctic.

METHODS

Field Sampling Procedures.

Samples for bryophyte-fossil analyses are often bulk samples taken from freshly cleaned natural exposures. Moss remains are commonly concentrated along bedding planes or as dispersed detritus and peat lenses in inorganic sediments. Care should be taken not to dry field samples completely because this causes the moss remains to become brittle and susceptible to damage before and during laboratory processing. If humified *Sphagnum* peats are completely dried no amount of soaking can restore the original texture, and dispersal of the fossil fragments in water becomes extremely difficult. When bulk samples are prescreened in the field to reduce the volume of the inorganic component, a preservative such as phenol should be added to the screened macrofossil remains to limit fungal growth. Alternatively, samples can be frozen.

Figure 3 *Peat monolith extracted from a "muskeg" peatland on Douglas Island, Juneau area, southeastern Alaska. Living* Sphagnum lindbergii *constitutes the top 5 cm of the monolith. The next 25 cm constitutes the aerobic acrotelm layer with well-preserved* Sphagnum *peat (Ingram, 1983; Verry, 1984), overlying the catotelm of more decomposed, anaerobic peat. (Photo by the author).*

Lake-sediment and peatland samples are obtained from cores or monoliths (Figure 3). Coring devices have been discussed most recently in Berglund (1986) for a wide variety of sediments and in Wright *et al.* (1984) and Janssens (1987) for Holocene peat deposits.

Laboratory methods and identification.
Bulk samples and subsamples of peat or lake sediment are wet-screened in the laboratory before fossil bryophytes are extracted. KOH solution, which is quite destructive to moss remains, should not be used to disperse the sample. Instead, a mild wetting agent, such as Aerosol OT (Fisher), is preferred. Boiling is rarely necessary. Dilute HCl can help to break up a slightly consolidated calcareous matrix and does little damage to mosses. A 300-μm mesh screen is sufficient to retain significant moss remains. Screened material can be stored in glycerine before extracting fossil fragments. Detailed techniques for volumetric measurement, identification, and quantification of *Sphagnum* species in Holocene peats are given by Janssens (1987, 1988). For leafy mosses other than *Sphagnum*, it is sufficient to detach the best-preserved leaves from the fossil fragment and mount both fragment and detached leaves under two separate coverslips on a microscope slide in Hoyer's solution (Schuster, 1966, p. 335). The specimens are then identified under a compound microscope with standard keys (Crum, 1984; Nyholm, 19541969; see also the list of manuals in Dickson, 1986), and with reference collections of slides and herbarium specimens. Dichotomous keys for the identification of fossil *Sphagnum* and *Drepanocladus* fragments are available in Janssens (1983b, 1987).

Presentation of results.
Differential breakage of fragments does not allow counts of stem or leaf remains to be performed easily (Watts and Winter, 1966; Birks, 1970; Rybníček, 1973; Janssens, 1983a). Abundance estimates are commonly based on volumetric estimates or on a combination of these estimates and tallies of fragments (Rybníček and Rybníčková, 1968; Aaby and Jacobson, 1979; Dickson, 1981; Janssens and Barnosky, 1985). For instance, Rybníček and Rybníčková (1968) commonly screen 100 mL to estimate moss fragments and other plant macrofossils volumetrically. Countable fragments are tallied. Both volumetric estimates (in %) and counts are converted to a 7-point scale of relative abundance.

Table 1 Quaternary mosses with more than 10 Quaternary records in Canada (Miller, 1980a; Janssens, 1983b; Janssens, 1984; Miller and Janssens, in prep.)

Scorpidium scorpioides [1,2]	54
Tomenthypnum nitens[2]	47
Drepanocladus exannulatus[1]	43
Calliergon giganteum[1,2]	42
Drepanocladus lycopodioides var. *brevifolius*[1,3]	32
Calliergon richardsonii[1,2]	28
Drepanocladus crassicostatus[1]	25
Drepanocladus revolvens[1,2]	25
Meesia triquetra[2]	21
Bryum pseudotriquetrum[2]	20
Aulacomnium palustre	20
Calliergon trifarium[1,2]	19
Drepanocladus fluitans[1]	19
Campylium stellatum[1,2]	18
Drepanocladus aduncus[1]	17
Hylocomium splendens	16
Drepanocladus sendtneri[1]	14
Aulacomnium turgidum[3]	14
Ditrichum flexicaule	11
Distichium capillaceum	10

[1] *belongs to the family Amblystegiaceae (brown mosses)*
[2] *rich-fen indicator*
[3] *subarctic-arctic species*

Sub-sampling of sediments with widely varying amounts of bryophyte material is more efficient when an exotic marker is added. Poppy seed, which has dispersal characteristics similar to bryophyte fragments, has proven useful (Janssens, 1983a). Only with the results of future studies on shoot densities, differential decomposition, and other taphonomically significant variables will actual counting of fragments be preferable to volumetric estimates or a combination of these estimates and fragment counts.

The most elaborate method of presenting results has been developed for Holocene autochthonous peats (Janssens, 1987, 1988). Standardized methods are essential for accurate quantitative inference (Birks, 1985). The first challenge in measuring the abundance of different species in a peat assemblage is the representation of *Sphagnum* species and other mosses on a common, comparative scale that reflects the volumetric contribution of each taxon to the total peat component. Janssens (1988, table 2) gives an example of such an assemblage and its quantification. The complexity of handling *Sphagnum* fragments is caused by the extensive fragmentation of their remains as compared with other mosses. No attempts have yet been made to compensate for selective decomposition (Clymo, 1965) and differential fragmentation (Rybníček and Rybníčková, 1968) among related species. Because of problems associated with quantification, qualitative aspects are still important, such as the presence or absence of certain "indicator" species (Rybníček, 1973). Techniques for quantitative inference are further discussed in the Paleoenvironmental Reconstruction section below and by Birks (1985).

TAPHONOMY

Study of the ontogeny of death assemblages is important for paleoenvironmental reconstruction. Selective decomposition (Steere, 1942; Clymo, 1965; Janssens, in prep.) and transport (Glaser, 1981; Miller, 1984; Warner and Barnett, 1986) are two important processes that alter the composition of fossil assemblages. Both processes affect even autochthonous peats as indicated by poor recovery of ubiquitous bog liverworts (Duckett and Clymo, 1988) and the representation of hummock and hollow species in single fossil assemblages (Janssens and Engstrom, in prep.). In detrital deposits, reworking of older deposits often leads to mixed assemblages that are characterized by distinctly different preservation of the components, heterogeneous assemblages, and differential fragment lengths (Janssens, 1981; Miller, 1984).

Etiolation and aquatic modifications are often encountered in bryophytes from terrestrial habitats that became dislodged and grew under water before burial (Dickson, 1973). The presence of aquatic modifications is used in paleoenvironmental reconstruction (Janssens and Glaser, 1986). Some of the post-burial history of fragments can be gleaned from the index of reliability devised by Dickson (1973), and elaborated by Janssens (1983a, 1987), Janssens and Zander (1980), Odgaard (1982), and Schweger and Janssens (1980).

PALEOENVIRONMENTAL RECONSTRUCTION

Distributional shifts were the first inferences made from fossil bryophyte records. With the development of better stratigraphical control and dating techniques, more significant qualitative paleoclimatic and paleoenvironmental inferences were made possible. Quantitative paleoenvironmental inference, suggested by models for paleoclimatology, has been developed only recently (Birks, 1985).

Commonly, Quaternary paleoenvironmental inference has been based on apparent climatic and environmental tolerances of individual species (Dickson, 1986) or on their value as syntaxonomic "indicators" for certain mire communities. H.H. Birks and Mathewes (1978) compared complementary macrofossil and pollen evidence in a paleoecological study of the late-glacial and early-Holocene development of Abernethy Forest, Scotland. The present distribution and edaphic preferences of *Calliergon trifarium* in Scotland were considered in the paleoenvironmental reconstruction of rich-fen communities based on bryophyte and vascular-plant macrofossils and pollen types of local origin. A similar approach was used by Webb and Moore (1982) in reconstructing stadial late-glacial paleoenvironments in southeastern Scotland. For example, bare-ground communities and late-snowmelt areas were suggested by the presence of *Salix herbacea* and *Polytrichastrum alpinum*. An exceptionally diverse Middle Weichselian moss assemblage from Jutland allowed Odgaard (1982) to postulate a treeless landscape dominated by herbs and dwarf shrubs. The climate was non-oceanic, and mean July temperatures were probably between 8° and 10°C. A different approach was taken by Rybníček (1973), who compared past and present mire communities in Czechoslovakia using a synecological approach. The presence of certain key species, rather than listing individual species, classified the fossil assemblages into predefined past community types. These predefined past community types are comparable to phytosociological units recognized in present-day mires. Paleoenvironmental reconstruction throughout a profile was based on a dated sere of community types that replaced each other through time.

Because of the absence of deep rooting systems, mosses are much more specific indicators of peatland habitat conditions than vascular plants (Janssens *et al.*, in press). Obviously only extant mosses that have been studied sufficiently in relation to environmental gradients can be used. Some of the most common and best-studied species, however, have such wide-ranging edaphic and climatic tolerances that they are of little indicator value in paleoenvironmental or paleoclimatic interpretations. *Hylocomium splendens* (Tamm, 1953) is an example. More detailed study of growth-forms and modifications of these common species can nevertheless indicate valuable paleoenvironmental aspects (Miller, 1984). The sympodial growth form of *Hylocomium splendens* remains of Sangamonan age from Hungry Creek in the Yukon Territory indicates the presence of a forest-floor community where presently no trees grow (Janssens, 1981; Hughes *et al.*, 1981). Aquatic modifications of *Scorpidium scorpioides* contribute to the understanding of the broader extent of marl pools in late-glacial-rich fens in contrast to the present limited occurrence of this habitat type (Janssens, 1981).

There is some evidence from phytocenological studies (Rybníček and Rybníčková, 1968) and from fossil records (Janssens *et al.*, 1986) that species represented in late-glacial assemblages may have had ecological requirements that differed from those in their extant populations. This problem is indicated by the presence of non-analogous assemblages (death assemblages without a modern analog: Birks and Birks, 1980), and sometimes can be explained by the possible existence of ecotypes (Gorham, 1956, in press; Green, 1968; Damman, 1977). The existence of different ecotypes in mosses (structurally identical populations only differentiated by edaphic or climatic requirements) has been suggested several times, *e.g.*, for *Sphagnum imbricatum* and *S. flavicomans* (Green, 1968; Tolonen and Tolonen, 1984; Flatberg, 1986). It is increasingly evident that there is a genetic basis for the habitat differentiation among populations of the same species growing in different regions (Longton, 1976, 1988; Dietert, 1980; Wyatt and Anderson, 1984; Sveinbjörnsson and Oechel, 1983; Shaw *et al.*, 1987; Shaw, 1987). However, the role of composition is most often still ignored. Presently, the most practical approach is to derive paleoecological inferences from the **total** known tolerance range of the species, including all putative ecotypes. Reference data sets should ideally encompass the entire distributional range of relevant species (Birks and Birks, 1980).

Another approach to paleoenvironmental reconstruction using macrofossils has been successfully employed by Wolfe (1978) with Tertiary leaf floras. It can be quite independent of taxonomic identification, but uses certain structural features, such as leaf outlines and margins, that differentiate along climatic gradients. Nothing has been attempted in this way with fossil bryophytes, but classification of bryophyte growth-forms (Gimingham and Birse, 1957; Gimingham and Smith, 1971) could possibly lead to a similar technique (Miller, 1980c; Janssens, 1981; Odgaard, 1981).

Old Crow Basin, detailed study. Janssens (1981) recognized six groups of fossil-moss assemblages for the Yukon refugium, some characterized by different growth-form spectra. These groups were formed by collating all fossil assemblages (65) from numerous sediment samples collected along exposures in the Old Crow Flats, northern Yukon Territory. The area was never glaciated and is the easternmost extremity of the large Beringian Pleistocene refugium (Hopkins *et al.*, 1982). The assemblage groups are illustrated by their growth-form spectra in Figure 4 along with a simplified composite section of the Old Crow River Quaternary exposures. The oldest assemblages are from Lower Lake clays and the thick alluvial deposit overlying them. These sediments are all older than 149 ka, the age of Old Crow tephra (Westgate, 1988), and, at some sites, the lake clays are known to be of Plio/Pleistocene age (Schweger, 1989). The alluvial section grades into an overlying paleosol, abruptly truncated by a disconformity (Morlan, 1980) and strongly disturbed by periglacial activity. The paleosol and associated autochthonous peats postdate the Happy Interval and probably belong to the latter part of the Boutellier Interval, a period of intermediate climate and less extensive glaciation in Beringia (65,000-30,000 years age; Hopkins, 1982; Schweger and Janssens, 1980). This period was followed by the last cold

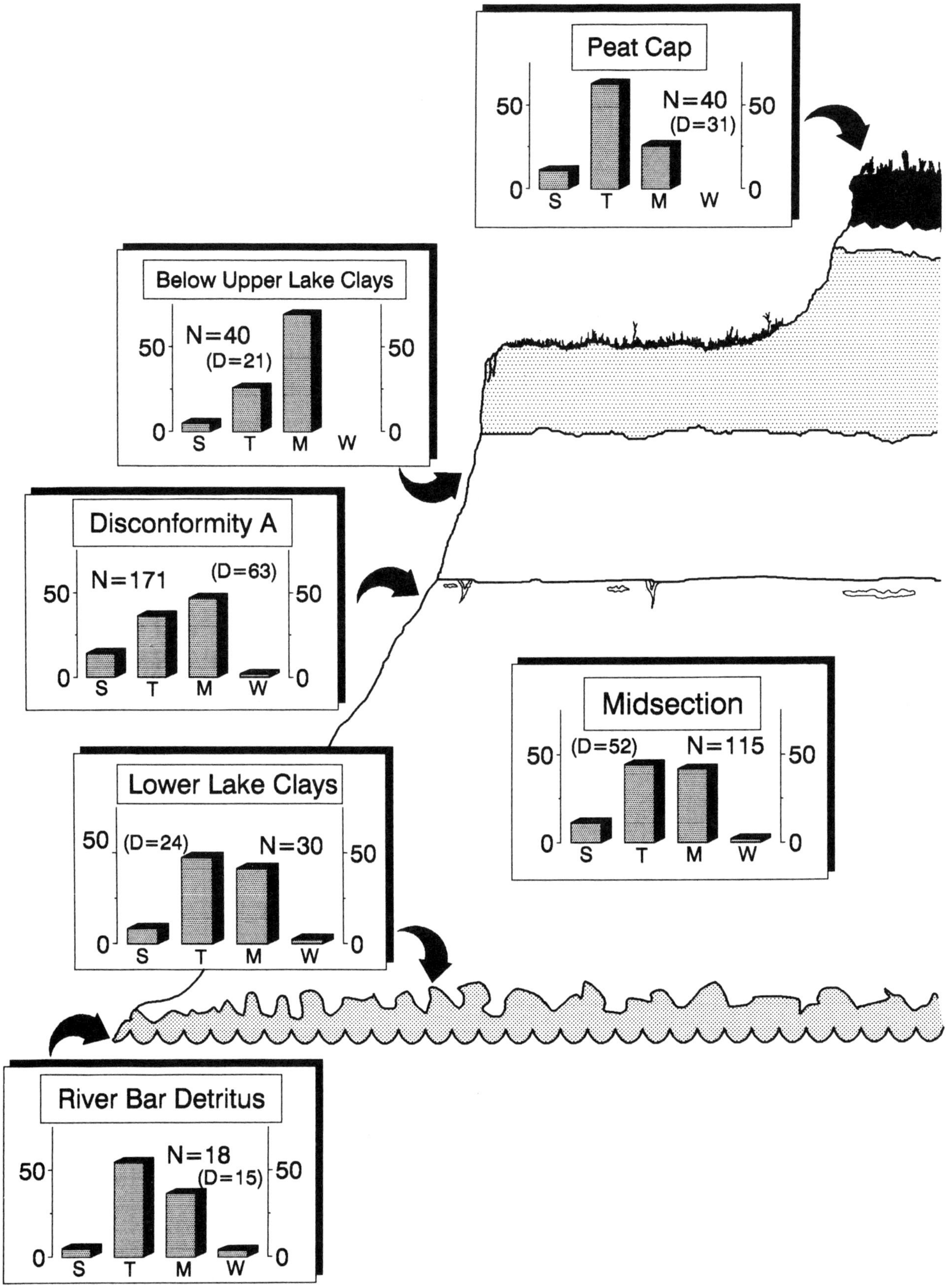

Figure 4 *Composite section along the Old Crow River in the Old Crow Flats basin and six groups of fossil-bryophyte assemblages, illustrated by their growth-form spectra. The Old Crow tephra (149 ka) is located just below Disconformity A. See text for discussion of chronostratigraphy and paleoenvironmental reconstruction.* ***N****, number of records;* ***D****, number of taxa;* ***S****, small turfs;* ***T****, tall turfs;* ***M****, mats; and* ***W****, wefts. (Compiled from Janssens, 1981; and Morlan and Matthews, 1978; see also Morison and Smith, 1987; and Schweger, 1989).*

and dry periglacial interval of the Pleistocene in unglaciated Beringia, the Duvanny Yar Interval (30,000-13,000 years ago). Locally, on the Old Crow Flats, the Duvanny Yar period is represented by the Upper Lake clays. The alluvial fossiliferous sediments below these sterile lake clays indicate the steep drying and cooling trend before the formation of the periglacial lakes. As shown in Figure 4, most Old Crow sections are capped by Holocene peat.

Six assemblage groups, illustrated by their growth-form spectra and species diversity in Figure 4, support the previously outlined Beringian paleoenvironmental reconstruction. All fossils were assigned a most common growth form (Janssens, 1981).

The habitat of a moss species is partially correlated with its strategy to obtain and retain the available water for growth (Birse, 1958; Gimingham and Smith, 1971; Longton, 1988). In Arctic or Antarctic regions, summers are short and favourable periods for growth are irregular. The growth form of a bryophyte will influence critically its survival since dense growth forms, such as small cushions, and turfs without branches are often effective in restricting water loss. The species with these growth forms are most commonly found in more exposed environments. Carpet-forming species (tall turfs with spreading branches such as *Sphagnum* spp.) and large cushions are found in swampy habitats provided with a permanent water supply. Tall turfs with erect branches and loose structure, and thus relatively little influence upon water conservation, often grow both in habitats with significant evaporation stress and on tall hummocks in more water-logged conditions. Often other mechanisms (for example cuticularization, and lamina and lamellae movement) also operate to restrict water loss. Most mat growers exhibit an extreme variability. Typically developed smooth and rough mats are never in contact with the water table, as is also the case with wefts (Tamm, 1953) that are mostly associated with forest-floor habitats. More robust forms approach carpets and are found in swampy habitats (for example, species of *Drepanocladus*).

In Figure 4, the seven different growth forms (tall turfs with erect branches, with spreading branches, with tomentum, short turfs, rough and smooth mats, and wefts) are grouped together in four classes (small turfs, tall turfs, mats, and wefts). The proportions of each class in the six assemblage groups are reflected in the height of the histogram bars. Proportions are calculated on the basis of the number of records for each fossil species in the assemblages. Species with two common growth forms are scored twice.

The Lower Lake Clay and Midsection assemblages consist mostly of detrital remains. Turfs are dominant over mats and wefts. A slightly higher proportion of mats is recorded in the more autochthonous set of assemblages associated with the paleosol below disconformity A (probably formed during the more moderate climate of the Boutellier Interval; Hopkins, 1982). This assemblage group has also the largest number of species (63) for the entire Old Crow Flats stratigraphy. In contrast, the sediments below the Upper Lake clays, which represent late Wisconsinan advance of the Laurentide ice sheet, have few species. Most of these assemblages consist of a few, transported, poorly preserved fossils. No weft-formers are evident and mats are dominant. High-energy fluvial habitats are suggested for the period before the formation of the upper glacial lakes. The growth-form spectrum of the Holocene peat cap is based on autochthonous

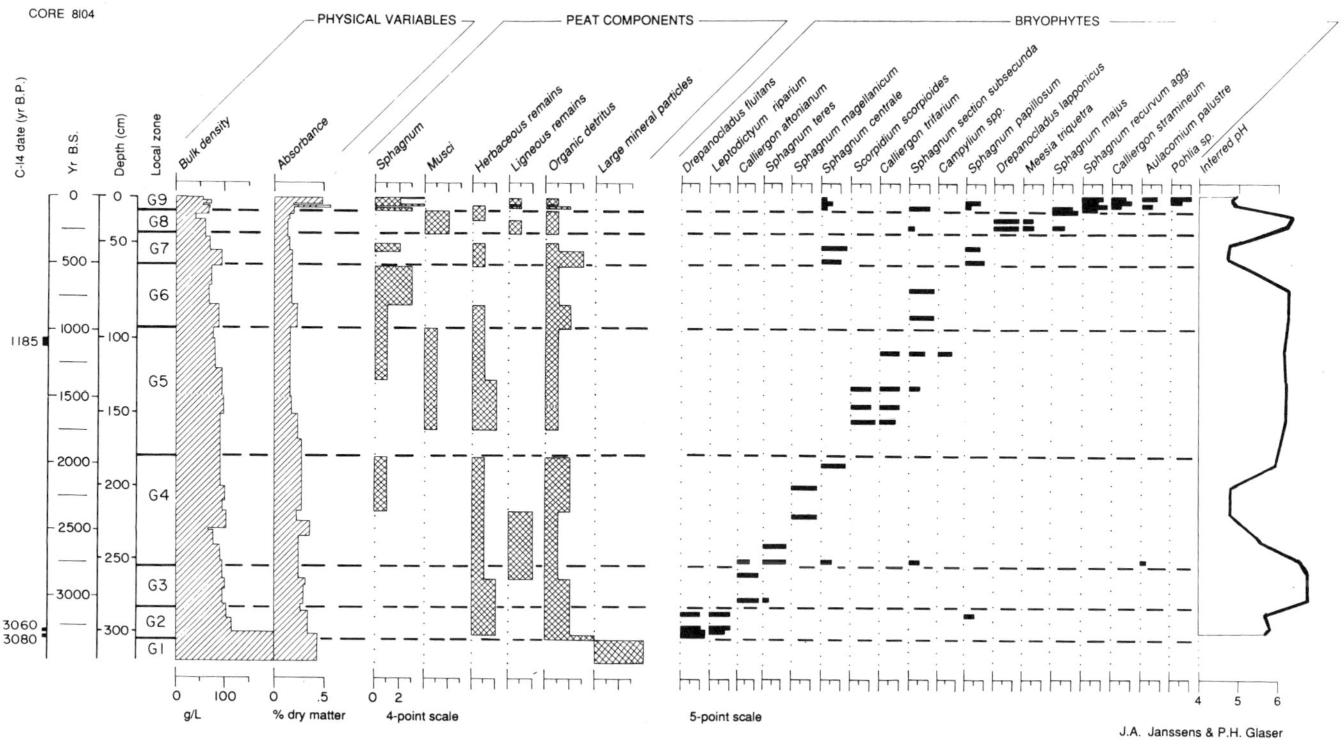

Figure 5 *Selected stratigraphic profiles versus calendar-year age for analyses from peat core 8104 of the Red Lake Peatland transect in northern Minnesota. The bryophyte profiles are discontinuous bar histograms at the level of each sample along a 5-point estimated volume scale (see Janssens (1987, 1988) for details). The inferred pH is estimated by weighted-averaging regression and calibration for each fossil-bryophyte assemblage using the FORTRAN program WACALIB. (From Janssens* et al., *in press.)*

peat samples, except for a few detrital samples immediately above the sterile glacio-lacustrine sediments.

Present-day detritus collected on river bars during the summer consists of material that was probably deposited during spring floods of the same year. Notwithstanding the different genesis of the material, the growth-form spectra of the peat cap and river-bar detritus are very similar. The bulk of the detrital river-bar moss fragments is probably derived from the easily eroded Holocene peat cap rather than the more consolidated Pleistocene exposures. Tall turfs, mainly *Sphagnum* and common peatland species, are dominant.

Other studies.
In contrast to the few studies mentioned above, most bryophyte fossils have been recovered from deposits of a single age or as allochthonous fragments in lake and alluvial deposits, representing only brief stratigraphic intervals. Some studies on lake and peatland profiles, however, have produced a continuous stratigraphic record at a single site, often encompassing several thousands of years (Birks, 1976; Miller, 1984; Anderson *et al.*, 1986; Miller and Futyma, 1987; Janssens *et al.*, in press).

Many individual bryophyte species have little value for edaphic or paleoclimatic inference (Miller, 1980b, and see previous discussion on ecotypes). However, stratigraphic boundaries based on changes of common peatland bryophyte communities can be established, dated, and correlated among nearby coring sites and can support wide-ranging paleoclimatic inferences (Janssens *et al.*, in press). The nearly synchronous start of the development of the extensive oligotrophic mire over the entire Red Lake Peatland watershed correlates well with a major climatic boundary between the prairie period and the mixed conifer-hardwood period established by pollen analysis and radiocarbon dating at about 2750 years B.P. This paleoenvironmental reconstruction is supported by a number of core analyses similar to the one illustrated by Figure 5. pH optima and tolerances for each bryophyte species are estimated using WACALIB, a computer program for weighted-averaging regression and calibration (Janssens and Birks, in prep.). It was developed initially for reconstructing paleoenvironmental gradients of pH based on diatom assemblages from

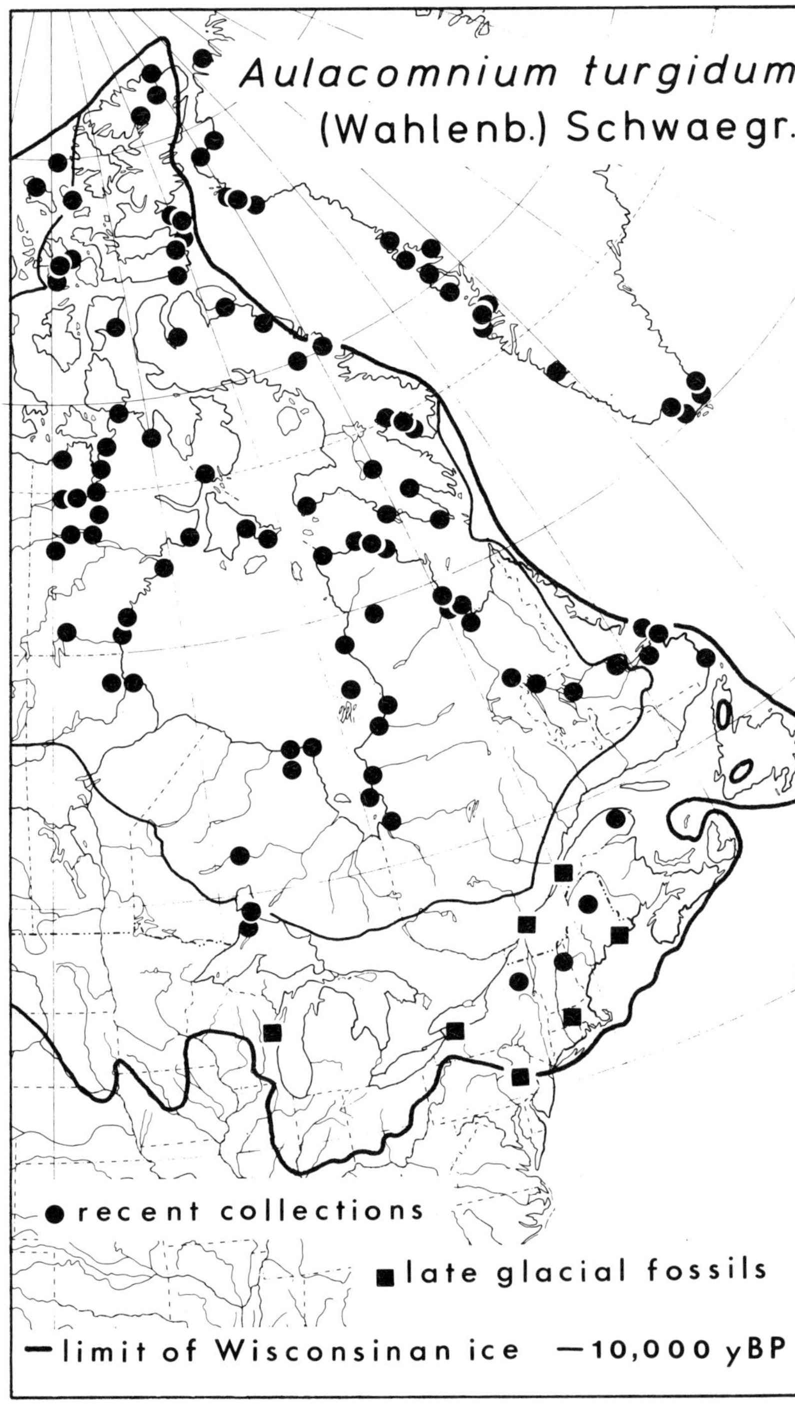

Figure 6 *Present and late-glacial distribution of* Aulacomnium turgidum *in eastern North America. (Modified from Miller, 1984).*

Figure 7 *Fragment of* Fontinalis flaccida *recovered from early postglacial lake sediments from Lake Hope Simpson, southeastern Labrador.* **(1)** *gametophore fragment, 12 ×;* **(2)** *rhizoid, 220 ×;* **(3)** *leaf apex, 220 ×;* **(4)** *medial leaf cells, 330 ×;* **(5)** *alar leaf cells, 360 ×;* **(6)** *leaf, 70 ×. (From Janssens* et al., *1986).*

lakes affected by acid rain (Birks, 1987; Birks *et al.*, in press). The advantages of using numerical techniques (Birks and Birks, 1980) for the reconstruction of past peatland environments are (1) the possibility of using **all** taxa, irrespective of the amount of quantitative data accumulated, (2) the explicit nature and repeatable techniques of the method (the WACALIB reference data set for fossil mosses of North America will be available soon from the author), and (3) the discovery of unsuspected problems and the formulation of new ecological theory (Gorham and Janssens, in prep.).

No detailed bryophyte developmental sequence, correlated with climatic phases during the entire Holocene, exists for North American peatlands (the present continental peatlands are mostly about 5000 years old). In Europe, however, major climatic periods (the late-glacial Alleröd and Younger Dryas, and the Holocene Pre-Boreal and Boreal) show an excellent correlation with changes in peatland communities (Rybníček and Rybníčková, 1968), and the principal developmental phases, according to the Central European phytosociocenological system. After bog (oligotrophic mire) developed, no fundamental changes took place, but only changes in the proportion of more hydrophilic and more xerophilic species. During the "Little Ice Age" (15th-18th century) the extent of mires also increased substantially as attested by early records quoted in Gorham (1953).

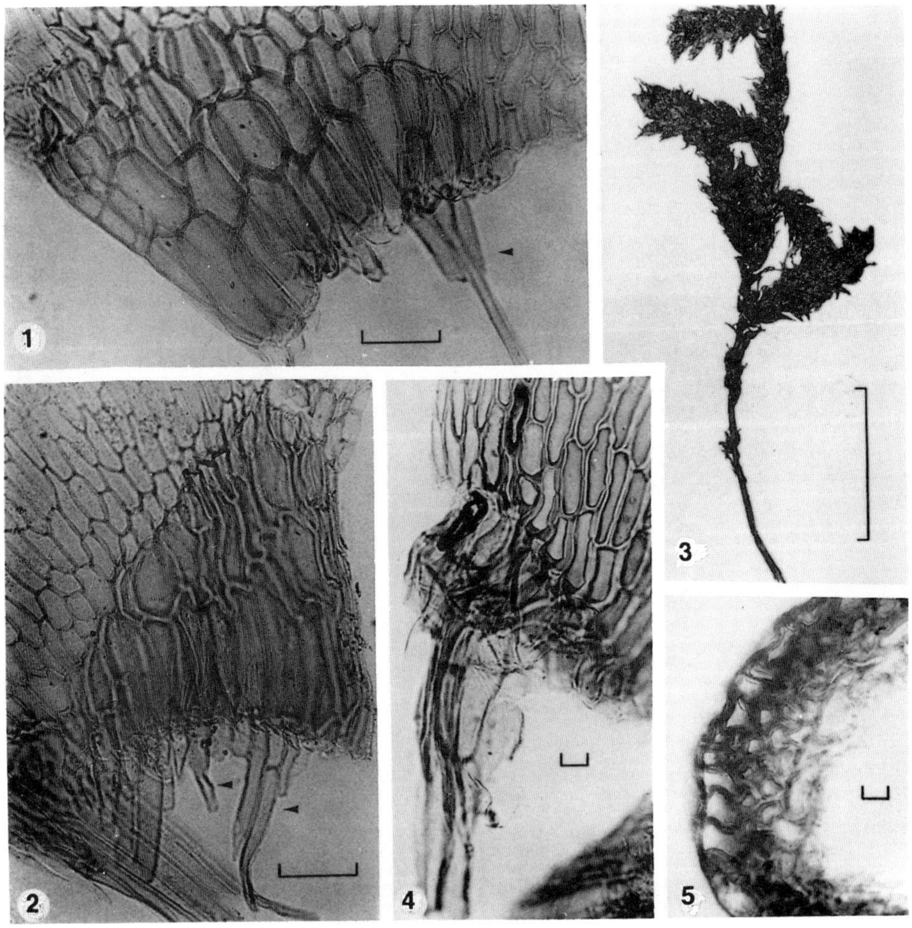

Figure 8 Drepanocladus minnesotensis *(=* D. aduncus *var.* kneiffii) *and* Neocalliergon integrifolium *(=* Scorpidium scorpioides). **(1-2)** D. minnesotensis*: alar cells of stem leaves with cortical cells of stem (at arrows) attached to leaf bases, scale = 50μm.* **(3-5)** Neocalliergon integrifolium. **(3)** *Plant, scale = 1 cm.* **(4)** *Alar region showing small group of enlarged cells at corner of leaf and enlarged cortical cells adhering to base of leaf, scale = 10μm.* **(5)** *Portion of transverse section of stem showing enlarged cortical cells, scale = 10μm. (From Miller, 1983).*

BRYOGEOGRAPHY, EVOLUTION, AND PUTATIVE EXTINCT SPECIES

Bryogeographical conclusions have always been a major part of all studies of Quaternary fossil bryophytes. Even the most recent Holocene records have distinct phytogeographical importance (Janssens and Glaser, 1986). Most bryogeographic conclusions have been phrased in the context of glacial or rich-fen relicts (Miller, 1984, 1987; Rybníček and Rybníčková, 1968; Rybníček, 1973; Dickson, 1973) and some have also had phylogenetic relevance (Janssens *et al.*, 1986). Several late-glacial localities in North America have records of species that do not extend as far south at the present: *Cinclidium latifolium* (Janssens, 1984); *Aulacomnium turgidum* (Figure 6; Miller, 1984); *Aulacomnium acuminatum* (Miller, 1973; LaSalle *et al.*, 1979); *Hypnum bambergeri* (Miller, 1976). As well, not all Pleistocene or early Holocene refugia were south of the main continental glaciers (Miller, 1987): *Fontinalis flaccida* (Figure 7) was identified from lake sediments in southeastern Labrador (Janssens *et al.*, 1986), and *Leptodontium flexifolium*, a southern United States and Central American species, was present during the mild mid-Wisconsinan Boutellier Interval in Beringia (Janssens and Zander, 1980).

Some of the species that are quite rare in present mire communities were more widespread during late Pleistocene and early Holocene time (Miller, 1987). For example, *Calliergon trifarium* formed thick peat layers throughout Europe (Rybníček and Rybníčková, 1968) and in the west-central watershed profiles of the Red Lake Peatland (Janssens *et al.*, in press). It has a considerably restricted distribution today. *Meesia triquetra* is commonly associated with species of *Sphagnum* section *Subsecunda* in fossil assemblages (Rybníček, 1973), but, at present, the two are rarely found together (Dickson, 1973). Rybníček (1973) suggests that this undoubtedly results from drainage in mires. However, the extensive expansion of oligotrophic mire communities restricted available habitat for species such as *M. triquetra*. These oligotrophic mire communities are only known from the Atlantic in Europe (Rybníček, 1973) and from after the mid-Holocene warm period in continental North America (Zoltai *et al.*, 1988) and there is no preserved pre-Quaternary record of similar extensive deposits. This also suggests that speciation in *Sphagnum* could be very recent compared with the usual concept of bryophyte evolutionary rates. Compilation of the fossil record of *Sphagnum* (excluding doubtful spore records) does not give any clear hint of evolutionary development. The pre-Quaternary record is extremely meagre, and the Mesozoic records seem to be contaminants (Arnold, 1932; Reissinger, 1938). The order Protosphagnales, created by Neĭburg (1958) to accommodate the Permian fossil genera *Junjagia, Vorcutannularia,* and *Protosphagnum,* appears unrelated to the Sphagnales. Apart from the superficial resemblance in leaf-cell dimorphism these form species are structurally very different from *Sphagnum* (*e.g.*, well-developed costa and border present). In addition, the cell differentiation seen in photographs of these leaf areolations (Neĭburg, 1958, 1960) could sometimes be an artefact of fossilization: compression fossils often show the sideways collapse of large-lumened cells (as in present-day *Funaria* and *Splachnum* species) suggestive of an areolation pattern seen in *Sphagnum*. Moreover, pores or fibrils, other characteristic structures of *Sphagnum* branch-leaf cells, are not evident, but septae are (Oostendorp, 1987). The resolution of this artefact problem is not insignificant: the placement of *Sphagnum* in the moss clade is secondarily supported by the character complex of the Protosphagnales (Mishler and Churchill, 1984).

Few species of Quaternary age have been described as extinct (Miller, 1984), and none of these is still recognized as extinct. Even the late Tertiary Beaufort Formation (Kuc, 1973, 1974; Kuc and Hills, 1971; Matthews *et al.*, in press) contains no extinct species. It contains, however, *Calliergon aftonianum* (Kuc, 1974b), initially described as a Quaternary extinct moss from Aftonian deposits in Iowa (Steere, 1942). It is now recognized as a common Quaternary fossil (Kuc, 1974b; Janssens *et al.*, in press). It is indistinguishable from extant *Calliergon* populations found in the Lost River Peatland in northern Minnesota (Janssens and Glaser, 1986; Glaser *et al.*, in press). The fossil specimens are most likely a modification of the common *C. giganteum* or *C. richardsonii* (see also Miller, 1984, 1987). Similar conclusions have been drawn by Miller about the identity of *Drepanocladus minnesotensis* Williams (= *D. aduncus* var. *kneiffii*) and *Neocalliergon integrifolium* Williams (= *Scorpidium scorpioides*), based on study of type specimens (Figure 8; Miller, 1983). The morphological aspects of Tertiary and Quaternary fossil bryophytes are amazingly similar to those of the extant flora (Janssens *et al.*, 1979; Grolle, 1985) and contrast sharply with the well-known fossils of Mesozoic age (Harris, 1939; Krassilov and Schuster, 1984; Schuster and Janssens, 1989) that are uniquely different from anything now living.

FUTURE DIRECTIONS

The record of fossil bryophytes, pre-Quaternary as well as Pleistocene and Holocene, has, to date, played only a minor role in phylogenetic and taxonomic studies. The absence or non-preservation of many characters, presently used extensively in the taxonomic study of extant populations, will always limit the value of fossil bryophytes for tracing phylogenetic lines. Alternatively, additional autecological study will enhance the capability of using fossil bryophytes for paleoenvironmental inference. Development of the new empirical and statistical procedures will finally establish quantitative bryophyte paleoecology in its own right among other Quaternary disciplines.

ACKNOWLEDGEMENTS

I thank Barry Warner, the editor of this series, for his valuable help and persistent encouragement to complete this review. I also thank Norton Miller for critical remarks and permission to republish illustrations from his papers, and H.J.B. Birks, H.A. Crum, J.H. Dickson, E. Gorham, J.V. Matthews, Jr., B.D. Mishler, and H.E. Wright, Jr. for comments on the manuscript. Some of the research mentioned here was supported by Natural Sciences and Engineering Research Council (Canada) to D.H. Vitt, by contracts from the Minnesota Department of Natural Resources and the United States Department of Energy to H.E. Wright, Jr. and by grants from the United States National Science Foundation (DEB7922142) and the A.W. Mellon Foundation to E. Gorham.

REFERENCES

Aaby, B. and Jacobson, J., 1979, Changes in biotic conditions and metal deposition in the last millennium as reflected in ombrotrophic peat in Draved Mose, Denmark: Danmarks Geologiske Undersogelse, Årbog 1978, p. 5-43.

Anderson, R.S., Davis, R.B., Miller, N.G. and Stuckenrath, R., 1986, History of late- and post-glacial vegetation and disturbance around Upper South Branch Pond, northern Maine: Canadian Journal of Botany, v. 64, p. 1977-1986.

Arnold, C.A., 1932, Microfossils from Greenland coal: Michigan Academy of Sciences, Arts and Letters, Papers, v. 15 (1931), p. 51-61.

Berglund, B.E., 1986, ed., Handbook of Holocene Palaeoecology and Palaeohydrology: John Wiley and Sons, New York, 869 p.

Birks, H.H. and Mathewes, R.W., 1978, Studies in the vegetational history of Scotland. V: Late Devensian and early Flandrian pollen and macrofossil stratigraphy at Abernethy Forest, Inverness-shire: New Phytologist, v. 80, p. 455-484.

Birks, H.J.B., 1970, Inwashed pollen spectra at Loch Fada, Isle of Skye: New Phytologist, v. 69, p. 807-820.

Birks, H.J.B., 1976, Late-Wisconsinan vegetational history at Wolf Creek, central Minnesota: Ecological Monographs, v. 46, p. 395-428.

Birks, H.J.B., 1982, Quaternary bryophyte palaeo-ecology, *in* Smith, A.J.E., ed., Bryophyte Ecology: Chapman and Hall, London, p. 473-490.

Birks, H.J.B., 1985, Recent and possible future mathematical developments in quantitative paleoecology: Palaeogeography, Palaeoclimatology, Palaeoecology, v. 50, p. 107-147.

Birks, H.J.B., 1987, Methods for pH-calibration and reconstruction from paleolimnological data: procedures, problems, potential techniques, *in* Proceedings of Surface Water Acidification Programme (SWAP), Mid-Term Review Conference, Bergen 22-26 June 1987, p. 370-380.

Birks, H.J.B. and Birks, H.H., 1980, Quaternary Palaeoecology: Edward Arnold Publishers, London, 289 p.

Birks, H.J.B., Line, J.M., Juggins, S., Stevenson, A.C. and ter Braak, C.J.F., in press, Diatoms and pH reconstruction: Royal Society of London, Philosophical Transactions, Series B, in press.

Birse, E.M., 1958, Ecological studies on growth-form in bryophytes. III. The relationship between growth-form of mosses and ground-water supply: Journal of Ecology, v. 46, p. 9-27.

Boros, A. and Jarai-Komlodi, M., 1975, An Atlas of Recent European Moss Spores: Akademiai Kiado, Budapest, 466 p.

Brassard, G.R., 1976, The mosses of northern Ellesmere Island, Arctic Canada III. New or additional records: The Bryologist, v. 79, p. 480-487.

Brassard, G.R. and Blake, W., Jr., 1978, An extensive subfossil deposit of the arctic moss *Aplodon wormskioldii*: Canadian Journal of Botany, v. 56, p. 1852-1859.

Clymo, R.S., 1965, Experiments on breakdown of *Sphagnum* in two bogs: Journal of Ecology, v. 53, p. 747-758.

Clymo, R.S. and Hayward, P.M., 1982, The ecology of *Sphagnum*, *in* Smith, A.J.E., ed., Bryophyte Ecology: Chapman and Hall, London, p. 229-289.

Conolly, A.P. and Dickson, J.H., 1969, A note on a Late-Weichselian *Splachnum* capsule from Scotland: New Phytologist, v. 68, p. 197-199.

Crandall-Stotler, B., 1984, Musci, hepatics and anthocerotes — an essay on analogues, *in* Schuster, R.M., ed., New Manual of Bryology, Volume 2: Hattori Botanical Laboratory, Nichinan, Japan, p. 1093-1129.

Crum, H.A., 1984, Sphagnopsida, Sphagnaceae, *in* North American Flora, Series II, Part 11: The New York Botanical Garden, 180 p.

Damman, A.W.H., 1977, Geographical changes in the vegetation pattern of raised bogs in the Bay of Fundy region of Maine and New Brunswick: Vegetatio, v. 35, p. 137-151.

de Vries, B. and Bird, C.D., 1965, Bryophyte subfossils of a late-glacial deposit from the Missouri Coteau, Saskatchewan: Canadian Journal of Botany, v. 43, p. 947-953.

Dickson, J.H., 1973, Bryophytes of the Pleistocene: Cambridge University Press, London, 256 p.

Dickson, J.H., 1981, Mosses from a Roman well at Abingdon: Journal of Bryology, v. 11, p. 559-560.

Dickson, J.H., 1986, Chapter 30: Bryophyte analysis, *in* Berglund, B.E., ed., Handbook of Holocene Paleoecology and Paleohydrology: John Wiley and Sons, New York, p. 627-643.

Dietert, M.F., 1980, The effect of temperature and photoperiod on the development of geographically isolated populations of *Funaria hygrometrica* and *Weissia controversa*: American Journal of Botany, v. 67, p. 369-380.

Duckett, J.G. and Clymo, R.S., 1988, Regeneration of bog liverworts: New Phytologist, v. 110, p. 119-127.

During, H.J., 1979, Life strategies of bryophytes: a preliminary review: Lindbergia, v. 5, p. 2-18.

Flatberg, K.I., 1986, Taxonomy, morphovariation, distribution and ecology of the *Sphagnum imbricatum* complex with main reference to Norway: Gunneria, v. 54, p. 1-118.

Gimingham, C.H. and Birse, E.M., 1957, Ecological studies on growth-form in bryophytes. I. Correlations between growth-form and habitat: Journal of Ecology, v. 45, p. 533-545.

Gimingham, C.H. and Smith, R.I.L., 1971, Growth form and water relations on mosses in the maritime antarctic: Bulletin of the British Antarctic Survey, v. 25, p. 1-21.

Glaser, P.H., 1981, Transport and deposition of leaves and seeds on tundra: a late-glacial analog: Arctic and Alpine Research, v. 13, p. 173-182.

Glaser, P.H., Janssens, J.A. and Siegel, D.I., in press, Responses of vegetation to chemical and hydrological gradients in the Lost River Peatland, northern Minnesota: Journal of Ecology, in press.

Gorham, E., 1953, Some early ideas concerning the nature, origin, and development of peatlands: Journal of Ecology, v. 41, p. 257-274.

Gorham, E., 1956, On the chemical composition of some waters from the Moor House Nature Reserve: Journal of Ecology, v. 44, p. 377-384.

Gorham, E., in press, Biotic impoverishment of northern peatlands, *in* Woodwell, G.M., ed., The Earth in Transition, Patterns and Processes of Biotic Impoverishment: Cambridge University Press, in press.

Green, B.H., 1968, Factors influencing the spacial and temporal distribution of *Sphagnum imbricatum* Hornsch. *ex* Russ. in the British Isles: Journal of Ecology, v. 56, p. 47-58.

Grolle, R., 1985, Monograph of *Frullania* in Baltic amber: Prace Muzeum Ziemi, Warszawa, v. 37, p. 87-100.

Harris, T.M., 1939, *Naiadita*, a fossil bryophyte with reproductive organs: Annales Bryologici, v. 12, p. 57-70.

Hinde, G.J., 1877, The glacial and interglacial strata of Scarboro' Heights, and other localities near Toronto: Canadian Journal of Science, N.S., v. 15, p. 388-413.

Hopkins, D.M., 1982, Aspects of the paleogeography of Beringia during the late Pleistocene, *in* Hopkins, D.M., Matthews, J.V., Jr., Schweger, C.E. and Young, S.B., eds., Paleoecology of Beringia: Academic Press, New York, p. 3-28.

Hopkins, D.M., Matthews, J.V., Jr., Schweger, C.E. and Young, S.B., 1982, Paleoecology of Beringia: Academic Press, New York, 489 p.

Horton, D.G., 1982, The evolutionary significance of surficial spore characters in the Bryidae: Journal of the Hattori Botanical Laboratory, v. 53, p. 99-105.

Hueber, F.M., 1961, *Hepaticites devonicus*, a new fossil liverwort from the Devonian of New York: Annals of the Missouri Botanical Garden, v. 48, p. 125-132.

Hughes, O.L., Harrington, C.R., Janssens, J.A., Matthews, J.V., Jr., Morlan, R.E., Rutter, R.W. and Schweger, C.E., 1981, Upper Pleistocene stratigraphy, paleoecology, and archaeology of the northern Yukon Interior, eastern Beringia, 1: Bonnet Plume Basin: Arctic, v. 34, p. 329-365.

Ingram, H.A.P., 1983, Hydrology, *in* Gore, A.J.P., ed., Mires: Swamp, Bog, Fen, and Moor: Elsevier Publishing Company, Amsterdam, General Studies, Ecosystems of the World 4A, p. 67-158.

Janssens, J.A., 1981, Subfossil Bryophytes in Eastern Beringia: their paleoenvironmental and paleogeographical significance, Ph.D. dissertation, University of Alberta, 450 p.

Janssens, J.A., 1983a, A quantitative method for stratigraphical analysis of bryophytes in Holocene peat: Journal of Ecology, v. 71, p. 189-196.

Janssens, J.A., 1983b, Past and extant distribution of *Drepanocladus* in North America, with notes on the differentiation of fossil fragments: Journal of the Hattori Botanical Laboratory, v. 54, p. 251-298.

Janssens, J.A., 1984, Quaternary fossil bryophytes in North America: new records: Lindbergia, v. 9, p. 137-151.

Janssens, J.A., 1987, Ecology of peatland bryophytes and paleoenvironmental reconstruction of peatlands using fossil bryophytes, Manual for Bryological Methods Workshop, Mainz, FRG, 17-23 July 1987, Update August 1987, 68 p.

Janssens, J.A., 1988, Fossil bryophytes and paleoenvironmental reconstruction of peatlands, *in* Glime, J.M., ed., Methods in Bryology, Proceedings of the Bryological Methods Workshop, Mainz: Hattori Botanical Laboratory, Nichinan, Japan, p. 299-306.

Janssens, J.A. and Baker, R.G., 1984, A full-glacial bryophyte assemblage from southeastern Iowa: Journal of Bryology, v. 13, p. 201-207.

Janssens, J.A. and Barnosky, C.W., 1985, Late-Pleistocene and Early-Holocene bryophytes from Battle Ground Lake, Washington, USA: Review of Palaeobotany and Palynology, v. 46, p. 97-116.

Janssens, J.A., Engstrom, D.R. and Horton, D.G., 1986, *Climacium dendroides* and *Fontinalis flaccida*: fossils of the early postglacial of Labrador: Lindbergia, v. 12, p. 67-72.

Janssens, J.A. and Glaser, P.H., 1986, The bryophyte flora and major peat-forming mosses at Red Lake Peatland, Minnesota: Canadian Journal of Botany, v. 64, p. 427-442.

Janssens, J.A., Hansen, B.C.S., Glaser, P.H. and Barnosky, C.W., in press, Development of a raised bog complex in northern Minnesota, *in* Wright, H.E., Jr., Coffin, B. and Aasing, N., eds., Patterned Peatlands of northern Minnesota, in press.

Janssens, J.A., Horton, D.G. and Basinger, J.F., 1979, *Aulacomnium heterostichoides* sp. nov., an Eocene moss from central British Columbia: Canadian Journal of Botany, v. 57, p. 2150-2161.

Janssens, J.A. and Zander, R.H., 1980, *Leptodontium flexifolium* and *Pseudocrossidium revolutum* as 60,000 year old subfossils from the Yukon Territory: The Bryologist, v. 83, p. 486-496.

Krassilov, B.A. and Schuster, R.M., 1984, Paleozoic and Mesozoic fossils, *in* Schuster, R.M., ed., A New Manual of Bryology: Hattori Botanical Laboratory, Obi, Japan, p. 1172-1193.

Kuc, M., 1973, Fossil flora of the Beaufort Formation, Meighen Island, NWT-Canada: ERA, Canadian-Polish Research Institute, Biology and Earth Sciences, Series A, No. 1, Toronto, Canada, 44 p.

Kuc, M., 1974a, The interglacial flora of Worth Point, western Banks Island: Geological Survey of Canada, Paper 74-1B, p. 227-231.

Kuc, M., 1974b, *Calliergon aftonianum* Steere in Late Tertiary and Pleistocene deposits of Canada: Geological Survey of Canada, Paper 74-24, 8 p.

Kuc, M. and Hills, L.V., 1971, Fossil mosses, Beaufort Formation (Tertiary), Northwestern Banks Island, Western Canadian Arctic: Canadian Journal of Botany, v. 49, p. 1089-1094.

LaSalle, P., Martineau, G. and Chauvin, L., 1979, Lits de bryophytes du Wisconsin moyen, Vallée-Jonction, Québec: Canadian Journal of Earth Sciences, v. 16, p. 593-598.

Longton, R.E., 1976, Reproductive biology and evolutionary potential in bryophytes: Journal of the Hattori Botanical Laboratory, v. 41, p. 205-223.

Longton, R.E., 1988, The biology of polar bryophytes and lichens: Cambridge University Press, Cambridge, 391 p.

Luken, J.O., 1984, Net ecosystem production in a subarctic peatland, Ph.D. dissertation, Duke University.

Mathewes, R.W., 1973, A palynological study of postglacial vegetation changes in the University Research Forest, southwestern British Columbia: Canadian Journal of Botany, v. 51, p. 2085-2103.

Matthews, J.V., Jr., 1987, Plant macrofossils from the Neogene Beaufort Formation on Banks and Meighen Islands, District of Franklin: Geological Survey of Canada, Paper 87-1A, p. 73-87.

Matthews, J.V., Jr., 1989, New information on the flora and age of the Beaufort Formation, Arctic Archipelago, and related Tertiary deposits in Alaska: Geological Survey of Canada, Paper 89-1D, p. 105-111.

Matthews, J.V., Jr., Ovenden, L.E. and Fyles, J.G., in press, Plant and insect fossils from the late Tertiary Beaufort Formation on Prince Patrick Island, Northwest Territories, in press.

Miller, H.A., 1982, Bryophyte evolution and geography: Biological Journal of the Linnean Society, v. 18, p. 145-196.

Miller, N.G., 1973, Late glacial plants and plant communities in northwestern New York State: Journal of the Arnold Arboretum, v. 54, p. 123-159.

Miller, N.G., 1976, Quaternary fossil bryophytes in North America: a synopsis of the record and some phytogeographic implications: Journal of the Hattori Botanical Laboratory, v. 41, p. 73-85.

Miller, N.G., 1980a, Quaternary fossil bryophytes in North America: catalog and annotated bibliography: Journal of the Hattori Botanical Laboratory, v. 47, p. 1-43.

Miller, N.G., 1980b, Fossil mosses in North America and their significance, *in* Taylor, R.J. and Leviton, A.L., eds., The Mosses of North America: American Association for the Advancement of Science, San Francisco, p. 9-36.

Miller, N.G., 1980c, Mosses as paleoecological indicators of lateglacial terrestrial environments: some North American studies: Torrey Botanical Club, Bulletin, v. 107, p. 373-391.

Miller, N.G., 1983, The identity of the Pleistocene mosses of *Drepanocladus minnesotensis* and *Neocalliergon integrifolium* (Amblystegiaceae): Brittonia, v. 35, p. 87-92.

Miller, N.G., 1984, Tertiary and Quaternary fossils, *in* Schuster, R.M., ed., A New Manual of Bryology: Hattori Botanical Laboratory, Obi, Japan, p. 1194-1232.

Miller, N.G., 1987, Phytogeography and paleoecology of a late Pleistocene moss assemblage from northern Vermont: Memoirs of the New York Botanical Garden, v. 45, p. 242-258.

Miller, N.G. and Futyma, R.P., 1987, Paleohydrological implications of Holocene peatland development in northern Michigan: Quaternary Research, v. 27, p. 297-311.

Mishler, B.D. and Churchill, S.P., 1984, A cladistic approach to the phylogeny of the "bryophytes": Brittonia, v. 36, p. 406-424.

Mishler, B.D. and Churchill, S.P., 1985, Cladistics and the land plants: a response to Robinson: Brittonia, v. 37, p. 282-285.

Morison, S.R. and Smith, C.A.S., 1987, eds., Guidebook to Quaternary Research in Yukon: XXII INQUA Congress, Ottawa, Canada.

Morlan, R.E., 1980, Taphonomy and archeology in the Upper Pleistocene of the northern Yukon Territory: a glimpse on the peopling of the new world: Archeological Survey of Canada, Paper 94, 398 p.

Morlan, R.E. and Matthews, J.V., Jr., 1978, New dates for Early Man: Geos, Winter 1978, p. 2-5.

Ne ĭburg, M.F., 1958, Permian true mosses of Angaraland: Journal of the Paleontological Society, India, v. 3, p. 22-29.

Ne ĭburg, M.F., 1960, Listostebelnye mkhi iz Permskikh otlozhenii Angaridy: Akademiya Nauk SSSR, Trudy Geologicheskogo Instituta, Vypusk 19, Moskva, 104 p.

Nyholm, E., 1954-1969, Illustrated Moss Flora of Fennoscandia: C.W.K. Gleerup, Lund, 799 p.

Odgaard, B.V., 1981, The Quaternary Bryoflora of Denmark. I: Species list: Danmark geologiske Undersogelse, Årbog 1980, p. 45-74.

Odgaard, B.V., 1982, A Middle Weichselian moss assemblage from Hirtshals, Denmark, and some remarks on the environment 47,000 B.P.: Danmark geologiske Undersogelse, Årbog 1981, p. 5-45.

Oostendorp, C., 1987, The bryophytes of the Palaeozoic and Mesozoic: Bryophytorum Bibliotheca, Band 34: J. Cramer, Berlin-Stuttgart, 112 p.

Pratt, W.H., 1876, Report on a geological examination of the section of the bluffs recently exposed by the C.R.I. & P.R.R.: Davenport Academy Natural Sciences, Proceedings, v. 1, p. 96-99.

Reissinger, A., 1938, Die "Pollenanalyse" ausgedehnt auf alle Sedimentgesteine der geologischen Vergangenheit: Palaeontographica, v. 84B, p. 1-20.

Rybníček, K., 1973, A comparison of the present and past mire communities of central Europe, *in* Birks, H.J.B. and West, R.G., eds., Quaternary Plant-Ecology: 14th Symposium of the British Ecological Society: Blackwell Scientific Publications, Oxford, p. 238-261.

Rybníček, K. and Rybníčková, E., 1968, The history of flora and vegetation on the Blato Mire in southeastern Bohemia, Czechoslovakia: Folia Geobotanica Phytotaxonomica Praha, v. 3, p. 117-142.

Scagel, R.F., Bandoni, R.J., Maze, J.R., Rouse, G.E., Schofield, W.B. and Stein, J.R., 1982, Nonvascular Plants: an evolutionary survey: Wadsworth, Belmont, California, 570 p.

Schofield, W.B., 1985, Introduction to bryology: MacMillan Publishing Co., New York, 431 p.

Schuster, R.M., 1966, The Hepaticae and Anthocerotae of North America, east of the hundredth Meridian: Columbia University Press, New York, v. 1.

Schuster, R.M. and Janssens, J.A., 1989, On *Diettertia*, an isolated Mesozoic member of the Jungermanniales: Review of Palaeobotany and Palynology, v. 57, p. 277-287.

Schweger, C.E., 1989, The Old Crow and Bluefish Basins, northern Yukon: development of the Quaternary history, *in* Carter, L.D., Hamilton, T.D. and Galloway, J.P., eds., Late Cenozoic History of the Interior Basins of Alaska and the Yukon: United States Geological Survey, Circular 102B, p. 30-33.

Schweger, C.E. and Janssens, J.A., 1980, Paleoecology of the Boutellier Non-glacial Interval, St. Elias Mountains, Yukon Territory, Canada: Arctic and Alpine Research, v. 12, p. 309-317.

Shaw, J., 1987, Evolution of heavy metal tolerance in bryophytes II. An ecological and experimental investigation of the "copper moss", *Scopelophila cataractae* (Pottiaceae): American Journal of Botany, v. 74, p. 813-821.

Shaw, J., Antonovics, J. and Anderson, L.E., 1987, Inter- and intraspecific variation of mosses in tolerance to copper and zinc: Evolution, v. 41, p. 1312-1325.

Steere, W.C., 1942, Pleistocene mosses from the Aftonian Interglacial deposits of Iowa: Michigan Academy of Science, Arts and Letters, Papers, v. 27, p. 75-104.

Stewart, T.G, 1981, The Holocene paleoenvironment of Clements Markham Inlet, Northern Ellesmere Island, N.W.T., Canada, M.Sc. dissertation, University of Alberta, 135 p.

Stewart, T.G. and England, J., 1983, Holocene sea-ice variations and paleoenvironmental change, northernmost Ellesmere Island, N.W.T., Canada: Arctic and Alpine Research, v. 15, p. 1-17.

Sveinbjörnsson, B. and Oechel, W.C., 1983, The effect of temperature preconditioning on the temperature sensitivity of net CO_2 flux in geographically diverse populations of the moss *Polytrichum commune*: Ecology, v. 64, p. 1100-1108.

Tamm, C.O., 1953, Growth, yield and nutrition in carpets of the forest moss *(Hylocomium splendens):* Meddelanden Fran Statens Skogsforskningsinstitut, Band 43 (1), 140 p.

Tolonen, K. and Tolonen, M., 1984, Late-glacial vegetational succession at four coastal sites in northeastern New England: Ecological and phytogeographical aspects: Annales Botanici Fennici, v. 21, p. 59-78.

Verry, E.S., 1984, Microtopography and water table fluctuation in a *Sphagnum* mire: Proceedings of the 7th International Peat Congress, Dublin, Ireland, 18-23 June 1985, Volume II.

Warner, B.G. and Barnett, P.J., 1986, Transport, sorting, and reworking of late Wisconsinan plant macrofossils from Lake Erie, Canada: Boreas, v. 15, p. 323-329.

Wasylikowa, K., 1986, Analysis of fossil fruits and seed, *in* Berglund, B.E., ed., Handbook of Holocene Palaeoecology and Palaeohydrology: John Wiley and Sons, New York, p. 571-590.

Watts, W.A. and Winter, T.C., 1966, Plant macrofossils from Kirchner Marsh, Minnesota — a paleoecological study: Geological Society of America, Bulletin, v. 77, p. 1339-1360.

Webb, J.A. and Moore, P.D., 1982, The late Devensian vegetational history of the Whitlaw Mosses, southeast Scotland: New Phytologist, v. 91, p. 341-398.

Westgate, J.A., 1988, Isothermal plateau fission track age of the late Pleistocene Old Crow tephra, Alaska: Geophysical Research Letters, v. 15, p. 376-379.

Wolfe, J.A., 1978, A paleobotanical interpretation of Tertiary climate in the Northern Hemisphere: American Scientist, v. 66, p. 694-703.

Wright, H.E., Jr., Mann, D.H. and Glaser, P.H., 1984, Piston corers for peat and lake sediment: Journal of Ecology, v. 65, p. 657-659.

Wyatt, R. and Anderson, L.E., 1984, Breeding systems in bryophytes, *in* Dyer A.F. and Duckett, J.G., eds, The Experimental Biology of Bryophytes: Experimental Botany, v. 19, p. 39-64.

Zoltai, S.C., Taylor, S., Jeglum, J.K., Mills, G.F. and Johnson, J.D., 1988, Wetlands of Boreal Canada, *in* Rubec, C.D.A., ed., Wetlands of Canada: Polyscience Publications, Montreal, Quebec, p. 97-154.

Originally published in
Geoscience Canada Volume 17 Number 1
(March 1990)

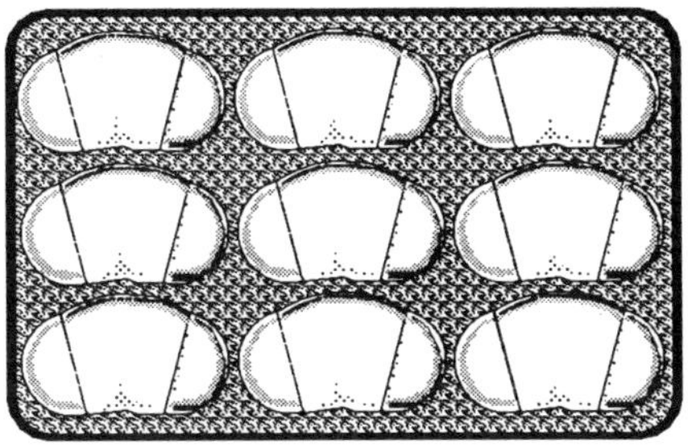

Palynology

Glen M. MacDonald
Department of Geography
McMaster University
Hamilton, Ontario L8S 4K1

SUMMARY

Palynology is the study of organic-walled microfossils which include pollen, spores, dinoflagellates, acritarchs, tasmantids, and chitinozoa. Quaternary palynology is concerned predominantly with the paleoecological analysis of fossil pollen and spores. Palynology is one of the most widely used technique in Quaternary paleoecology. Fossil pollen and spores (palynomorphs) can be found in a wide variety of deposits. Recovery of the fossils from sediment samples is relatively easy. Transmitted light microscopy is sufficient for routine identification and counting of palynomorphs. Their occurrence in large quantities makes fossil pollen and spores amenable to numerical and statistical analysis. Interpretation of fossil samples in North America is aided by a large data-base of modern pollen and spore samples taken from various depositional environments and different vegetation types. Despite the long history and wide application of fossil pollen analysis in Quaternary studies, significant methodological and interpretive problems remain to be resolved. In this paper, I outline the basic principles and methods of Quaternary palynological analysis and review some representative examples of contemporary research in Canada.

INTRODUCTION

To the layman, the study of fossil pollen and spores may appear esoteric at best and a singular waste of time and money at worst! However, palynology is one of the most widely used research tools in Quaternary studies (Edwards, 1983). A quick tabulation indicates that fossil pollen analysis was incorporated in approximately 25% of all research papers published in English and French language Quaternary journals in 1986. Indeed, palynology has a long history as an integral part of Quaternary research. Although descriptions of fossil pollen assemblages were made in the last century, the Swedish geologist Lennart von Post is recognized as the father of quantitative pollen analysis. In a paper first published in 1916 (von Post, 1967), he put forward the concept of presenting pollen data as percentages of the sum of pollen grains in a sediment sample plotted against stratigraphic position. This method remains the standard technique used by palynologists. Furthermore, von Post demonstrated the usefulness of pollen analysis for environmental reconstruction and stratigraphic correlation by showing that pollen diagrams from the same region were similar to one another but were different from pollen diagrams of other areas.

From the time of von Post to the present, Quaternary pollen analysis has been predicated upon the following general principles (Birks and Birks, 1980): (1) Pollen and spores are produced in large numbers during the natural reproductive cycle of many plants; (2) The proportion of pollen and spores released in the environment depends on the number of parent plants and therefore reflects vegetation composition; (3) Most of the grains never fulfill their reproductive function and some of these are deposited in environments where they may be preserved as fossils; (4) The grains can be extracted from sediment and identified to taxonomic levels ranging from family to species; (5) Pollen and spores from different statigraphic levels provide information on vegetation at specific periods in the past; (6) Palynological data from different sites provide information on differences in vegetation at the sites.

Several important methodological advances have been made in Quaternary palynology since von Post's time. Szafer (1935) introduced the widely used technique of presenting palynological data from many sites in the form of isopollen maps. Radiocarbon dating was developed in the late 1940s and is now routinely used to provide direct chronological control of pollen stratigraphies. Mathematical transfer functions were developed to provide quantitative estimates of plant species abundances (Davis, 1963) and climatic conditions (Webb and Bryson, 1972). Methods for the calculation of pollen concentrations and accumulations were introduced (Davis and Deevey, 1964; Maher, 1972) and are now widely applied.

HISTORICAL DEVELOPMENT IN CANADA

Auer (1927a,b, 1930) provided the first Quaternary pollen diagrams and systematic analysis of fossil pollen records in Canada through the examination of peat deposits, in Ontario, Quebec, New Brunswick and Nova Scotia. A pollen percentage diagram from peat sections along the north shore of the Saint Lawrence was published by Bowman (1931). The first pollen diagrams from Labrador were constructed from peat deposits by Wenner (1947). The initial palynological research in much of British Columbia, the Yukon and Alberta was conducted by Hansen (1940, 1949, 1953). The first radiocarbon-dated pollen diagrams from Canada were reported from Quebec by Potzger and Courtemanche (1954). Quaternary palynological research in arctic Canada began with the work of Terasmae (1956) on Banks Island. Kupsch (1960) published the first pollen diagram from Saskatchewan while Ritchie (1964) published the first diagram from Manitoba. Early paleoecological work on the fossil pollen record of Quaternary marine sediment was conducted by Piper *et al.* (1978) on material from the Grand Banks.

The tempo of Quaternary palynological research in Canada has continued to increase into the 1980s. Much of the current research is conducted using lake sediment cores and almost all published diagrams include radiocarbon dates or other forms of absolute chronological control. Palynological laboratories have been established by Archaeology, Biology, Geography, and Geology departments in universities and museums. Several laboratories are operated by the federal and provincial governments. Despite the continuing research in Quaternary palynology, pollen records remain sparse from large sections of Canada (Hills and Sangster, 1980; Bryant and Holloway, 1985; Ritchie, 1987). Most Quaternary palynological records reported from Canada are Holocene in age. However, notable Pleistocene records have been recovered from sites in British Columbia (summarized by Clague and MacDonald, in press), the western interior and arctic (summarized by Ritchie, 1985), southern Ontario (Terasmae, 1960; Berti, 1975; Karrow and Warner, 1984), and eastern Canada (summarized by T.W. Anderson, 1985). Most Pleistocene assemblages are from earlier non-glacial intervals. Pollen records from the height of the late Wisconsinan glaciation are available from unglaciated portions of the Queen Charlotte Islands (Warner *et al.*, 1982), the Yukon (Rampton, 1971; Cwynar, 1982) and Baffin Island (Short and Andrews, 1988). Additional palynological data on conditions during the last glacial maximum have been obtained from marine sediments off the east and west coasts (summarized by Heusser, 1985; de Vernal and Hillaire-Marcel, 1987) and from the Devon Island Ice Cap (McAndrews, 1984).

PRODUCTION, DEPOSITION AND PRESERVATION

Pollen grains are the male gametophytes of seed plants. The spores examined by Quaternary palynologists are the immature gametophytic generation of mosses, ferns and fern allies. The pollen and spores are produced in tetrads in the anthers of angiosperms, the male cones of gymnosperms, or the sporangia of lower plants. The grains consist of a living cytoplasm which is surrounded by intine composed of cellulose, pectins, callose, proteins, polysaccharides and antigens, and a resistant exine composed of complex polymers known as sporopollenin (Heslop-Harrison, 1968). The polymers of sporopollenin have a basic formula $(C_{90}H_{142}O_{36})^n$ and the substance is largely inert. The size of pollen grains and spores ranges from less than 10μm to over 200μm.

In Canada, pollen and spores are generally released by plants in the spring and summer (Bassett *et al.*, 1978). Production of spores and pollen varies for different species. Anemophilous (wind pollinated) plants depend on wind to disperse pollen and produce large numbers of grains. A lodgepole pine *(Pinus contorta)* tree may release as many as 21 billion grains in one season (Critchfield, 1985). In contrast, cleistogamous (closed flowering) and entomophilous (insect pollinated) species may produce only a few hundred thousand pollen grains or less. The amount of pollen produced by an individual plant often varies significantly from year to year (S.T. Andersen, 1974).

The majority of grains, particularly for anemophilous species, never fulfill their reproductive function and are deposited on the ground or in water. The cytoplasm of the grains dies shortly after release. The intine and exine decay rapidly through oxidation and biological activity. However, when deposited in anoxic environments, or where the accumulation of enclosing sediment is rapid, the exine is extremely resistant to decay and palynomorphs may be preserved indefinitely. Pollen and spores have been recovered from marine and lacustrine sediments, peats, soil horizons, tufas, cave deposits, and glacial ice. The degree of resistance to decay varies from species to species and appears to be correlated with the amount of sporopollenin in the exine (Sangster and Dale, 1964; Havinga, 1967, 1984).

The pollen and spore catchment area (Figure 1) of a deposit depends on a number of factors (Jacobson and Bradshaw, 1981). Pollen may be transported and deposited either through aerial fallout, surface wash, or stream and river flow. Recent mathematical modelling and empirical studies of airborne pollen deposition (Bradshaw and Webb, 1985; Prentice, 1985) suggest that the source area for lakes increases with lake size and is inversely related to pollen grain mass. Rivers and streams may transport significant amounts of pollen long distances into lakes and the ocean (McAndrews and Power, 1973; Bonny, 1976; Heusser, 1978). The pollen deposited in peat includes both regional fallout and pollen deposited directly from plants growing on the bog surface (Janssen, 1984).

Pollen and spores have a specific gravity of 1.4 to 1.5 (Flenley, 1971) and in suspension behave like silt-size particles. Generally, the grains are deposited in low-energy environments within a matrix of fine sediments. However, due to differences in size, shape and density of grains, there appear to be differences in the hydrodynamic efficiency of pollen and spores from different species (Hopkins, 1950). Significant variation in the deposition and redeposition of pollen and spores in oceans, lakes, and bogs has been documented. Pine pollen becomes increasingly dominant seaward in marine deposits off the west coast of North America (Heusser, 1978). The increase in the dominance of *Pinus* may be a result of superior hydrodynamic efficiency (Heusser, 1978). Davis *et al.* (1971) and Davis and Brubaker (1973) showed that in thermally stratified lakes, heavy spheroidal grains such as oak *(Quercus)* were evenly distributed in the sediments. Pollen grains of pine and herbs were deposited in greater numbers at the lake edges. In addition, the pollen of plants which grew along the edges of the lakes was most abundant in sediments along the periphery. Davis and Brubaker (1973) suggested that relatively small differences in density and settling velocity determined whether the grains would sink through the thermo-

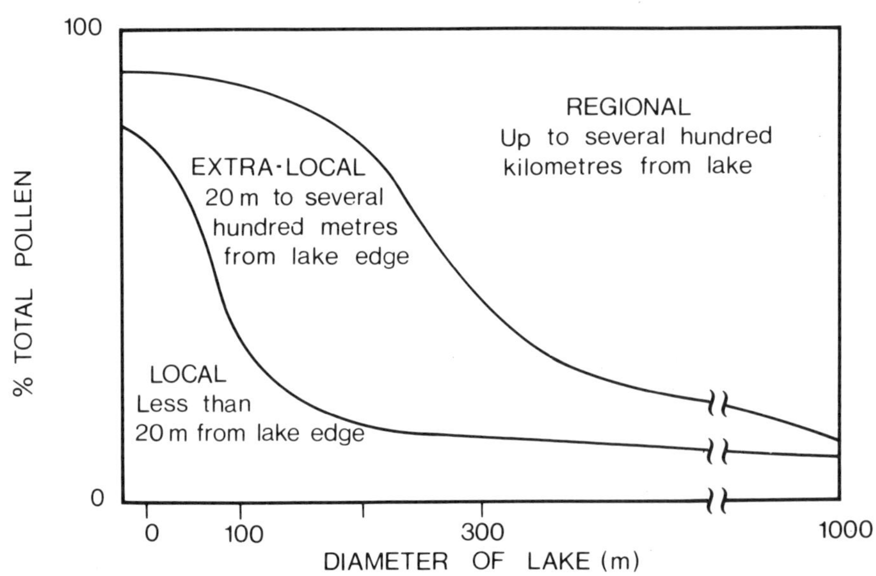

Figure 1 *Hypothetical relationship between pollen source area and lake size for lake basins that lack inflowing streams. (After Jacobson and Bradshaw, 1981).*

cline and be deposited or would remain suspended and moved to the periphery of the lakes. Lake bathymetry can exert control on the overall depositional pattern of pollen sized particles (Davis *et al.*, 1971; Lehman, 1975). Changes in lake bathymetry can lead to significant variation in the accumulation rates of pollen and spores at specific sites on the lake bottom. This effect is referred to as sediment focussing and can lead to difficulties in interpreting pollen accumulation data (Davis *et al.*, 1984). Following deposition, bioturbation can redistribute pollen vertically by as much as 15 cm in small lakes (Davis, 1974). Rowley and Rowley (1956) demonstrated that small spheroidal pollen grains could percolate downward in peat deposits while the movement of larger grains was more restricted.

FIELD AND LABORATORY TECHNIQUES

Many of the fondest reminiscences of Quaternary palynologists revolve around sampling expeditions. Although Quaternary palynomorphs have been recovered from a number of sediment types, the principal sources remain lake and bog deposits. Samples from sediment exposures may be taken at selected intervals during the logging of the sections. Sampling ice-caps and the deep sea sediments requires relatively specialized and sophisticated equipment. Platforms for sampling sediment from extant lakes have included watercraft, float planes, and winter ice. Samples of the surface sediments of lakes may be obtained using a variety of cup samplers and gravity corers (Aaby and Digerfeldt, 1986). The most widely used device for obtaining cores of lake sediments is the modified Livingstone piston sampler (Wright *et al.*, 1984). The piston action preserves the finest laminations and sedimentary structures. The Livingstone sampler is driven into the sediment and recovered using a series of metal core-rods, making it difficult to use in deep lakes ($>$50m). Corers driven by compressed air (Mackereth, 1958) or cables and weights (Aaby and Digerfeldt, 1986) have been devised for use in deep water. Peats are often sampled using chamber corers which are driven into the sediment and then rotated to fill a sample chamber. The two most common peat corers are the Hiller and modified versions of Russian samplers (Jowsey, 1966). The Russian sampler has the advantage of preserving stratigraphy. In addition, a new version of the Livingstone corer has been developed to sample peats (Wright *et al.*, 1984).

The concentrations of pollen and spores in sediment samples may range from $<$1 grain per mL in loess, cave and marine sediments to $>$100,000 grains per mL in lake sediments from the temperate forest region. However, the grains must be concentrated from the sediment before analysis. The resistant nature of the exine allows palynologists to use chemical reagents to remove most other mineral and organic matter from the samples. A standard processing sequence is as follows:

(1) Sub-sampling of a standard volume of sediment. Often 1 mL of material is used.
(2) Addition of a known quantity of exotic palynomorphs (Stockmarr, 1971). The exotic palynomorphs are used to calculate the volumetric concentrations of pollen and spores in the sediment samples.
(3) Removal of carbonates using 10% HCl.
(4) Removal of humic acids using 10% KOH in a hot water bath.
(5) Removal of pyrite using concentrated HNO_3.
(6) Removal of silica using 50% HF in a hot water bath.
(7) Removal of cellulose using an "acetolysis" treatment (9 parts $(CH_3CO)_2O$ to 1 part H_2SO_4 preceeded and followed by washes in CH_3COOH).

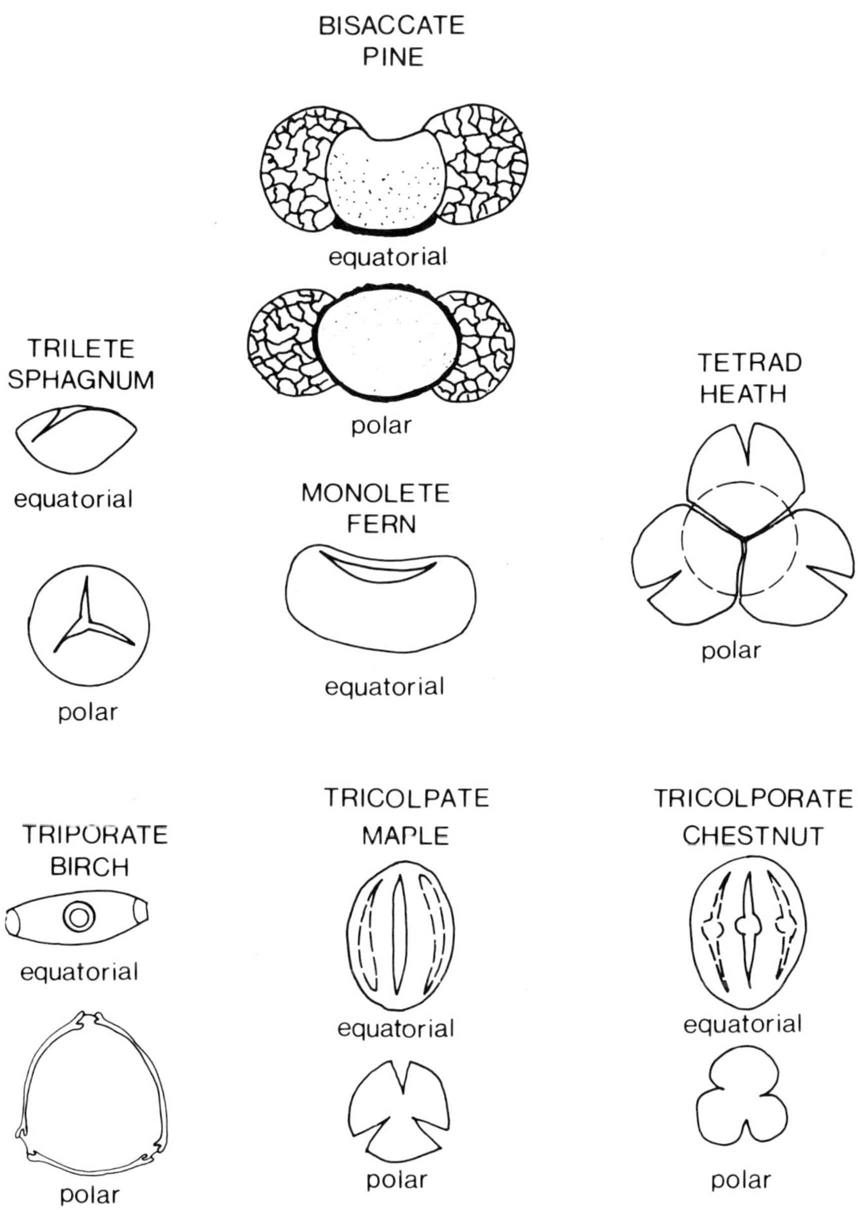

Figure 2 *Selected examples of shapes and aperture arrangements of pollen and spores.*

(8) Dehydration using $(CH_3)_3COH$.
(9) Staining with safranin or basic fuchsin.
(10) Storage of prepared material and mounting on microscopic slides with 2000 cs silicone oil or glycerine (refractive index 1.55).

Sieving may be used to remove large particles of peat or coarse clastic sediments. Micro-sieving techniques have been developed to remove small ($<7\mu m$) mineral and organic particles (Cwynar *et al.*, 1979). Flotation methods may be used also to remove heavy minerals. The exact sequence of chemical treatment varies for different labs and different samples. Full details are available from Faegri and Iversen (1975) and Berglund and Ralska-Jasiewiczowa (1986).

Taxonomic keys for the identification of modern pollen and spores from Canada are available from Kapp (1969), McAndrews *et al.* (1973), and Bassett *et al.* (1978). Most palynological identification and counting is done using transmitted light microscopy. Typically, a magnification of 400× is used to identify and count samples, while 1000× magnification is used for critical determinations. Pollen and spores may be differentiated by a suite of morphological characteristics (Figures 2 to 4). The most important characteristics are shape, size, aperture arrangement, exine structure, and surface ornamentation.

Most palynomorphs are spheroidal, but may be significantly compressed or extended along the polar axis. Pollen from many gymnosperms is bisaccate in form, consisting of a central body (colpus) with two wing-like bladders (sacci). Some plants, including cattails *(Typha)* and heaths (Ericales), release pollen in which the tetrad form is retained.

Apertures are apparent on most grains. On spores, the aperture often takes the form of a monolete or trilete scar on the proximal face of the grain. Angiosperms possess apertures which are pores (pori), furrows (colpi), or a combination of both (colporate grains).

Three important structural subdivisions are recognized for the exine (Figure 3). These are the tectum, the columellae, and the foot layer. The three structures are often referred to collectively as the ektexine (Faegri and Iversen, 1975). The tectum may be complete, perforated, or absent. Surface ornamentation may be present on the tectum or it may be featureless. The columellae may form surface ornamentation on the foot layer.

Many pollen taxa cannot be satisfactorily differentiated to the specific or generic level using transmitted light microscopy. Electron scanning microscopy (Figure 4) may be useful for some determinations (Bassett *et al.*, 1978). This technique is not practical for the large counts (~300-1000 grains per sample) required for quantitative palynology. Numerical taxonomic techniques based upon precise measurements of one or more grain dimensions have proven useful in separating morphologically similar species such as *Picea glauca* and *P. mariana* (white and black spruce) (Birks and Peglar, 1980). However, natural variation commonly occurs in the size of pollen grains from the same species (Ives, 1977; Birks, 1978). In addition, the size of palynomorphs can be affected by chemical preparation techniques and slide mounting mediums (Andersen, 1960, 1978).

Palynological data from sediment sequences are most often presented as percentages, volumetric concentrations, or accumulation rate diagrams (Figure 5). The concentrations of palynomorphs are usually calculated using the exotic pollen method (Stockmarr, 1971). Volumetric concentrations may be converted to pollen accumulation rates when a reliable estimate of sediment deposition rates is available (Maher, 1972). The advantage of pollen accumulation rates is that the record of a plant taxon is not influenced by variations in the abundance of other taxa, but accumulation rates are prone to variation and anomalies caused by variations in sedimentation (Davis *et al.*, 1984). Independent chronological control for pollen records is often provided by

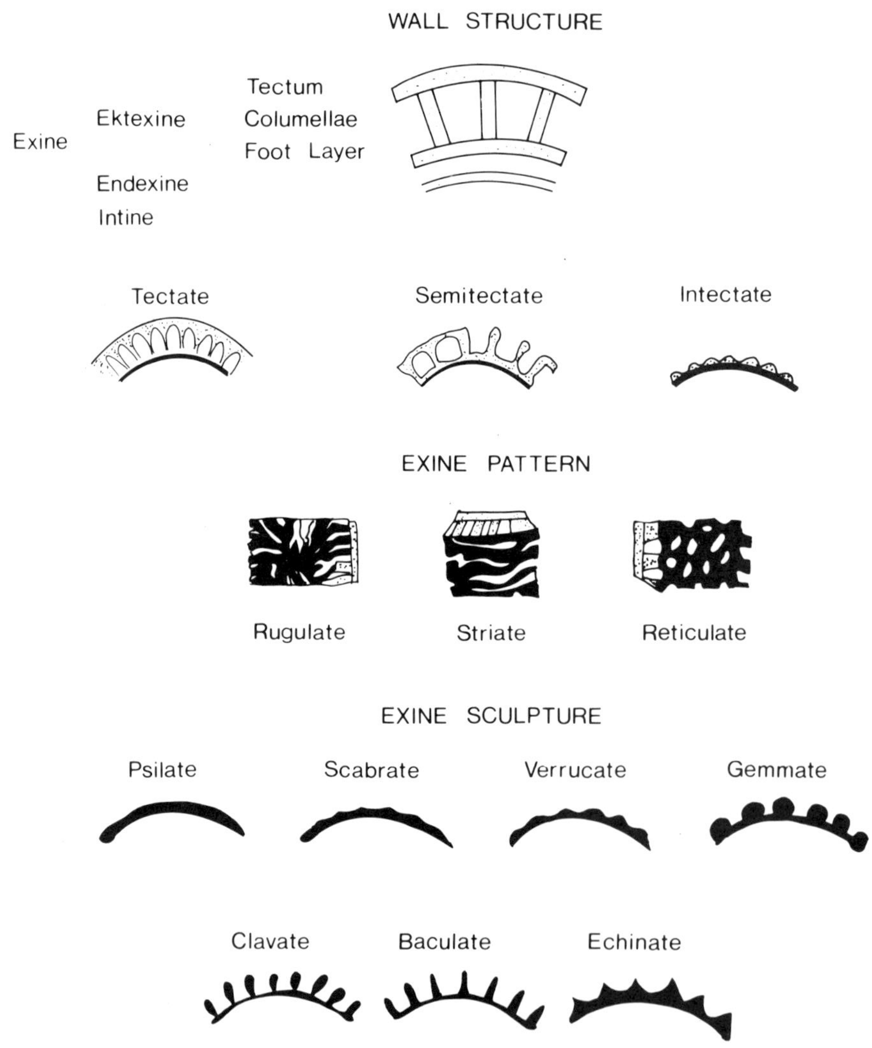

Figure 3 *Pollen wall structure and sculptural elements.*

radiocarbon dating. Chronological control may also be provided by short-lived isotopes such as ^{210}Pb (Oldfield and Appleby, 1984), paleomagnetism (Thompson, 1986), tephrochronology (Einarsson, 1986), and varve counts of annually laminated sediments (Saarnisto, 1986).

Pollen diagrams are often divided into zones that contain stratigraphically conterminous samples with pollen assemblages that are internally similar and dissimilar from the assemblages of adjacent zones (Figure 5). The zones are generally based upon percentage data and correspond most closely to the assemblage biozone defined by the International Stratigraphic Code (Hedberg, 1976). The local assemblage zones described from individual sections may be combined to form composite regional zones. A variety of numerical techniques are available for the zonation and comparison of pollen diagrams (Birks and Gordon, 1985). Some workers question the usefulness of zones and suggest that they are often artificial constructs which serve to obscure the importance of transition periods in pollen records (Walker, 1982). Others suggest that zones are useful for ease of discussion and formal stratigraphic description (Birks and Gordon, 1985).

Non-stratigraphic pollen data are often presented in tabular form. The spatial distribution of pollen taxa from surface samples or from temporally synchronous levels in chronologically correlated cores may also be presented as isopollen maps (Figure 6) and their variants (MacDonald and Waters, 1987).

SELECTED APPLICATIONS OF QUATERNARY PALYNOLOGY

Palynology is an invaluable research tool for a wide variety of Quaternary studies. Chronostratigraphic correlation, paleoecology, paleoclimatology and archaeology are arguably the four sub-disciplines of Quaternary research where palynology is of the greatest importance.

Chronostratigraphic Correlation. Pollen and spores from Quaternary deposits have long been used as chronostratigraphic tools. The use of fossil pollen for chronostratigraphy requires a regional master stratigraphy. If absolute dates are to be assigned on the basis of pollen assemblages, then the regional stratigraphy must be reliably dated by independent means. In southwestern Ontario, deposits that contain pollen assemblages dominated by *Picea* can be assigned an age of approximately 13,000 to 10,000 years B.P. (^{14}C years **B**efore **P**resent) on the basis of the regional pollen stratigraphy (McAndrews, 1981). Chronostratigraphic applications of pollen spectra are based upon the assumption that vegetation changes are regionally synchronous. Figure 7 illustrates the asynchronous nature of the pine *(Pinus)* pollen record in western Canada (MacDonald and Cwynar, 1985). Pine pollen is an important component of pollen assemblages throughout the Holocene in central Alberta. Pine does not appear in pollen records of northwestern Alberta until 5500 years B.P. In central Yukon, pine does not appear until ~500 years B.P. Clearly, the asynchronous nature of the pine record precludes the use of pine pollen as a reliable indicator fossil for broad-scale chronostratigraphic correlation in the western interior of Canada. Fossil pollen mapping for eastern North America indicates that most taxa have asynchronous records (Davis, 1981b). However, the distances at which the asynchronous nature of the record becomes significant may vary from a few hundred to several thousand kilometres. As reliable absolute dating techniques have developed, less emphasis has been placed on the chronostratigraphic use of the Holocene pollen record. Indeed, much research has centered on documenting and attempting to explain the asynchronous nature of vegetation change reflected in the pollen stratigraphy. Pollen-based chronostratigraphic correlations remain important for Pleistocene deposits that are often impossible to date using the radiocarbon technique.

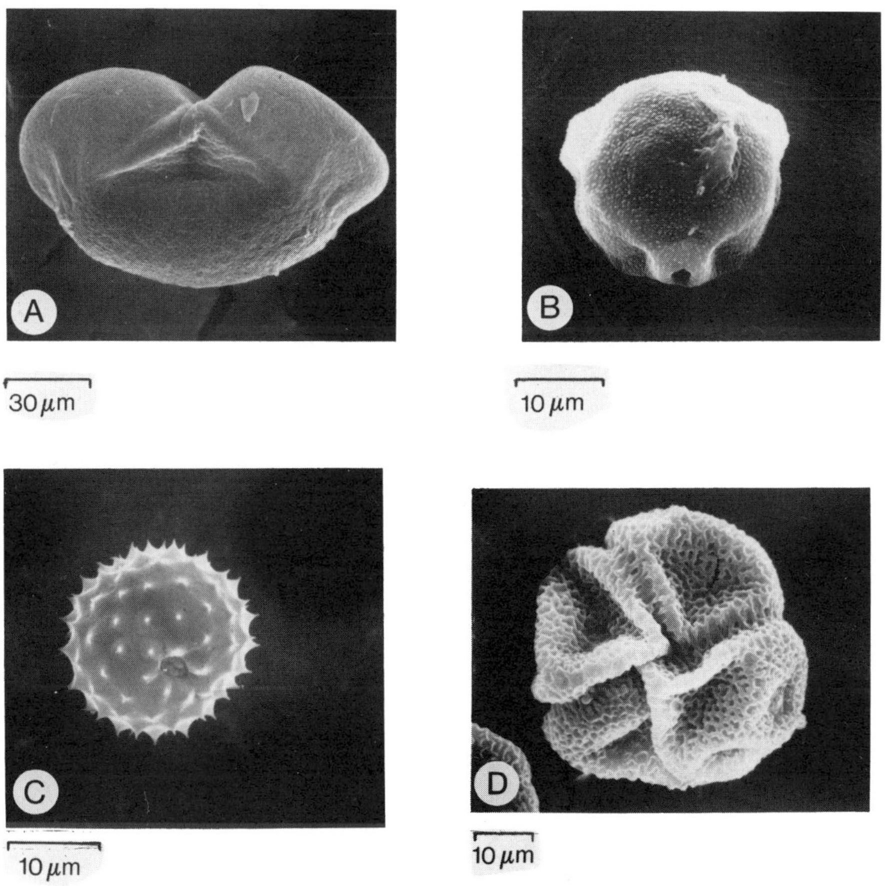

Figure 4 *Scanning electron micrographs of selected pollen types:*
(a) *White spruce* (Picea glauca), *a bisaccate grain.*
(b) *Dwarf birch* (Betula glandulosa), *a triporate grain with microscabrate sculpturing.*
(c) *Giant ragweed* (Ambrosia trifida), *a tricolporate grain with echinate sculpturing.*
(d) *Cattail* (Typha latifolia), *a tetrad with reticulate sculpturing.*
Scanning electron micrographs reproduced from Bassett et al. *(1978) by permission of the Biosystematics Research Centre.*

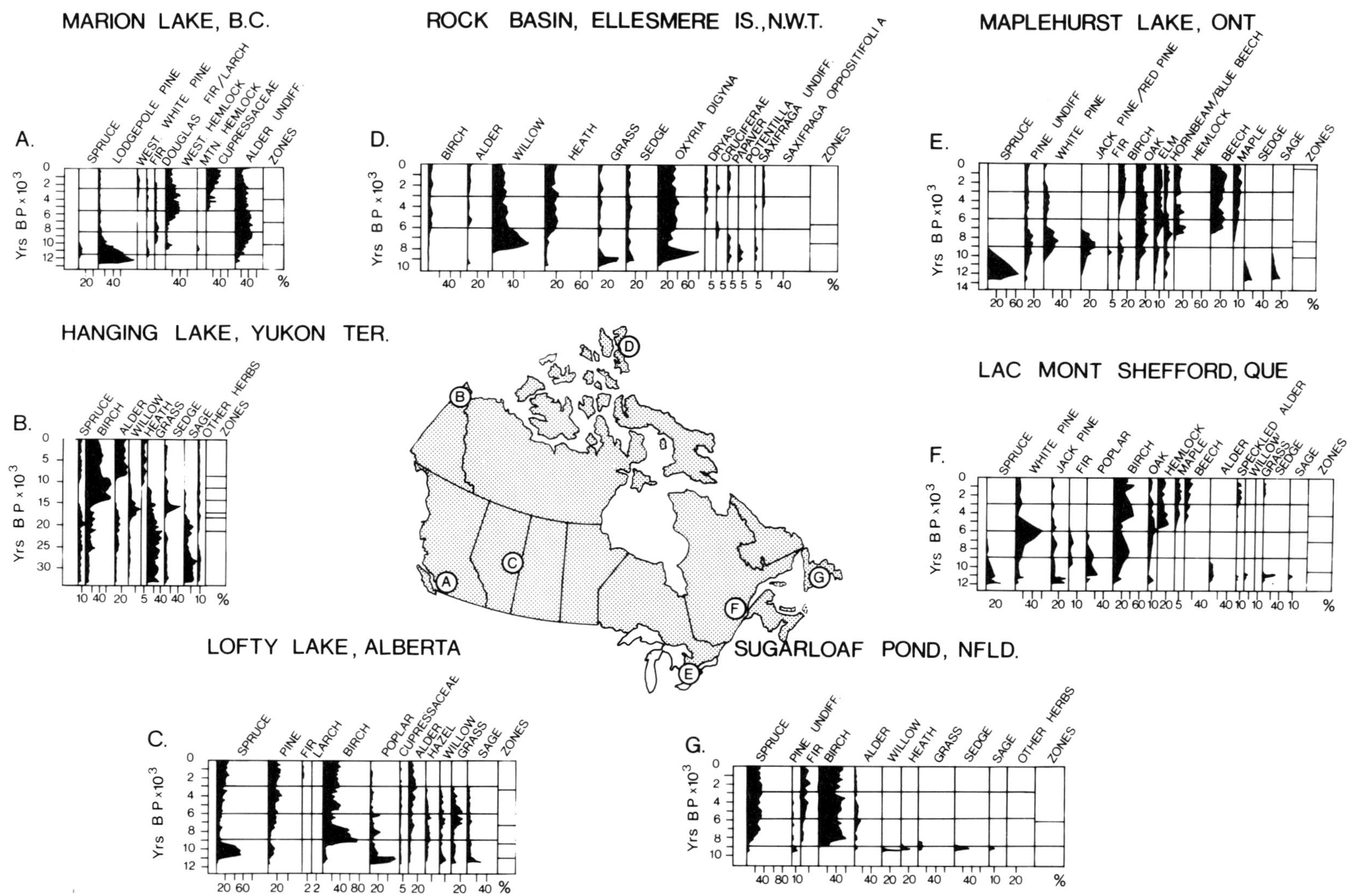

Figure 5 *Representative pollen percentage diagrams from Canada. Sources are:* **(a)** *Mathewes (1973),* **(b)** *Cwynar (1982),* **(c)** *Lichti-Federovich (1970),* **(d)** *Hyvarinen (1985),* **(e)** *Mott and Farley-Gill (1978),* **(f)** *Richard (1978), and* **(g)** *Macpherson (1982).*

Paleoecology.
Fossil pollen and spores provide long-term records of vegetation that are unavailable from any other source. Perhaps the most powerful technique for reconstructing vegetation history from pollen records is the comparative approach (Birks and Birks, 1980). Fossil assemblages are compared with modern samples from known vegetation types in order to aid in the interpretation of the fossils by discovering if past vegetation has a modern counterpart. In many cases, the comparison of the fossil and modern spectra is done using numerical ordination procedures (*e.g.*, Ritchie and Yarranton, 1978; Liu and Lam, 1985; MacDonald and Ritchie, 1986) or direct mathematical distance measures (Overpeck *et al.*, 1985). Examples of the numerical comparison of fossil and modern spectra are available from a number of sites in the western interior of Canada (Ritchie and Yarranton, 1978; Ritchie, 1977; MacDonald, 1983; MacDonald and Ritchie, 1986; MacDonald, 1987). These studies suggest that early Holocene vegetation of the western interior of Canada lacks a modern counterpart in the region. During the mid-Holocene, the grassland boundary and the northern limits of boreal forest were both located significantly north of their modern positions, presumably due to climatic change. Between approximately 5000 and 3500 years B.P. the modern vegetation was established. Deducing the factors that produced early Holocene vegetation which lacks modern regional counterparts remains an important and sometimes contentious field of research for Quaternary palynologists (Prentice, 1983).

Fossil pollen can be used to provide

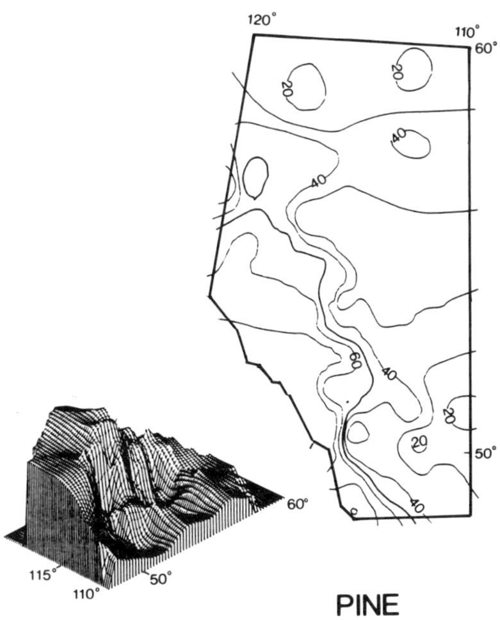

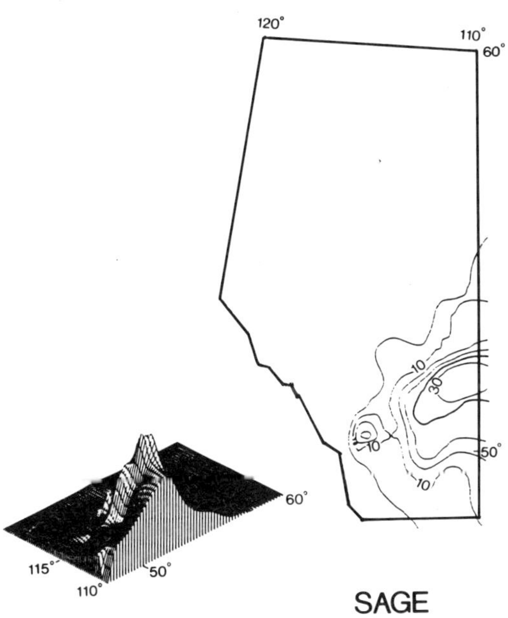

Figure 6 *Isopollen maps and simulated 3-D surfaces for the percentages of pine* (Pinus), *birch* (Betula) *and sage* (Artemisia) *pollen in modern lake sediments from Alberta. (After MacDonald and Waters, 1987). The maps indicate a correlation between vegetation zones and pollen deposition. High pine percentages typify the subalpine forest. Birch is at a maximum at boreal forest sites. Sage is important in the grassland.*

records of vegetation disturbance and response which span millenia. Particular interest has focussed on the use of fossil pollen and charcoal to examine the role of fire in forested ecosystems. Charcoal particles are retained during the chemical processing of sediments for fossil pollen and spores. The abundance of charcoal may be estimated using transmitted light microscopy (Tolonen, 1986; Patterson *et al.*, 1987). Cwynar (1978) examined the charcoal, pollen and geochemistry of a five hundred year long section of laminated sediment from a small lake in Algonquin Park, Ontario. He deduced that fires had occurred approximately every 80 years in the lake drainage basin. Greene (1982) used time series analysis to statistically examine Holocene fire record preserved in the fossil pollen and charcoal from a small lake in southwest Nova Scotia. He suggested that fire was a much more important agent of vegetation disturbance prior to 6000 years B.P. than it is in the modern forests of the region.

The epidemiology of plant pathogens is an intriguing application of palynology. The North American chestnut *(Castanea dentata)* was decimated by the accidental introduction of the fungal pathogen at the turn of this century. The decline of chestnut trees is clearly reflected in the declines in chestnut pollen in recent deposits (T.W. Anderson, 1974). Fossil pollen records from eastern North America provide evidence that hemlock trees *(Tsuga canadensis)* underwent a similar decline at approximately 4850

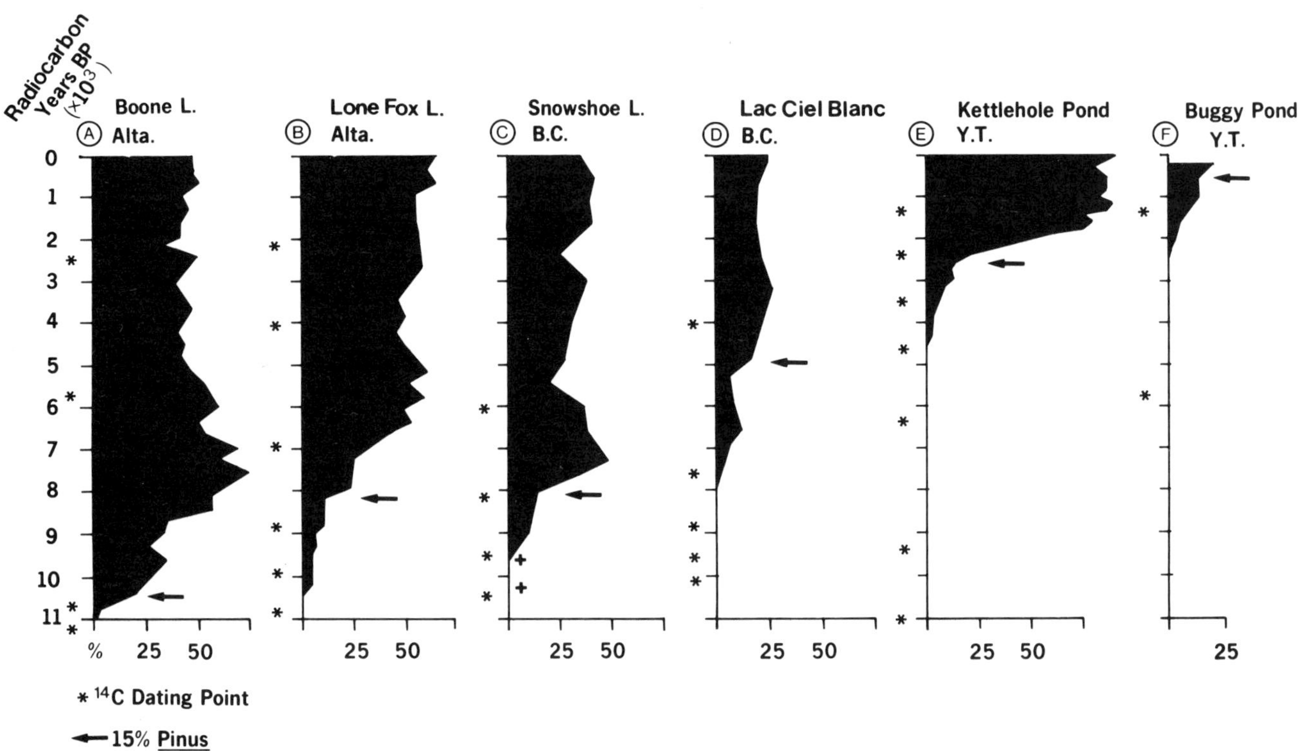

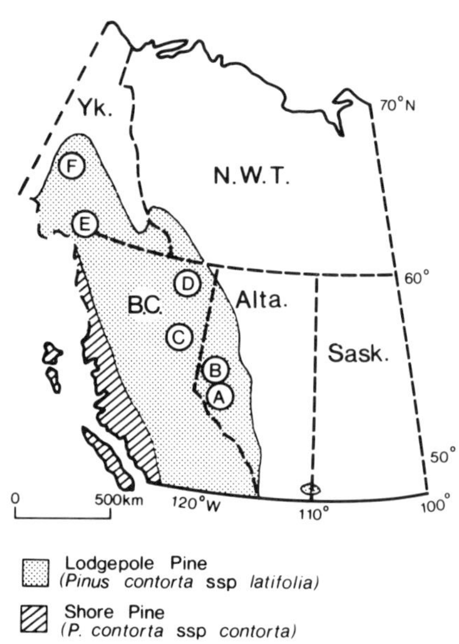

Figure 7 *Pine* (Pinus) *pollen percentages for sites from the western interior of Canada. (After MacDonald and Cwynar, 1985). The diagrams demonstrate asynchroneity in the stratigraphic record of pine. The asynchroneity reflects the slow northward spread of pine from a southern refugium following the close of the last glaciation.*

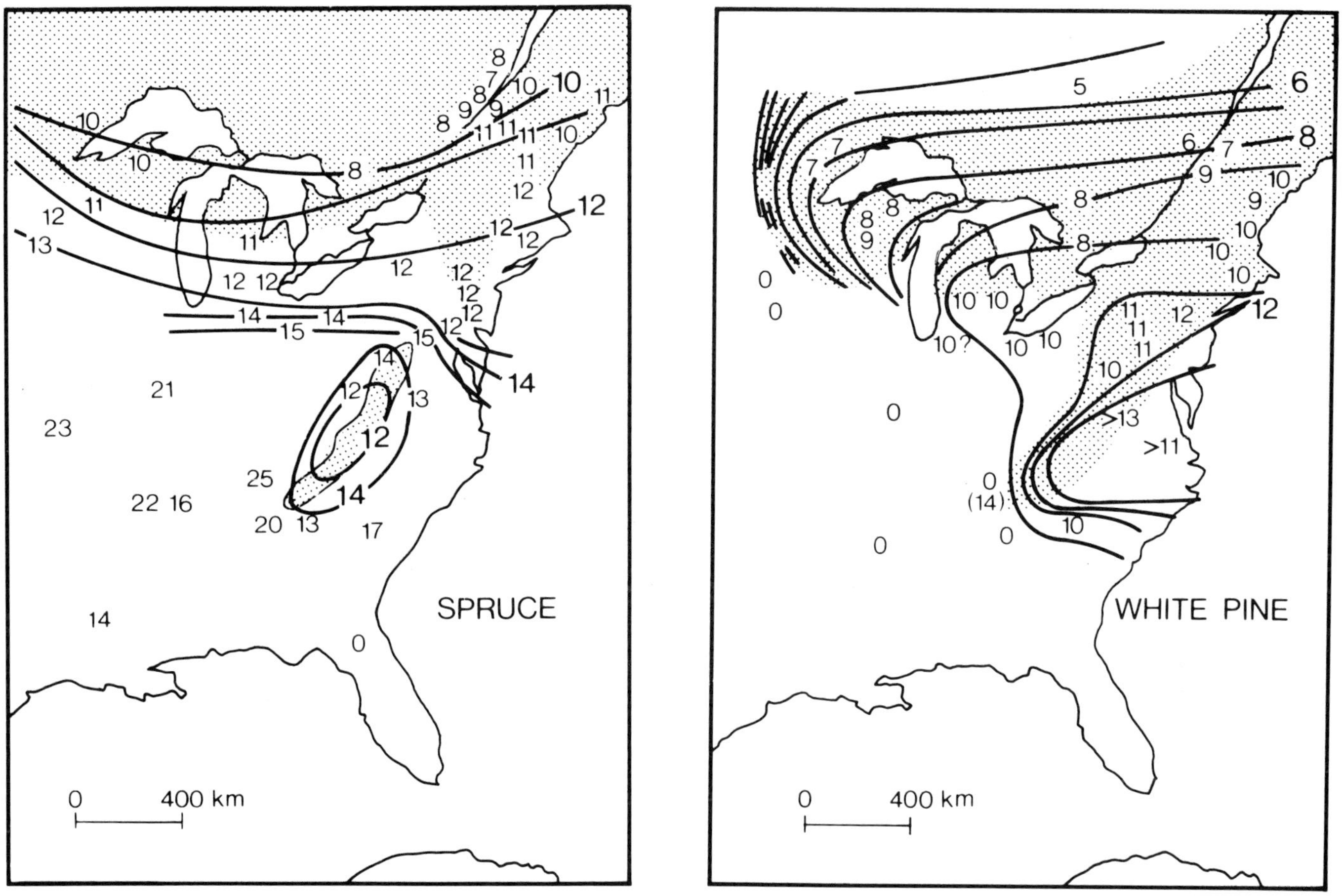

Figure 8 *Isochrone maps depicting the northward spread of spruce* (Picea spp.) *and white pine* (Pinus strobus) *during late Quaternary time. Each isochrone line depicts the range limits of the taxa at thousand year intervals. (After Davis, 1981b).*

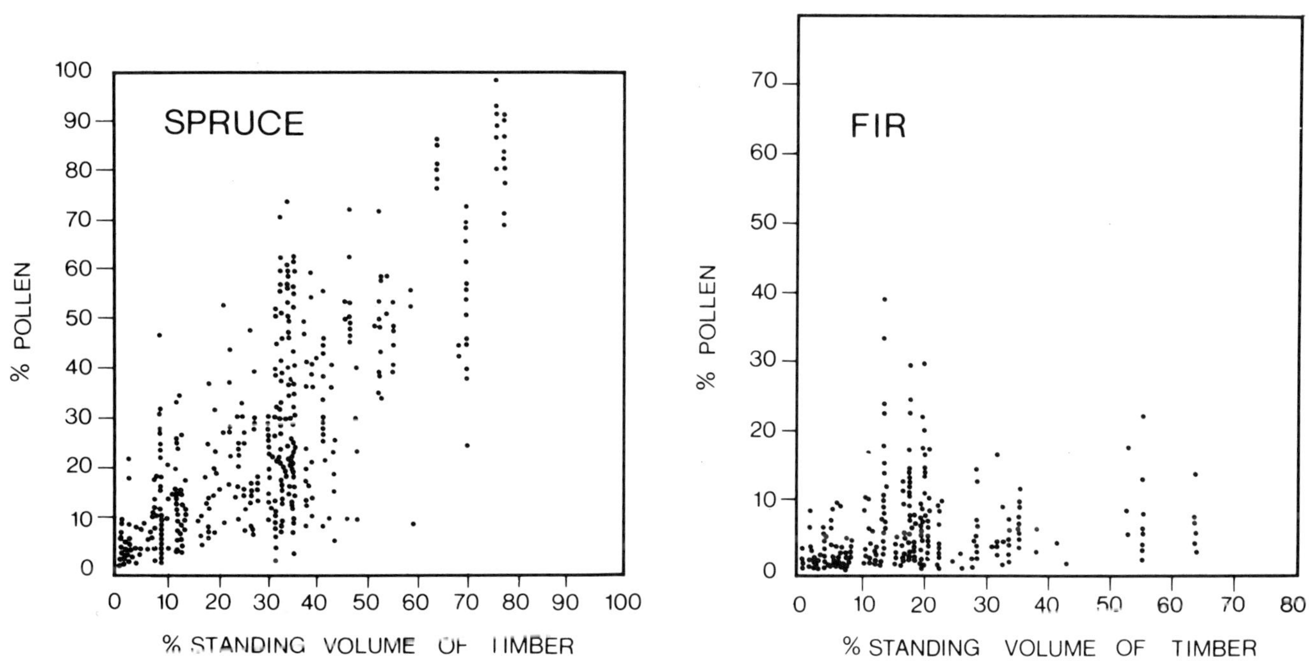

Figure 9 *The relationship between the percentages of spruce* (Picea) *and fir* (Abies) *pollen in modern sediments and the percentage of the spruce and fir trees in adjacent forests. (After Delcourt* et al., *1984). A good relationship is suggested for spruce. However, the relationship between fir pollen percentages and abundance in the vegetation is less clear.*

years B.P. (Davis, 1981a; Allison *et al.*, 1986). Davis (1981a) interpreted this decline as the impact of a virulent pathogen. The pollen record indicates that hemlock populations recovered approximately 1900 years after the decline, possibly due to the development of an immunity to the pathogen (Davis, 1981a; Allison *et al.*, 1986). It is impossible, however, to determine what pathogenic agent affected the hemlock populations.

Fossil pollen can also be used as a long-term record of individual plant species and genera. Maps depicting the northward movement of plant populations into Canada following the end of the last glacial period have been produced from pollen accumulation and percentage data (*e.g.*, Davis, 1981b; Webb *et al.*, 1983; Ritchie and MacDonald, 1986; Delcourt and Delcourt, 1987)(Figure 8). These maps indicate that many plant species found in association today had very different migrational histories during the late Pleistocene and Holocene. Study of the migrational patterns provides important insights into the long-term patterns of community development and the response of individual plant species to environmental change (Davis, 1981b).

Interest has focussed on the use of the fossil pollen record to test hypotheses developed by population ecologists and geneticists. Studies in Japan, the United States and England have used pollen accumulation data to estimate the doubling time of local tree populations as they increased following the end of the Pleistocene (Tsukada, 1982; Tsukada and Sugita, 1982; Bennett, 1983, 1988). The estimates of doubling times for tree species range from 31 to 462 years and appear to correspond well with the available observational data on tree population doubling times (Bennett, 1986). A recent study in western Canada used the fossil pollen record of lodgepole pine *(Pinus contorta* ssp. *latifolia)* to test hypotheses about the relations between plant population history and genetic variation (Cwynar and MacDonald, 1987). This study concluded that population residence time as determined from the pollen record was positively correlated with the genetic diversity of pine populations.

Table 1 R-values for important pollen taxa from the Inuvik area, N.W.T. (Ritchie, 1984). In this region, alder is greatly over-represented in the pollen record, the heaths are greatly under-represented, and juniper is equally represented.

Pollen Taxon	R-value
Spruce	0.5
Larch	0.2
Poplar	0.6
Birch	3.0
Alder	11.8
Willow	0.6
Juniper	1.0
Heath	0.1
Sage	5.0
Grass	3.4
Sedge	2.2

Both migration maps and tree species population estimates made from fossil pollen data are prone to significant uncertainties. It is often unclear from most pollen records if low amounts of pollen from a particular species indicate that the species is present near the site in low numbers, or if the species is present some distance from the site and the pollen is derived from long distance transport (Bennett, 1985). The sampling area represented by the pollen record may vary from several hectares to $>1000 km^2$ depending on the size of the lake and the pollen production and dispersal capabilities of the species under study. It is unknown whether the relation between plant population size and representation in the pollen record is linear. A non-linear relation between plant population and pollen accumulation rates could confound estimates of doubling time. Sediment focussing leads to increases and decreases in pollen accumulation rates which may invalidate the population doubling rates calculated from the pollen record.

Due to inter-specific variation in pollen production and preservation, the abundance of a plant species in the fossil record may not be directly proportional to the abundance of the species in the vegetation (Fagerlind, 1952). The relationship between modern pollen deposition and the abundances of contributing trees (Figure 9) has been examined to provide correction factors or "R-values" (Davis, 1963) which allow fossil pollen data to be used to estimate the past abundance of plants. R-values are calculated as:

$$\text{R-value for species } \mathbf{a} = \frac{\text{\% of species } \mathbf{a} \text{ in pollen spectra}}{\text{\% of species } \mathbf{a} \text{ in vegetation}}$$

Taxa that are over-represented in the pollen record have high values while under-represented taxa have low R-values (Table 1). Both linear and non-linear relations between pollen abundances and tree abundances have been used to approximate R-values (Webb *et al.*, 1981; Prentice and Webb, 1986). Past vegetation abundance can be estimated:

$$\text{\% of species } \mathbf{a} \text{ in past vegetation} = \frac{\text{\% of species } \mathbf{a} \text{ in pollen spectra}}{\text{R-value}}.$$

The use of R-values to calculate past plant species abundances is based upon the assumption that the R-value for the species has not changed through time. This assumption is uncertain when past vegetation lacks a modern counterpart.

Paleoclimatology.

Two approaches are used to estimate past climatic conditions from pollen data. The indicator species approach requires knowledge of the climatic limits of one or more plant species. The presence or absence of the species in the fossil record is used to infer past climatic conditions. In northwestern Canada, poplar and aspen *(Populus)*, cattail *(Typha latifolia)*, and bog myrtle *(Myrica gale)* do not occur north of the present treeline. Ritchie *et al.* (1983) found evidence in the pollen record that all three of these species were present at sites beyond the modern treeline during the period 11,000 to 6,000 years B.P. (Figure 10). This was interpreted as evidence for warmer summer conditions during the early Holocene in northwestern Canada.

The assemblage approach (Birks and Birks, 1980) assumes that climate serves to control the proportions of plant species in vegetation assemblages. Multivariate numerical analysis is used to determine the modern relationship between pollen proportions and climatic variables (Figure 11). A mathematical transfer function is derived from the modern pollen-climate relationships and used to estimate past climatic conditions from fossil pollen spectra:

$$\text{Climate Estimate} = \text{Transfer Function} \times \text{Pollen Proportions}.$$

Detailed discussion of transfer function methodology is presented by Howe and Webb (1983).

Mathewes and Heusser (1981) constructed mathematical transfer functions from modern pollen and climate data from the Pacific Northwest. They applied transfer functions to the fossil pollen record from Marion Lake, British Columbia (Figure 5) and produced estimates of mean July temperature and mean annual precipitation. The results suggest that the early Holocene climate was significantly warmer during July and drier than present (Figure 12).

A number of factors may give rise to unreliable estimates of past climate from transfer functions. The two most troubling assumptions of the transfer function approach are:

(1) past and present vegetation condi-

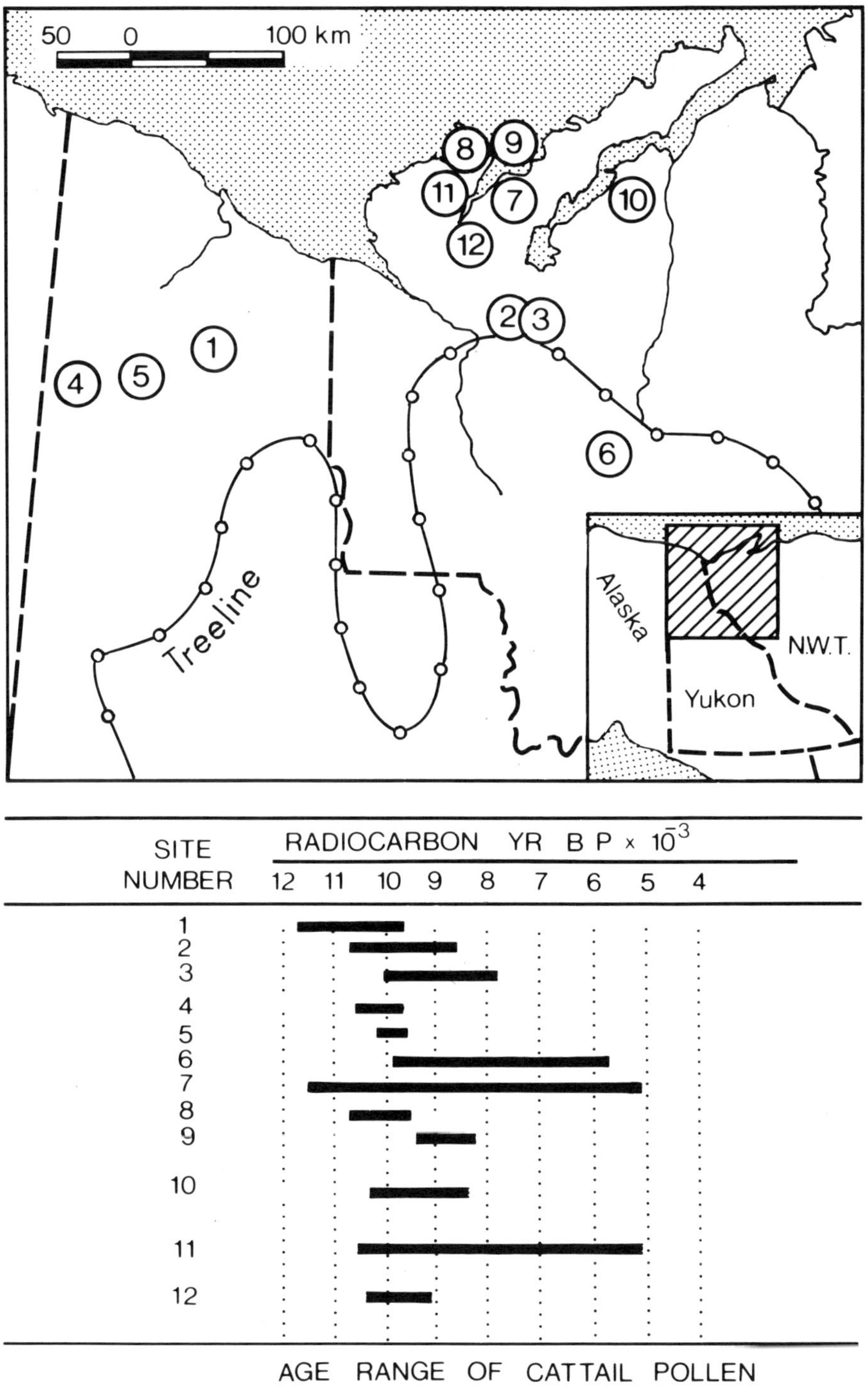

Figure 10 *Evidence from northwestern Canada for the northward extension of cattail* (Typha latifolia) *during the early to mid Holocene. (After Ritchie, 1984).*

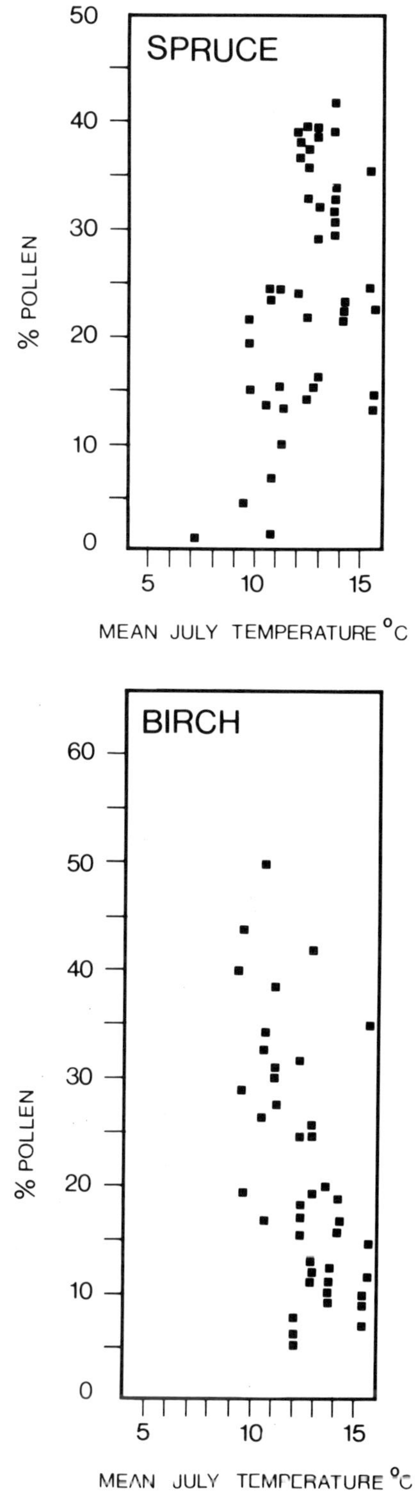

Figure 11 *The relationship between the percentages of spruce* (Picea) *and birch* (Betula) *pollen in modern lakes sediments and mean July temperature. (After Kay and Andrews, 1983).*

tions reflect states of equilibrium between climate and vegetation; and (2) the past vegetation-climate relationships have analogues in the present relationships between climate and vegetation. A number of factors including climatic conditions which were unique to the past, lags between climatic change and vegetation response, and the effect of progressive soil development on vegetation may produce spurious estimates of past climate using the transfer functions. However, the disequilibrium conditions caused by the above factors will only produce significant distortion of cli-

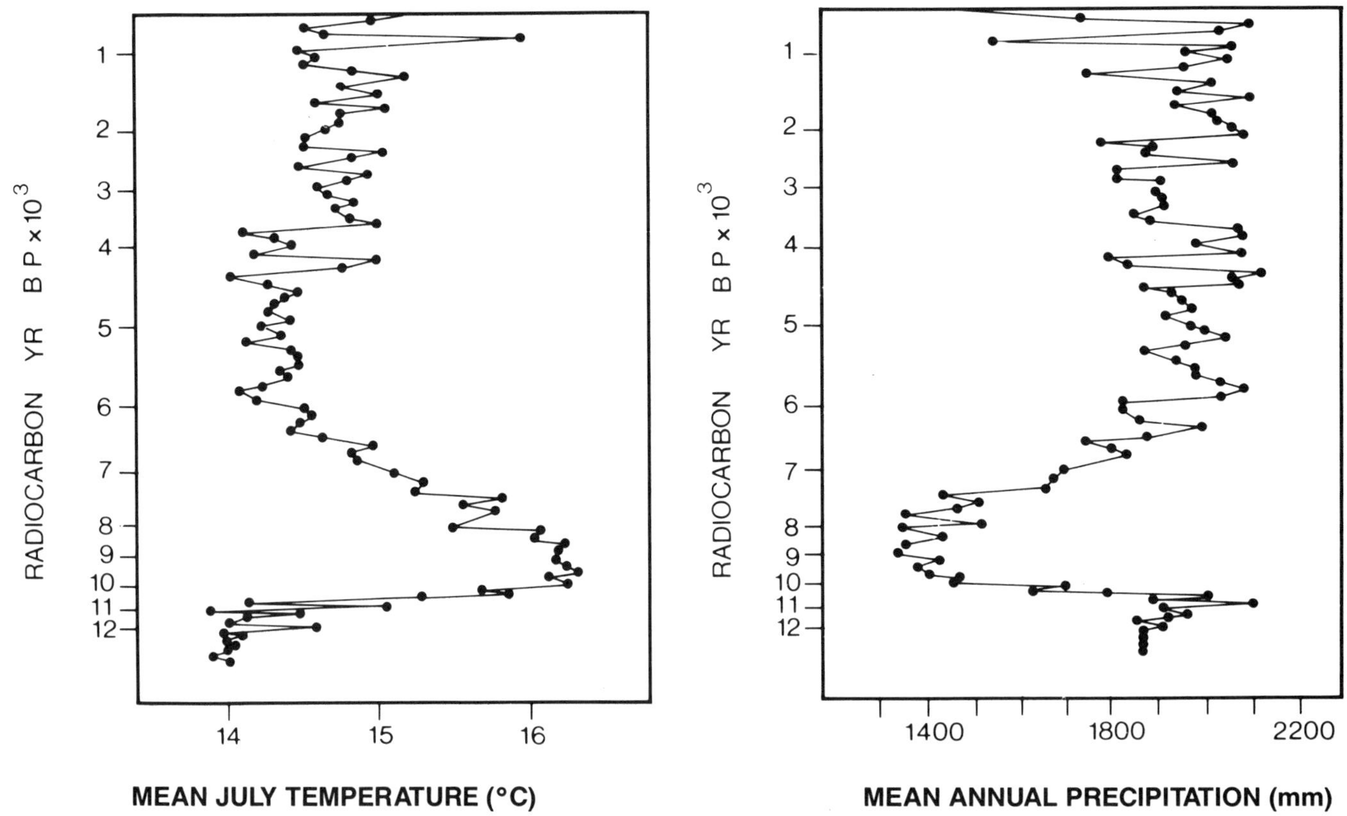

Figure 12 *A reconstruction of post-glacial temperature and precipitation based on pollen-climate transfer functions from the Pacific Northwest applied to the fossil pollen record from Marion Lake, B.C. (After Mathewes and Heusser, 1981).*

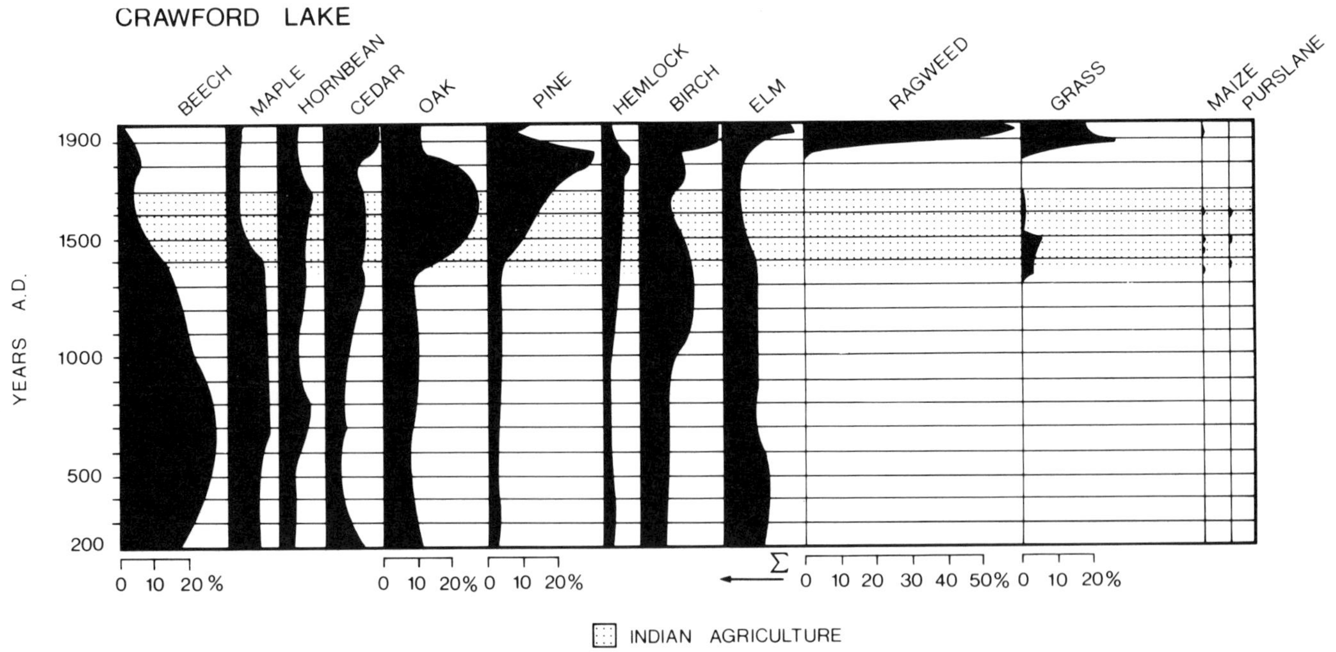

Figure 13 *The fossil pollen record from Crawford Lake, Ontario, highlighting the period of Indian agriculture. (After McAndrews, 1976).*

matic estimates if the lag time between vegetation change and climate are large compared with the rate of climatic change (Prentice, 1986; Webb, 1986). In addition, climatic change involves simultaneous variation in a number of variables, and the simplifying assumptions made during the estimation of mean summer temperatures or annual precipitation may be unreasonable given the complex nature of the climate and vegetation-climate relationships (Bryson, 1985). Comparison of climatic estimates from the pollen record with estimates from independant sources such as other fossil groups or paleoclimatic models provides a potential test of pollen-climate transfer functions.

Archaeology.

Pollen analysis has been widely used in archaeology to provide chronological control and paleoenvironmental information. In a classic study, Byrne and McAndrews (1975) recovered purslane *(Portulaca oleracea)* pollen from the annually laminated sediments of Crawford Lake. They were able to demonstrate that purslane was native to North America and was likely an important Indian pot-herb (Figure 13). Detailed pollen analysis indicated that relatively large amounts of maize *(Zea mais)* pollen were deposited in Crawford Lake between 1400 AD and 1700 AD (McAndrews, 1976). This suggested that an Indian settlement had been located near the lake. Archaeological investigations have uncovered an Iroquois village of approximately 500 people which was inhabited during the period of maize and purslane pollen deposition (Finlayson *et al.*, 1973).

Recently, Hebda and Mathewes (1984) compared the pollen record of cedar *(Thuja)* pollen and the age range of native artifacts used for massive woodworking at a number of sites in coastal British Columbia and Washington. They found that cedar migrated northward along the coast, spreading from central Washington to northern British Columbia between 9000 and 2500 years B.P. A clear relationship was revealed between the appearance of cedar and the first occurrence in the archaeological record of massive woodworking tools. Hebda and Mathewes suggested that the development of massive woodworking technology was environmentally constrained until large cedar trees became available.

FUTURE DIRECTIONS

A number of research directions will continue to entice Quaternary palynologists in Canada. Large areas of the country, particularly in the arctic, are unknown in terms of Quaternary pollen stratigraphy and vegetation history. Important data on both Pleistocene and Holocene palynology remains to be collected. The usefulness of fossil pollen for plant population studies will serve to provide exciting research avenues for palynologists and to attract a new audience of population ecologists and population geneticists. Palynology has the potential to provide empirical tests of plant population and genetic hypotheses and perhaps contribute new theory (Edwards, 1983; Birks, 1985). Quaternary palynology will continue to play a key role in paleoclimatology. In particular, the fossil pollen record provides one of the best sources of data to test the large-scale paleoclimatic models which are now being developed (Bartlein *et al.*, 1986). Archaeological studies will long benefit from the paleoecological insights which the fossil pollen record can contribute to the understanding of cultural traditions and development. It is safe to say that palynology will continue to play a fundamental role in Quaternary ecology.

ACKNOWLEDGEMENTS

Dr. C.W. Crompton and the Biosystematics Research Centre kindly provided the SEM photomicrographs used here. I thank R.J. Mott, B.G. Warner, and T. Webb III for many useful comments and suggestions on an earlier version of this manuscript. Ongoing palynological research at McMaster is supported by Natural Sciences and Engineering Research Council.

REFERENCES

Aaby, B. and Digerfeldt, G., 1986, Sampling techniques for lakes and bogs, *in* Berglund, B.E., ed., Handbook of Holocene Palaeoecology and Palaeohydrology: John Wiley and Sons, Chichester, p. 181-194.

Allison, T.D., Moeller, R.E. and Davis, M.B., 1986, Pollen in laminated sediments provides evidence for a mid-Holocene forest pathogen outbreak: Ecology, v. 67, p. 1101-1105.

Andersen, S.T., 1960, Silicone oil as a mounting medium for pollen grains: Danmarks Geologiske Undersogelse, Series IV, 4, 22 p.

Andersen, S.T., 1974, Wind conditions and pollen deposition in a mixed deciduous forest. II. seasonal and annual pollen deposition: Grana, v. 14, p. 64-77.

Andersen, S.T., 1978, On the size of *Corylus avellana* L. pollen mounted in silicone oil: Grana, v. 17, p. 5-13.

Anderson, T.W., 1974, The chestnut pollen decline as a time horizon in lake sediments in eastern North America: Canadian Journal of Earth Sciences, v. 11, p. 678-685.

Anderson, T.W.,1985, Late-Quaternary pollen records from eastern Ontario, Quebec, and Atlantic Canada, *in* Bryant, V.M. and Holloway, R.G., eds., Pollen Records of Late-Quaternary North American Sediments: American Association of Stratigraphic Palynologists, Dallas, p. 281- 326.

Auer, V., 1927a, Botany of interglacial peat beds of Moose River Basin: Geological Survey of Canada Summary Report, Part-C, p. 45-47.

Auer, V., 1927b, Stratigraphical and morphological investigations of peat bogs of southeastern Canada: Communications ex Instituto Quaestionum Forestalium Finlandiae Editae, v. 12, p. 1-62.

Auer, V., 1930, Peat bogs in southeastern Canada: Geological Survey of Canada, Memoir 162, p. 1-32.

Bartlein, P.J., Prentice, I.C. and Webb, T., III, 1986, Climatic response surfaces from pollen data for some eastern North American taxa: Journal of Biogeography, v. 13, p. 35-57.

Bassett, I.J., Crompton, C.W. and Parmelee, J.A., 1978, An atlas of airborne pollen grains and common fungus spores of Canada: Canada Department of Agriculture, Monograph 18, 321 p.

Bennett, K.D., 1983, Postglacial expansion of forest tree populations in Norfolk, UK: Nature, v. 303, p. 164-167.

Bennett, K.D., 1985, The spread of *Fagus grandifolia* across eastern North America the last 18,000 years: Journal of Biogeography, v. 12, p. 147-164.

Bennett, K.D., 1986, The rate of spread and population increase of forest trees during the postglacial: Royal Society of London, Philosophical Transactions, v. 314B, p. 523-531.

Bennett, K.D., 1988, Holocene geographic spread and population expansion of *Fagus grandifolia* in Ontario, Canada: Journal of Ecology, v. 76, p. 547-557.

Berglund, B.E. and Ralska-Jasiewiczowa, M., 1986, Pollen analysis and pollen diagrams, *in* Berglund, B.E., ed., Handbook of Holocene Palaeoecology and Palaeohydrology: John Wiley and Sons, Chichester, p. 455-484.

Berti, A.A., 1975, Paleobotany of Wisconsinan interstadials, eastern Great Lakes region, North America: Quaternary Research, v. 5, p. 591-519.

Birks, H.J.B., 1978, Geographic variation of *Picea abies* (L.) Karsten pollen in Europe: Grana, v. 17, p. 149-160.

Birks, H.J.B., 1985, Recent and possible future mathematical developments in quantitative palaeoecology: Palaeogeography, Palaeoclimatology, Palaeoecology, v. 50, p. 107-147.

Birks, H.J.B. and Birks, H.H., 1980, Quaternary Palaeoecology: Edward Arnold, London, 289 p.

Birks, H.J.B. and Gordon, A.D., 1985, Numerical Methods in Quaternary Pollen Analysis: Academic Press, London, 317 p.

Birks, H.J.B. and Peglar, S.M., 1980, Identification of *Picea* pollen of late Quaternary age in eastern North America: a numerical approach: Canadian Journal of Botany, v. 58, p. 2043-2058.

Bonny, A.P., 1976, Recruitment of pollen to the seston and sediment of some English Lake District lakes: Journal of Ecology, v. 64, p. 859-887.

Bowman, P.W., 1931, Study of a peat bog near the Matamek River, Quebec, Canada by the method of pollen analysis: Ecology, v. 12, p. 694-708.

Bradshaw, R.H.W. and Webb, T., III, 1985, Relationships between contemporary pollen and vegetation data from Wisconsin and Michigan, U.S.A.: Ecology, v. 66, p. 721-737.

Bryant, V.M. and Holloway, R.G., 1985, eds., Pollen Records of late-Quaternary North American Sediments: American Association of Stratigraphic Palynologists, Dallas, 423 p.

Bryson, R.A., 1985, On climatic analogs in paleoclimatic reconstruction, Quaternary Research, v. 23, p. 275-286.

Byrne, R. and McAndrews, J.H., 1975, Precolumbine purslane (*Portulaca oleracea* L.) in the New World: Nature, v. 253, p. 726-727.

Clague, J.J. and MacDonald, G.M., 1989, Paleoecology and paleoclimatology (Canadian Cordillera), *in* Fulton, R.J., ed., Quaternary Geology of Canada and Greenland: Geological Survey of Canada, Geology of Canada, no. 1, *in* Chapter 1, p. 70-74. [Also Geological Society of America, The Geology of North America, v. K-1]

Critchfield, W.B., 1985, The late Quaternary history of lodgepole and jack pines: Canadian Journal of Forestry Research, v. 15, p. 749-772.

Cwynar, L.C., 1978, Recent history of fire and vegetation from laminated sediment of Greenleaf Lake, Algonquin Park, Ontario: Canadian Journal of Botany, v. 56, p. 10-21.

Cwynar, L.C., 1982, A late-Quaternary vegetation from Hanging Lake, Northern Yukon: Ecological Monographs, v. 52, p. 1-24.

Cwynar, L.C., Burden, E. and McAndrews, J.C., 1979, An inexpensive sieving method for concentrating pollen and spores from fine-grained sediments: Canadian Journal of Earth Sciences, v. 16, p. 1115-1120.

Cwynar, L.C. and MacDonald, G.M., 1987, Geographical variation of lodgepole pine in relation to population history: American Naturalist, v. 129, p. 463-469.

Davis, M.B., 1963, On the theory of pollen analysis: American Journal of Science, v. 261, p. 897-912.

Davis, M.B., 1981a, Outbreaks of forest pathogens in forest history: Proceedings of the of the IV International Palynological Conference, Birbal Sahni Institute of Paleobotany, v. 3, p. 216-227.

Davis, M.B., 1981b, Quaternary History and the Stability of Forest Communities, *in* West, D.C., Shugart, H.H. and Botkin, D.B., eds., Forest Succession: Springer-Verlag, New York, p. 132-153.

Davis, M.B. and Brubaker, L.B., 1973, Differential sedimentation of pollen grains in lakes: Limnology and Oceanography, v. 18, p. 635-646.

Davis, M.B., Brubaker, L.B. and Beiswenger, J.M., 1971, Pollen grains in lake sediments: pollen percentages in surface sediments from southern Michigan: Quaternary Research, v. 1, p. 450-467.

Davis, M.B. and Deevey, E.S., 1964, Pollen accumulation rates: estimates from late-glacial sediment of Rogers Lake: Science, v. 145, p. 1293-1295.

Davis, M.B., Moeller, R.E. and Ford, J., 1984, Sediment focusing and pollen analysis, *in* Haworth, E.Y. and Lund, J.W.G., eds., Lake Sediments and Environmental History: Leicester University Press, Leicester, p. 261-293.

Davis, R.B., 1974, Stratigraphical effects of tubificids in profundal lake sediments: Limnology and Oceanography, v. 19, p. 466-488.

Delcourt, P.A. and Delcourt, H.R., 1987, Long-term Forest Dynamics of the Temperate Zone: Springer-Verlag, New York, 439 p.

Delcourt, P.A., Delcourt, H.R. and Webb, T., 1984, Atlas of Mapped Distributions of Dominance and Modern Pollen Percentages for Important Tree Taxa of Eastern North America: American Association of Stratigraphic Palynologists, Dallas, 131 p.

de Vernal, A. and Hillaire-Marcel, C., 1987, Éléments d'une climatostratigraphie du Pléistocène moyen et tardif dans l'est du Canada par l'analyse palynologique et isotopique du forage 84-030-003, mer du Labrador: Canadian Journal of Earth Sciences, v. 24, p. 1886-1912.

Einarsson, Th., 1986, Tephrochronology, *in* Berglund, B.E., ed., Handbook of Holocene Palaeoecology and Palaeohydrology: John Wiley and Sons, Chichester, p. 329-342.

Edwards, K.J., 1983. Quaternary palynology: consideration of a discipline: Progress in Physical Geography, v. 7, p. 113-125.

Faegri, K. and Iversen, J., 1975, Textbook of Pollen Analysis (Third Edition): Blackwell Scientific Publications, Oxford, 295 p.

Fagerlind, F., 1952, The real signification of pollen diagrams: Botaniska Notiser, v. 1952, p. 185-224.

Finlayson, W.D., Byrne, A.R. and McAndrews, J.H., 1973, Iroquoian settlement and subsistence patterns near Crawford Lake, Ontario: Canadian Archaeological Association Bulletin, v. 5, p. 134-136.

Flenley, J.R., 1971, Measurements of the specific gravity of the pollen exine: Pollen et Spores, v. 2, p. 179-186.

Greene, D.G., 1982, Fire and stability in the postglacial forests of southwest Nova Scotia: Journal of Biogeography, v. 9, p. 29-40.

Hansen, H.P., 1940, Paleoecology of two peat bogs in southwestern British Columbia: American Journal of Botany, v. 27, p. 144-149.

Hansen, H.P., 1949, Postglacial forests in south central Alberta, Canada: American Journal of Botany, v. 36, p. 54-65.

Hansen, H.P., 1953, Postglacial forests in the Yukon Territory and Alaska: American Journal of Science, v. 251, p. 505-542.

Havinga, A.J., 1967, Palynology and pollen preservation: Review of Palaeobotany and Palynology, v. 2, p. 81-98.

Havinga, A.J., 1984, A 20-year experimental investigation into the differential corrosion susceptibility of pollen and spores in various soil types: Pollen et Spores, v. 26, p. 541-558.

Hebda, R.J. and Mathewes, R.W., 1984, Holocene history of cedar and native Indians of the North America Pacific Coast: Science, v. 225, p. 711-713.

Hedberg, H.D., 1976, ed., International Stratigraphic Guide: a Guide to Stratigraphic Classification, Terminology and Procedure: John Wiley and Sons, London, 200 p.

Heslop-Harrison, J., 1968, Pollen wall development: Science, v. 161, p. 237.

Heusser, L.E., 1978, Spores and pollen in the marine realm, *in* Haq, B.L. and Boersma, A., eds., Introduction to Marine Micropaleontology: Elsevier Science Publishers, New York, p. 327-339.

Heusser, L.E., 1985, Quaternary palynology of marine sediments in the northeast Pacific, northwest Atlantic, and Gulf of Mexico, *in* Bryant, V.M., Jr. and Holloway, R.G., eds., Pollen Records of Late-Quaternary North American Sediments: American Association of Stratigraphic Palynologists, Dallas, p. 385-403.
Hills, L.V. and Sangster, E.V., 1980, A review of paleobotanical studies dealing with the last 20,000 years; Alaska, Canada, Greenland: Syllogeus, v. 26, p. 73-224.
Hopkins, J.S., 1950, Differential floatation and deposition of coniferous and deciduous tree pollen: Ecology, v. 31, p. 633-641.
Howe, S.E. and Webb, T., 1983, Calibrating pollen data in climatic terms: improving the methods: Quaternary Science Reviews, v. 2, p. 17-51.
Hyvarinen, H., 1985, Holocene pollen stratigraphy of Baird Inlet, east-central Ellesmere Island, arctic Canada: Boreas, v. 14, p. 19-32.
Ives, J.W., 1977, Pollen separation of three North American birches: Arctic and Alpine Research, v. 9, p. 73-80.
Jacobson, G.L. and Bradshaw, R.H.W., 1981, The selection of sites for paleovegetational studies: Quaternary Research, v. 16, p. 80-96.
Janssen, C.R., 1984, Modern pollen assemblages and vegetation in the Myrtle Lake peatland, Minnesota: Ecological Monographs, v. 54, p. 213-252.
Jowsey, P.C., 1966, An improved peat sampler: New Phytologist, v. 65, p. 245-248.
Kapp, R.C., 1969, How to Know Pollen and Spores: Wm. Brown, Dubuque, 249 p.
Karrow, P.F. and Warner, B.G., 1984, A subsurface middle Wisconsin interstadial site at Waterloo, Ontario: Boreas, v. 13, p. 67-85.
Kay, P.A. and Andrews, J.T., 1983, Re-evaluation of pollen-climate transfer functions in Keewatin, northern Canada: Annals of the Association of American Geographers, v. 73, p. 550-559.
Kupsch, W.D., 1960, Radiocarbon-dated organic sediment near Herbert, Saskatchewan: American Journal of Science, v. 258, p. 282-292.
Lehman, J.T., 1975, Reconstruction of the rate of accumulation of lake sediment: the effect of sediment focusing: Quaternary Research, v. 5, p. 541-550.
Lichti-Federovich, S., 1970, A pollen stratigraphy of a dated section of late-Pleistocene lake sediment from central Alberta: Canadian Journal of Earth Sciences, v. 7, p. 938-945.
Liu, K-B. and Lam, N. S-N., 1985, Paleovegetation reconstruction based on modern and fossil pollen data: an application of discriminant analysis: Annals of the Association of American Geographers, v. 75, p. 115-130.
MacDonald, G.M., 1983, Holocene vegetation history of the upper Natla River area, Northwest Territories, Canada: Arctic and Alpine Research, v. 15, p. 169-180.
MacDonald, G.M., 1987, Postglacial development of the subalpine-boreal transition forest of western Canada: Journal of Ecology, v. 75, p. 303-320.
MacDonald, G.M. and Cwynar, L.C., 1985, A fossil pollen based reconstruction of the late Quaternary history of lodgepole pine (*Pinus contorta* ssp. *latifolia)* in the western interior of Canada: Canadian Journal of Forestry Research, v. 15, p. 1039-1044.
MacDonald, G.M. and Ritchie, J.C., 1986, Modern pollen surface samples and the interpretation of postglacial vegetation development in the western interior of Canada: New Phytologist, v. 103, p. 245.
MacDonald, G.M. and Waters, N.M., 1987, An evaluation of automated mapping algorithms for the analysis of Quaternary pollen data: Review of Palaeobotany and Palynology, v. 51, p. 289-307.
Mackereth, F.J.H., 1958, A portable core sampler for lake deposits: Limnology and Oceanography, v. 3, p. 181-191.
Macpherson, J.B., 1982, Postglacial vegetation history of the eastern Avalon Peninsula, Newfoundland, and Holocene climatic change along the eastern Canadian seaboard: Géographie physique et Quaternaire, v. 36, p. 175-196.
Maher, L.J., Jr., 1972, Absolute pollen diagram of Redrock Lake, Boulder County, Colorado: Quaternary Research, v. 2, 531-553.
Mathewes, R.W., 1973, A palynological study of postglacial vegetation changes in the University Research Forest, southwestern British Columbia: Canadian Journal of Botany, v. 51, p. 2085-2103.
Mathewes, R.W. and Heusser, L.E., 1981, A 12,000 year palynological record of temperature and precipitation trends in southwestern British Columbia: Canadian Journal of Botany, v. 59, p. 707-710.
McAndrews, J.H., 1976, Fossil history of Man's impact on the Canadian flora: an example from southern Ontario: Canadian Botanical Association, Bulletin Supplement, v. 9, p. 1-6.
McAndrews, J.H., 1981, Late Quaternary climate of Ontario: temperature trends from the fossil pollen record, *in* Mahaney, W.C., ed., Quaternary Paleoclimate: Geo Abstracts, Norwich, p. 319-333.
McAndrews, J.H., 1984, Pollen analysis of the 1973 ice core from Devon Island Glacier, Canada: Quaternary Research, v. 22, p. 68-76.
McAndrews, J.H., Berti, A.A. and Norris, G., 1973, Key to the Quaternary Pollen and Spores of the Great Lakes Region: Royal Ontario Museum, Toronto, 61 p.
McAndrews, J.H. and Power, D.M., 1973, Palynology of the Great Lakes: the surface sediments of Lake Ontario: Canadian Journal of Earth Sciences, v. 10, p. 777-792.
Mott, R.J. and Farley-Gill, L.D., 1978, A late-Quaternary pollen profile from Woodstock, Ontario: Canadian Journal of Earth Sciences, v. 15, p. 1101-1111.
Oldfield, F. and Appleby, P.G., 1984, Empirical testing of ^{210}Pb-dating models for lake sediments, *in* Haworth, E.Y. and Lund, J.W.G., eds., Lake Sediments and Environmental History: Leicester University Press, Leicester, p. 93-124.
Overpeck, J.T., Prentice, I.C. and Webb, T., 1985, Quantitative interpretation of fossil pollen spectra: dissimilarity coefficients and the method of modern analogs: Quaternary Research, v. 23, p. 87-108.
Patterson, W.A., III, Edwards, K.J. and Maguire, D.J., 1987, Microscopic charcoal as a fossil indicator of fire: Quaternary Science Reviews, v. 6, p. 3-23.
Piper, D.J., Mudie, P.J., Aksu, A.E. and Hill, P.R., 1978, Late Quaternary sedimentation 50°N, north-east Newfoundland Shelf: Géographie physique et Quaternaire, v. 32, p. 321-332.
Potzger, J.E. and Courtemanche, A., 1954, A radiocarbon date of peat from James Bay, Quebec: Science, v. 119, p. 908-909.
Prentice, I.C., 1983, Postglacial climatic change: vegetation dynamics and the pollen record: Progress in Physical Geography, v. 7, p. 273-286.
Prentice, I.C., 1985, Pollen representation, source area and basin size: towards a unified theory of pollen analysis: Quaternary Research, v. 23, p. 76-86.
Prentice, I.C., 1986, Vegetation responses to past climatic variation: Vegetatio, v. 67, p. 131-141.
Prentice, I.C. and Webb, T., III, 1986, Pollen percentages, tree abundances and the Fagerlind effect: Journal of Quaternary Science, v. 1, p. 35-43.
Rampton, V.N., 1971, Late Quaternary vegetational and climatic history of the Snag-Klutlan area, southwestern Yukon Territory, Canada: Geological Society of America, Bulletin, v. 82, p. 959-978.
Richard, P.J.H., 1978, Histoire tardiglaciaire et postglaciaire de la végétation au Mont Shefford, Québec: Géographie physique et Quaternaire, v. 32, p. 81-93.

Ritchie, J.C., 1964, Contributions to the Holocene paleoecology of west-central Canada, I. the Riding Mountain area: Canadian Journal of Botany, v. 42, p. 182-196.
Ritchie, J.C., 1977, The modern and late Quaternary vegetation of the Campbell-Dolomite Uplands near Inuvik, N.W.T., Canada: Ecological Monographs, v. 47, p. 401-423.
Ritchie, J.C., 1984, Past and Present Vegetation of the Far Northwest of Canada: University of Toronto Press, Toronto, 251 p.
Ritchie, J.C., 1985, Quaternary pollen records from the western interior and the arctic of Canada, *in* Bryant, V.M. and Holloway, R.G., eds., Pollen Records of Late-Quaternary North American Sediments: American Association of Stratigraphic Palynologists, Dallas, p. 327-352.
Ritchie, J.C., 1987, Postglacial Vegetation of Canada: Cambridge University Press, Cambridge, 178 p.
Ritchie, J.C., Cwynar, L.C. and Spear, R. W., 1983, Evidence from northwest Canada for an early Holocene Milankovitch thermal maximum: Nature, v. 305, p. 126-127.
Ritchie, J.C. and MacDonald, G.M., 1986, The postglacial spread of white spruce: Journal of Biogeography, v. 13, p. 527-540.
Ritchie, J.C. and Yarranton, G.A., 1978, The late-Quaternary history of the boreal forest of central Canada: Journal of Ecology, v. 66, p. 199-212.
Rowley, J.R. and Rowley, J., 1956, Vertical migration of spherical and aspherical pollen in a *Sphagnum* bog: Proceedings of the Minnesota Academy of Science, v. 24, p. 29-30.
Saarnisto, M., 1986, Annually laminated lake sediments, *in* Berglund, B.E., ed., Handbook of Holocene Palaeoecology and Palaeohydrology: John Wiley and Sons, Chichester, p. 343-370.
Sangster, A.G. and Dale, H.M., 1964, Pollen grain preservation and the underrepresentated species in fossil pollen spectra: Canadian Journal of Botany, v. 42, p. 437-449.
Short, S.K. and Andrews, J.T., 1988, A sixteen thousand year old organic deposit, northern Baffin Island, N.W.T., Canada: palynology and significance: Géographie physique et Quaternaire, v. 42, p. 75-82.
Stockmarr, J., 1971, Tablets with spores used in absolute pollen analysis: Pollen et Spores, v. 13, p. 615-621.
Szafer, W., 1935, The significance of isopollen lines for the investigation of the geographical distribution of trees in the post-glacial period: Polska Akademia Umiejetnosci, Wydzial Matematycznoprzyrodiczy, Bulletin International, S.B., p. 235-239.
Terasmae, J., 1956, Palynological study of Pleistocene deposits on Banks Island, Northwest Territories, Canada: Science, v. 123, p. 801-802.
Terasmae, J., 1960, Contributions to Canadian palynology II, part 2: a palynological study of the Pleistocene interglacial beds at Toronto, Ontario: Geological Survey of Canada, Bulletin 56, p. 1-22.
Thompson, R., 1986, Palaeomagnetic dating, *in* Berglund, B.E., ed., Handbook of Holocene Palaeoecology and Palaeohydrology: John Wiley and Sons, Chichester, p. 313-327.
Tolonen, K, 1986, Charred particle analysis, *in* Berglund, B.E., ed., Handbook of Holocene Palaeoecology and Palaeohydrology: John Wiley and Sons, Chichester, p. 485-496.
Tsukada, M, 1982, *Cryptomeria japonica:* glacial refugia and late- glacial and post-glacial migration: Ecology, v. 63, p. 1091-1105.
Tsukada, M. and Sugita, S., 1982, Late Quaternary dynamics of pollen influx at mineral lake, Washington: Botanical Magazine of Tokyo, v. 95, p. 401-418.
Von Post, L., 1967, Forest tree pollen in south Swedish peat bog deposits: Pollen et Spores, v. 9, p. 375-401.
Walker, D., 1982, Vegetation's fourth dimension: New Phytologist, v. 90, p. 419-429.
Warner, B.G., Mathewes, R.W. and Clague, J.J., 1982, Ice-free conditions on the Queen Charlotte Islands, British Columbia, at the height of late Wisconsin glaciation: Science, v. 218, p. 675-677.
Webb, T., III, 1986, Is vegetation in equilibrium with climate? How to interpret late-Quaternary pollen data: Vegetatio, v. 67, p. 75-91.
Webb, T., III and Bryson, R.A., 1972, Late- and post-glacial climatic change in the northern Midwest, U.S.A.: quantitative estimates derived from fossil pollen spectra by multivariate statistical analysis: Quaternary Research, v. 2, p. 70-115.
Webb, T., III, Howe, S.E., Bradshaw, R.H.W. and Heide, K.M., 1981, Estimating plant abundances from pollen percentages: the use of regression analysis: Review of Palaeobotany and Palynology, v. 34, p. 269-300.
Webb, T., III, Richard, P.J.H. and Mott, R.J., 1983, A mapped history of Holocene vegetation in southern Quebec: Syllogeous, v. 49, p. 273-336.
Wenner, C.G., 1947, Pollen diagrams from Labrador: Geografiska Annaler, v. 29, p. 137-373.
Wright, H.E., Jr., Mann, D.H. and Glaser, P.H., 1984, Piston corers for peat and lake sediments: Ecology, v. 65, p. 657-659.

Originally published in
Geoscience Canada Volume 15 Number 1
(March 1988)

Plant Macrofossils

Barry G. Warner
Department of Earth Sciences
and Quaternary Sciences Institute
University of Waterloo
Waterloo, Ontario N2L 3G1

SUMMARY

Plant macrofossils consist of reproductive and vegetative plant parts visible to the unaided eye. Lakes and peatlands are the most common repositories for the preservation of Quaternary plant macrofossils; however, natural tar seeps, calcareous nodules, spring deposits, packrat middens, and fossil dung are important sources of plant remains also. In Canada, the first plant macrofossils were collected in the mid-1800s, with a brief period of activity around 1900, before the beginning of modern Quaternary plant macrofossil analyses around 1960. Taphonomic studies indicate that plant macrofossils can be transported, sorted and redeposited long distances, which complicates paleoecological reconstructions. Plant macrofossils are valuable in geological studies of environmental history and climatic change; however, the use of plant macrofossils for biostratigraphic correlation must await further research. Four broad phytogeographic patterns are recognizable on the basis of Quaternary plant macrofossil records in Canada, and plant macrofossils can be used to trace the evolutionary origins of the modern flora.

INTRODUCTION

Almost everyone has observed plant macrofossils in the making, be it leaves floating on a lake in the autumn, once forested lands flooded by dams and reservoirs, drifted plant debris around river inlets and tidal marshes, or perhaps some have seen the woodrat in the twilight hours busily gathering food and material for its nest. All of these processes contribute to accumulations of plant parts that may be preserved for many years or even millenia.

Macroscopic plant remains were discovered in Quaternary deposits over 200 years ago (Mannion, 1986a). Subsequently, the identification of plants by their macroscopic remains has grown into an important and useful branch of Quaternary ecology. In contrast with many pollen and spores, macrofossils are more readily identified to the species.

A macrofossil is any plant part which is visible to the naked eye and which becomes incorporated into a sedimentary sequence or geological deposit. Identifiable plant macrofossils often include reproductive structures such as seeds, bracts, cones, sporangia, megaspores, and complete inflorescences or the vegetative parts of plants such as wood, twigs, leaves, and rhizomes. Macroscopic remains of bryophytes, algae, and fungi are common, but details of their role in Quaternary ecology are dealt with in other papers of this series.

This paper reviews potential sources for plant macrofossils, field sampling techniques, and laboratory processing procedures. The role of plant macrofossils as indicators of Quaternary environments and biostratigraphy, and their use in assessing the phytogeographic and evolutionary origins of the modern flora is discussed. Plant macrofossils have important implications for archeology and paleoethnobotany, but their use in these fields is not considered here. The geographic focus is Canada. Recent reviews give details on the use of plant macrofossils primarily in Europe (Dickson, 1970; Watts, 1978; Birks, 1980; Wasylikowa, 1986; Grosse-Brauckmann, 1986; and Mannion, 1986a,b,c).

HISTORICAL DEVELOPMENT

The first observations of macroscopic plant remains were made by workmen clearing and digging drainage ditches in peatlands for agricultural expansion in the early 1800s in northern Europe (Barber, 1981). In Canada, the first notice of plant macrofossils in peatlands was not until Ganong's (1891, 1897) investigations in New Brunswick.

The recording of plant macrofossils from Quaternary geological formations in Canada began with Dawson's paper in 1857. He reported fossil wood, leaves and marine algae in calcareous concretions from what today are known as Champlain Sea deposits near Ottawa (Figure 1).

Plant macrofossils did not achieve much impact in geology until the 1870s and 1880s, though Dawson (1857, 1866) in Canada, Lesquereaux (1859) in America, and Heer (1862) in Britain made new discoveries in the intervening period. Real interest in plant macrofossils from geological deposits began in 1870 in Sweden with Alfred G. Nathorst (1873), and subsequently Andersson (1888, *vide* Tralau, 1963) in Scandinavia, Weber (1891, *vide* Tralau, 1963) in Germany, and Reid (1892, *vide* Tralau, 1963) in Britain.

Figure 1 *Purported* Populus balsamifera *leaves in concretions from Green Creek, Ottawa reported by Dawson in 1866 (Redpath Museum # 7407). These specimens were figures 2 and 3 in his 1866 paper.*

Figure 2 ***(above)*** *Fossil middens of* Neotoma cinerea *from Interior British Columbia. Such middens serve as urination perches making them hard, dark and shiny, thereby preserving a varied assortment of plant macrofossils. These specimens are about 1110 years old (Hebda* et al., *1990). They contain* Pseudotsuga menziesii *needles,* Juniperus scopulorum *branches, and Rosaceae thorns.*

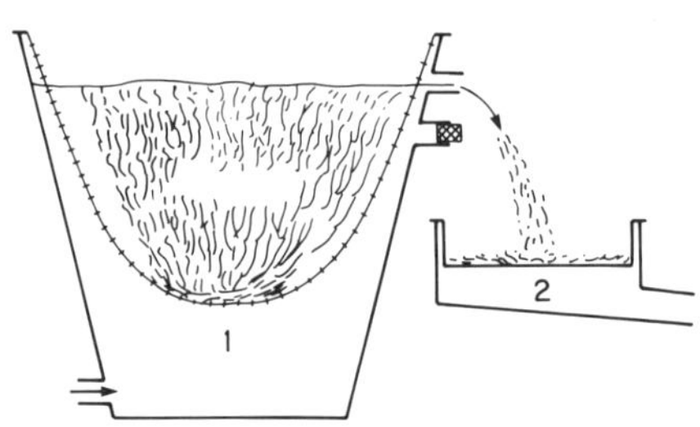

Figure 3 ***(above)*** *The earliest sieve apparatus for extraction of plant macrofossils designed by Andersson (1892). A hose is attached to the inflow spout at the base, the water is circulated in the bowl and net* ***(1)****, the fine debris floats off and is caught on a screen* ***(2)*** *(after Backman, 1965). Some laboratories in Scandinavia continue to use this apparatus today.*

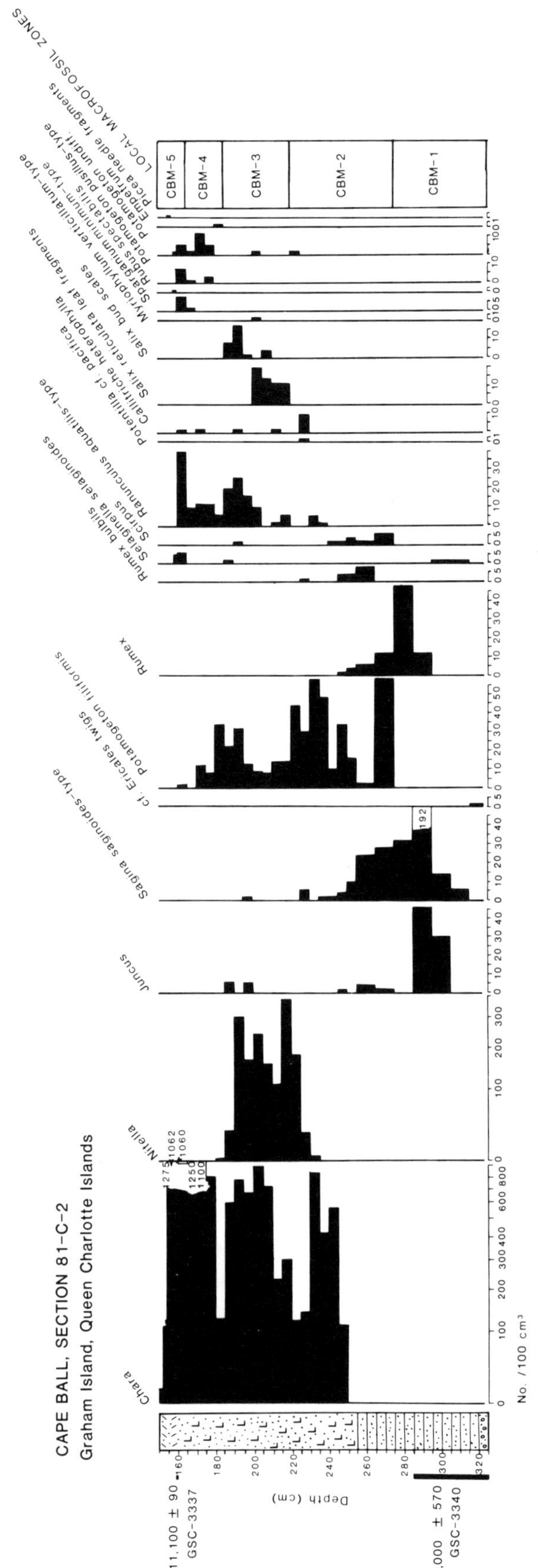

Figure 4 ***(right)*** *Representative plant macrofossil diagram from a late Quaternary site on eastern Graham Island, British Columbia (from Warner, 1984b). Five local macrofossil zones can be recognized. Generally, the sequence suggests wet sand habitats along a pond margin with* Juncus, Sagina saginoides*-type, and* Rumex, *followed by an increase in wetland and submerged vegetation such as* Chara, Nitella, Potamogeton filiformis, *and* Ranunculus aquatilis*-type, which ended with development of emergent aquatic and marsh communities with* Sparganium minimum*-type,* Scirpus, *and* Potamogeton pusillus*-type.*

Hollick (1892) and Knowlton (1896) in the United States and Dawson and Penhallow (1890) in Canada made the first contributions in North America. These early studies concentrated simply on compiling species lists with little regard given to the stratigraphic position of the fossils. Consequently, it is difficult to ascertain the age of these early plant records as many stratigraphic descriptions are incomplete or inaccurate, and most sites long since have changed or disappeared. Perhaps the most significant contributions made during this early period of investigation are the remarkably accurate paleoenvironmental reconstructions (Warner, 1986).

With the advent of pollen analysis around 1920, attention was diverted from plant macrofossils in North America. Nonetheless, in Europe, Knud Jessen maintained the tradition of plant macrofossil analysis, and successfully demonstrated the merits of combining the stratigraphic and chronological precision of pollen analysis with the taxonomic detail of plant macrofossils (Jessen and Milthers, 1928). One paper on plant macrofossils from an interstadial deposit near Vancouver, British Columbia was published during this period (Draycott, 1948). This was the only paper until the first papers appeared from the present period of plant macrofossil investigations (Terasmae, 1959; Terasmae and Fyles, 1959; Schofield and Robinson, 1960; Ritchie and DeVries, 1964).

SAMPLING PROCEDURES AND LABORATORY METHODS

Abundant and well-preserved plant macrofossils occur in any natural environment where aerobic activity has been minimal, accumulation of enclosing materials was rapid, and deposition was in a low-energy environment. Lakes and peatlands provide the most common repositories for plant macrofossils. Plant macrofossils are most valuable when analysis can be performed in conjunction with other paleoecological indicators and on samples in stratigraphic sequence from cores, monoliths or open faces. Inter- and post-glacial deltaic, alluvial, colluvial and diamicton deposits provide other opportunities to collect stratigraphic plant macrofossil samples.

Less conventional, non-stratigraphic sources for plant macrofossils include calcareous nodules such as those in post-glacial Champlain Sea deposits near Ottawa or purported occurrences in the Sangamonian Don Formation at Toronto (Penhallow, 1904).

Packrats, or woodrats *(Neotoma)* build bulky dens in caves, cliffs and river canyons in British Columbia and throughout the United States. Their waste piles or middens usually contain a rich assortment of plant fragments which can be preserved for thousands of years in the arid scrublands and deserts (Hebda *et al.*, 1990; Figure 2). Fossil ground squirrel, mammoth and giant ground sloth dung contain plant remains (*i.e.*, Mead *et al.*, 1986). In the United States, calcareous spring deposits and tar seeps are important sources of plant macrofossils (Haynes, 1985). Among the most spectacular plant macrofossil records are purported 9000 year old seeds that germinated after being thawed and planted (Porsild *et al.*, 1967). Volcanic ashfalls may contribute to well-preserved beds of plant macrofossils (Burnham and Spicer, 1986).

Storage and extraction of plant macrofossils in the laboratory varies with the nature of the sample. Wet field samples should be frozen or refrigerated, but they can be stored unrefrigerated for several months if sealed tightly to prevent dessication. Dry field samples should be kept dry until processing. Wet samples are preferred over dry ones as the wet macrofossils are less readily damaged during processing and analysis.

Extraction of fossils is relatively simple. Washing on a screen (usually 250 μm mesh) with a gentle jet of warm tap water sometimes after soaking overnight in water is usually all that is required for the majority of samples (Figure 3). A nest of sieves of different mesh sizes may be used when large volumes are processed. Chemicals should be avoided if at all possible, and when stubborn samples are involved, treatment for a few minutes in a mild detergent, a weak base (*i.e.*, KOH) in the case of peats, or weak acids (*i.e.*, HCl) in the case of calcareous sediments, can be used. Treatment with HNO_3 and H_2O_2 as advocated by Wasylikowa (1986) is not recommended because these chemicals can degrade plant remains. Submergence in an ultrasonic bath for about one minute to break down samples with no apparent damage or alteration to the macrofossils is preferred over chemical treatments.

The retained residue is sorted in water under a binocular microscope (magnification 8-40 ×). All identifiable remains are stored in 95% ethanol and glycerine in stoppered vials or in glycerine in plastic boxes. One or two drops of phenol can be added to prevent fungal growth.

The identification of plant macrofossils is based on comparison with modern reference collections. There are a number of publications containing photographs and drawings of fossil specimens. Montgomery (1977) is a valuable aid for southern Canadian floras. Delcourt *et al.* (1979) give a comprehensive bibliography of seed atlases and other publications useful for plant macrofossil identification.

Plant macrofossil data are most conveniently presented on a graph, with relative frequencies or macrofossil concentrations on the abscissa and depth on the ordinate axis (Figure 4). For leaf or stem fragments, or other plant remains that are difficult to count accurately, their presence can be recorded by an abundance scale. In some cases, tables or species lists are adequate.

TAPHONOMY AND INTERPRETATION OF PLANT MACROFOSSIL DATA

Generally, it is assumed that plant macrofossils are deposited not far from the parent plant, thereby making them valuable paleoecological indicators. However, in the case of many environments, the initial place of rest is temporary, and the plant remains can be transported some distance before being fossilized at the study site.

Plant macrofossils are the end product of processes that lead to their incorporation and preservation in the sediment. Consequently, the plant fossil assemblage represents only a fraction of the total flora that lived at the time of deposition. Knowledge of the factors that influence the accumulation of potential plant macrofossils in sediments can be understood best by observing modern processes. Such factors include natural biotic dispersal mechanisms and seed productivity, dynamics of stream and river hydrology, origin of storm flotsam accumulations on beaches or winter ice, or woodrat foraging habits.

Most macrofossil assemblages in Canada have been deposited in or by water. Numerous studies have demonstrated the capabilities of streams and rivers entering lakes to carry potential plant macrofossils great distances (Drake and Burrows, 1980; Hill, 1981; Holyoak, 1984; Spicer and Wolfe, 1987). Plant remains in montane lakes are particularly conducive to long-distance transport as streams may carry remains from high elevations to lowland lakes. Differentiation of distant *versus* local fossil components is nearly impossible, and perhaps the best way is simply to be aware of retransport possibilities, most often reflected by the sedimentology of the enclosing deposits. Small lakes and ponds lacking surface feeder streams most likely contain locally derived plant remains.

Potential plant macrofossils are affected by both water and biological sorting during transport and deposition (Spicer, 1980). Consequently, sorting can lead to the preservation of biased assemblages and misleading paleoecological inferences. Plant remains reflect the hydrodynamic characteristics of the site, in that low-energy environments where settling velocity is slow, contain small or more delicate plant parts and high energy environments tend to yield the more resistant seeds, cones and wood (Spicer, 1980, 1981; Collinson, 1983; Ferguson, 1985; Spicer and Wolfe, 1987). Consideration of transport and sorting processes aided in the interpretation of plant macrofossils in a late Quaternary deltaic sequence in southern Ontario as these processes probably contributed to a mixed assemblage of plant species which do not live together in the same vegetation zone today (Warner and Barnett, 1986). However, only sorting and little, if any, transport was apparent from macrofossils in another deltaic deposit in northern Ontario (Bajc *et al.*, 1986). Problems of interpretation are compounded with subtill organic deposits where the physiography of the pre-glacial landscape is often unknown. Retransport of plant macrofossils may produce incongruent pollen and plant macrofossil assemblages, though in his early studies in the Yukon Territory, Matthews (1975) considered only the pollen to be retransported.

Wind can contribute to transport and sorting by regulating the size or type of leaves that may fall on a lake, by carrying seeds that are adapted to wind dispersal, or by blowing plant macrofossils over flat snow and ice covered landscapes (Spicer, 1981; Glaser, 1981; Ferguson, 1985).

Reworking of deposits may mix macrofossils that are older than their contemporary assemblages. A recent study of plant macrofossils in alluvial deposits in the Mississippi River Valley demonstrated the dating and interpretation problems of the older components in such assemblages when only younger components fortuitously were dated (Givens and Givens, 1987). Matthews *et al.* (1986) discussed the possibilities of mixed Tertiary and Quaternary fossils in deposits in Arctic Canada. One site in southern Ontario revealed that a discordant radiocarbon date on wood was probably due to reworking and incorporation of the older wood into a much younger plant assemblage (Warner and Barnett, 1986). Karrow *et al.* (1977) described what are thought to be Middle

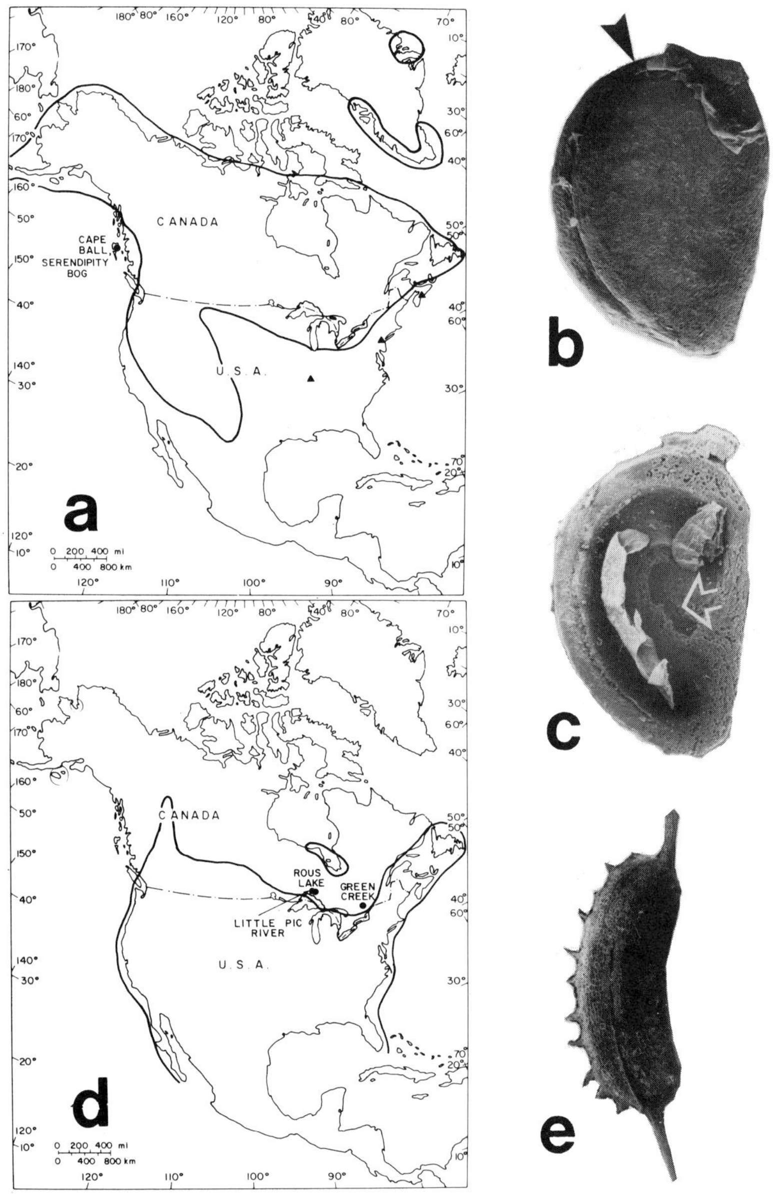

Figure 5 **(a)** *Modern distribution of* Potamogeton filiformis *in North America. (After Mason, 1957; Porsild and Cody, 1980; Packer, 1983; Brayshaw, 1985). The dots (new records) and triangles (from Watts, 1979; Mott* et al.*, 1982; Jackson and Miller, 1983) indicate Late Quaternary fossil localities. Fossil endocarps of* P. filiformis, **(b)** *arrow indicates saddle between beak and lid of endocarp,* **(c)** *arrow indicates characteristic cavity in endocarp loop.* **(d)** *Modern distribution of* Zannichellia palustris *in North America. (After Porsild and Cody, 1980; Riley and MacKay, 1980; Packer, 1983; Brayshaw, 1985). Dots indicate Late Quaternary range extensions.* **(e)** *Fossil seed of* Z. palustris. *Scale bars equal 1 mm.*

Wisconsinan plant remains in a Late Pleistocene esker deposit, suggesting retransport in glacier meltwaters. It seems that multiple radiocarbon dates on various materials will become necessary to ascertain the representativeness of plant macrofossils where retransport is likely. In at least one instance, Warner *et al.* (1987) obtained several radiocarbon dates on a Holocene macrofossil assemblage preserved in alluvial sediments and found good consistency among the dates, which suggests that reworking is not always a problem. Other obvious examples of reworking is the presence of wood in till or peat balls in ice-contact deposits (Heusser, 1973; Berti, 1975).

Further studies are needed to understand the origin of plant macrofossils in other less conventional deposits. The genesis of calcareous concretions such as those at Green Creek remains a mystery (Gadd, 1980).

Some of the most useful taphonomic studies involve surface samples from lakes (Birks, 1973; Dunwiddie, 1987) and peatlands (GreatRex, 1983). For example, among submerged aquatics, *Najas flexilis* seeds are over-represented because they produce prolific quantities annually, while perennial species such as *Brasenia schreberi* produce far fewer seeds in relation to its cover abundance (Birks, 1973). In contrast to lakes, where seeds can be transported around the basin, seeds in surface peats generally originate from plants within one metre of the sampling point, except for wind-dispersed propagules (GreatRex, 1983).

BIOSTRATIGRAPHY

The most valuable use of plant macrofossils in Quaternary geology is in the reconstruction of local environments and the history of past climates. Interglacial and interstadial plant macrofossils can be exceptionally difficult to interpret. It appears that the majority have been reworked or transported to some degree, and dating control may be poor. An *in situ* sub-till peat deposited in a shallow backwater channel on the Queen Charlotte Islands, British Columbia is an exception. Plant macrofossils reveal gradual hydroseral development from open water, through a shallow peatland and open fen phase with *Hippuris vulgaris, Veronica scutellata* and *Menyanthes trifoliata,* which eventually transformed into a *Sphagnum* peatland with *Selaginella selaginoides* (Warner *et al.*, 1984a). This sequence of peat development closely resembles Holocene developmental processes in the same region (Warner, 1984b), suggesting that the timing and ordering of peat development were similar during Middle Wisconsinan time in this region.

By virtue of their aquatic habitat, macrofossils of submerged aquatic macrophytes perhaps are the most common elements in lacustrine fossil plant assemblages. Extensive waterbodies in the past, such as pluvial and proglacial lakes, meltwater channels, and small ponds probably provided an efficient dispersal network and diverse habitats for colonization. The rapidity with which aquatic macrophytes colonize newly deglaciated areas is often thought to be due to efficient dispersal agents, most probably water. *Potamogeton filiformis* (Figure 5a) and *Zannichellia palustris* (Figure 5b) are two examples of aquatic macrophytes that spread rapidly into previously glaciated regions, and subseqently have undergone range contractions within the last 10,000 years in Canada. Two possible hypotheses may explain the previous more widespread distributions. The first suggests that local extinctions and range contractions reflect climatic responses to colder temperatures, and the second proposes that natural ontogenetic processes changed living conditions in the aquatic environment which lead to the demise and local extinction of certain aquatic species.

In view of the cosmopolitan distribution of *P. filiformis* and *Z. palustris* in habitats ranging from intertidal estuaries to alpine water bodies, climate does not have any apparent direct influence on distributional boundaries. Both of these plants achieve their best growth in shallow water bodies with high pH and conductivities. The amount of free carbon dioxide in the water available for photosynthesis is limited. *P. filiformis* and *Z. palustris* would be placed at a competitive advantage over other aquatic macrophytes not capable of using low amounts of free carbon dioxide or bicarbonate as their carbon sources for photosynthesis. Other factors such as water depth, substrates, turbidity, and dispersibility undoubtedly played a significant role in regulating the past distributions

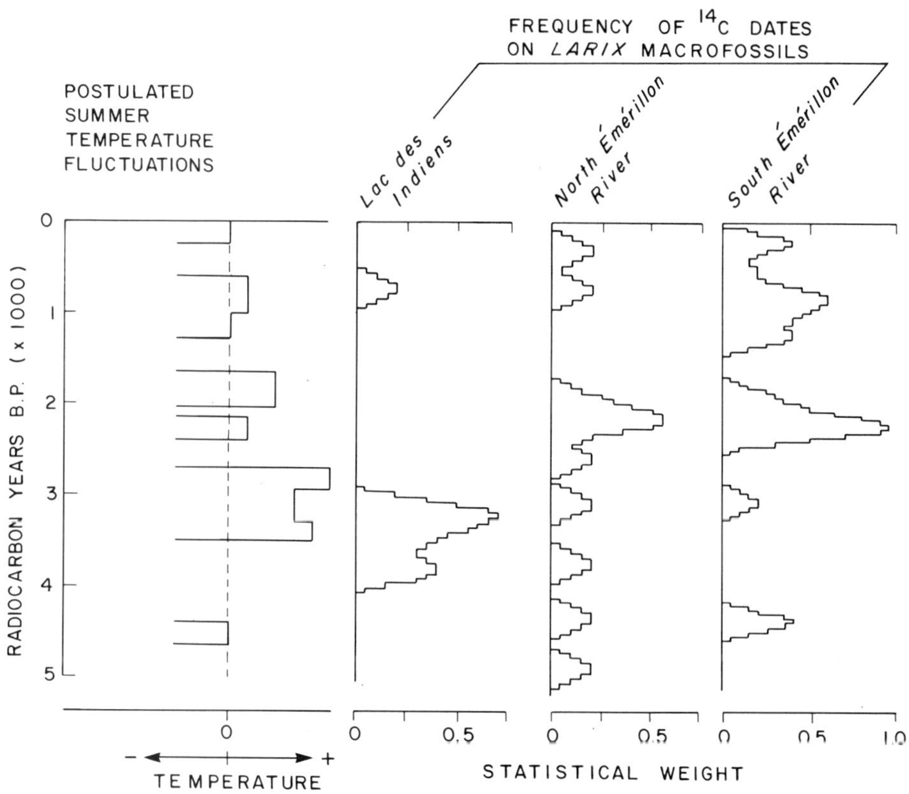

Figure 6 *The frequency of radiocarbon dates on* Larix *macrofossils north of its modern range indicate two periods of northward treeline expansion in the past 5000 years in the area of Nouveau-Québec. (Adapted from Gagnon and Payette, 1981).*

of certain submerged aquatic macrophytes. Much more ecological information is needed before aquatic macrophytes can be used as paleoclimatic indicators (Warner, 1984a) although they seem to have much potential in paleolimnological reconstructions (Birks *et al.*, 1976).

The first biostratigraphic schemes were developed on the basis of plant macrofossils which gave rise to the Aspen, Pine, Oak, and Alder zones in Scandinavia (Steenstrup, 1841, *vide* Mangerud *et al.*, 1974). Subsequently, European biostratigraphic schemes have been replaced largely by pollen chronologies with plant macrofossils playing a subordinate role. The use of plant macrofossils for biostratigraphic correlations in Canada is hampered simply by the paucity of studied sites covering a wide geographic and time range. The Sangamonian deposits at Toronto, Ontario contain a number of "southern" plant species whose modern ranges have reached the southern fringes of the Great Lakes region only since mid-Holocene time and are not known from any other older Quaternary deposit in North America (Kerr-Lawson, 1985; Warner, unpublished). *Najas guadalupensis* has been recovered from the interglacial deposits at Toronto and from purported interglacial deposits in the James Bay Lowlands (Lichti-Federovich, 1974). These are the only two records for this taxon in Canada. Can we assume temporal contemporaneity for the two deposits? Probably this would be a misleading conclusion in view of the reliability of aquatic macrophytes as paleoecological indicators.

Macrofossil studies on the north slope of Alaska have proven useful in distinguishing Quaternary and Tertiary deposits (Matthews, 1987b,c). The Tertiary deposits contain a number of Old World plant taxa such as *Epipremnum* and *Aracispermum* not reported previously from North America. In light of such ancient phytogeographic connections during Tertiary time, some plant macrofossils may prove useful as biostratigraphic indicators to differentiate Pliocene and Pleistocene deposits.

The study by Gagnon and Payette (1981) is one of the finest examples of the use of plant macrofossils in reconstructing past climates (Figure 6). Over 700 *Larix* wood specimens, on which 67 radiocarbon dates were obtained, were collected near treeline in the area of Nouveau Québec. *Larix* reached the

Figure 7 *Selected representative plant macrofossils.* **(a)** Najas guadalupensis sensu lato *seed,* **(b)** Dryas integrifolia *leaf (upper surface),* **(c)** Saxifraga oppositifolia *leaves,* **(d)** Selaginella densa *megaspore,* **(e)** Hypericum scouleri *seed,* **(f)** Elaeagnus commutata *seed,* **(g)** Taraxacum *cf.* lacerum *seed, and* **(h)** Dryas drummondii *leaf (under surface). Scale bar equals 1 mm except for (d) which is 0.5 mm.*

area around 4500 years B.P. Subsequently between 3500-2700 and between 2000-1600 years B.P., it underwent two periods of northward expansion up to 5 km north of its present northern limit. Reproduction by only seed makes *Larix* a good paleoclimatic indicator in contrast with other northern trees such as *Picea* which can reproduce vegetatively in spite of climatically severe growing seasons. Jackson (1989) used macrofossils to trace the movement of alpine treelines in the Adirondack Mountains of New York.

Numerous studies have demonstrated the merits of combining pollen and plant macrofossil analyses especially on lake and peatland deposits (*e.g.*, Karrow *et al.*, 1975; Hughes *et al.*, 1981; Mott *et al.*, 1981; Warner *et al.*, 1984b; Schwert *et al.*, 1985; Baker *et al.*, 1986, 1989). Under certain circumstances the enclosing sediments such as deltaic or alluvial sequences may be too coarse for the preservation of pollen and only plant macrofossils may be preserved in paleosols, calcareous concretions, mammal dung, or packrat middens. Plant macrofossils should be studied concurrently with pollen if at all possible to avoid biased or incomplete paleoenvironmental reconstructions.

PHYTOGEOGRAPHY

As a rough approximation, about 2.5% of Canada's modern flora has been found in the Quaternary fossil record. Phytogeographic extinctions are not known in Canada, and those reported by Penhallow are referrable to modern taxa (Warner, 1986). Possible Quaternary extinctions occurred in the United States (Beals and Melhorn, 1961; Blackwell, 1982).

Published plant macrofossil records exist for no less than 100 sites in Canada. Based on published records and on some of my unpublished work, four broad Quaternary phytogeographic patterns are recognizable. I have not included the early reports of Dawson and Penhallow because these records need verification (see references in Warner, 1986).

Selected representatives in each group are:

(a) Northern range extensions during the penultimate, Sangamonian Interglacial:
Taxa in this group are from sites in ***Ontario:***
Chamaecyparis thyoides
Elatine triandra var. *americana*
Najas guadalupensis sensu lato (Figure 7a)

At present, these taxa are restricted to the eastern and southeastern United States. Some taxa extended as far north as the James Bay Lowlands, which suggests a major contracted range since the last interglaciation. The presence of *Chamaecyparis thyoides* at Toronto during Sangamonian time (Kerr-Lawson, 1985) suggests an old inland origin for the Coastal Plain-Great Lakes disjunction that may have persisted south of the ice limit during later advances.

(b) Middle Wisconsinan disjunctions or range extensions:
Taxa showing this distribution include:
Southern Ontario
Dryas integrifolia (Figure 7b)
Betula glandulosa
Saxifraga oppositifolia (Figure 7c)
Alaska and Yukon Territory
Alisma gramineum
Scutellaria galericulata
Najas flexilis
Alnus incana (A. rugosa)

The presence of *Saxifraga oppositifolia* and *Dryas integrifolia* in southern Ontario and in southern Quebec (Matthews *et al.*, 1987; Warner *et al.*, 1987) suggests that the Arctic-Alpine or Holarctic-Montane disjunctions were probably in place since Middle Wisconsinan time. These taxa likely grew in ice marginal zones moving north-south several times with glacier positions before assuming modern distributions. Some components such as *Dryas integrifolia* maintained the Arctic-Alpine disjunction on into early Holocene time and presumably to the present along the north shore of Lake Superior (Bajc *et al.*, 1986).

(c) Late Wisconsinan disjunctions or range extensions:
Coastal British Columbia
Selaginella densa (Figure 7d)
Hypericum scouleri (Figure 7e)
Potamogeton filiformis
Salix cf. *pedicellaris*
Sagina cf. *saginoides*
Dryas drummondii
Ontario and Southern Quebec
Elaeagnus commutata (Figure 7f)
Thalictrum venulosum
Taraxacum cf. *lacerum* (Figure 7g)
Shepherdia canadensis
Dryas integrifolia
Salix herbacea
Sibbaldia procumbens
Oxyria digyna
Carex bigelowii
Armeria maritima

Late Wisconsinan environments characteristically contained mixtures of plants which at present have temperate, boreal and subarctic distributions. During this period, Cordilleran plants such as *Dryas drummondii* and prairie-type plants such as *Shepherdia canadensis, Elaeagnus commutata,* and *Thalictrum* cf. *venulosum* were in eastern Canada (Miller and Thompson, 1979; Karrow and Warner, 1988). This suggests that the Eastern North American-Western North American disjunction was in place by this time. Similarly, plants of modern subarctic and arctic affinities such as *Dryas integrifolia, Oxyria digyna* and *Carex bigelowii* grew much farther south indicating that the Arctic-Alpine disjunction continued on into Late Wisconsinan time (*e.g.*, Birks, 1976; Watts, 1979; Mott *et al.*, 1981; Baker *et al.*, 1986).

Northern interior and central interior elements, such as *Salix* cf. *pedicellaris, Selaginella densa* and *Dryas drummondii* extended to the coastal regions of British Columbia. Northern *(Sagina* cf. *saginoides)* and southern *(Hypericum scouleri)* plants extended beyond their modern boundaries to the Queen Charlotte Islands on the west coast (Warner, 1984b).

(d) Holocene range extensions:
Yukon
Larix laricina
Northern Quebec
Larix laricina
Ontario
Dryas drummondii (Figure 7h)
Pinus strobus
cf. *Zizania palustris*
Zannichellia palustris

Generally, the range extensions represent comparatively short distributional shifts on the order of 100 km or less. An exception is *Dryas drummondii* which represents persistence of the Western North American disjuncts in northern Ontario into and throughout Holocene time (Bajc *et al.*, 1986).

EVOLUTIONARY CHANGES AND PALEOECOLOGY

The dramatic changes in ice extent, sea-level positions, presence of pro-glacial lakes, and climate during Quaternary time almost certainly pushed plants to their limits of environmental adaptation. Plant macrofossils have been measured for morphological changes as a means of assessing the evolution of different biotypes. In Europe, *Dulichium arundinaceum* and *Menyanthes trifoliata* have shown some degree of change (Birks, 1980); however, *Menyanthes* shows no morphological change during Quaternary time in Japan (Kokawa, 1960).

Fossil cones of *Picea* from coastal British Columbia were used to assess the degree of change within the last 10,000 years (Warner and Chmielewski, 1987). This study indicated that, in contrast with today, past spruce populations probably consisted of more introgressive populations of *P. sitchensis* and *P. glauca* that extended farther west before the two species assumed more or less independent distributions during Holocene time.

Some fascinating areas of research involving plant macrofossils which have received little or no attention in Quaternary botany lie in the areas of paleobiochemistry, past plant-animal interactions, mycology, and plant ultrastructure. One study on fossil *Picea* needles and wood from Michigan ranked the rates of degradation of various compounds through geological time which proved useful for studying processes of coalification (Meyers *et al.*, 1980).

Obvious examples of plant-animal interactions are beaver-gnawed *Picea, Pinus* and *Juniperus* branches (Aalto *et al.*, 1983), insect trails on logs (Figure 8), or evidence of herbivory including chewed or eaten seeds (Figure 9). Such studies may provide important clues to co-evolution, herbivory and mimicry, or the role of animals as dispersal agents and pollinators.

Bark, leaf and other plant tissues often possess fragmentary fungal remains, thereby providing an additional means of tracing the origin, interrelationships and evolution of major fungal groups throughout Quaternary time.

Heightened awareness by Quaternary botanists and their communication with experts in allied fields will lead to new and exciting aspects of Quaternary plant macrofossil analyses.

ACKNOWLEDGEMENTS

Critiques of the manuscript were kindly provided by T.W. Anderson, R.G. Baker, and H.J. Kubiw. N. Bahar drafted the figures and R. Etmanski patiently typed drafts of the manuscript. Ongoing plant macrofossil studies which contributed to this paper are supported by an Operating Grant from the Natural Sciences and Engineering Research Council of Canada.

REFERENCES

Aalto, M., Donner, J., Hirvas, H. and Niemelä, H., 1983, Majavan jyrsimiä pesä- ja patopuita moreenin alla Vimpelissä: Geologi, v. 35, p. 137-139.

Andersson, G., 1892, Om metoden för växtpaleontologisk undersökning af torfmossar: Geologiska Förhandlingar, v. 14, p. 165-175, 506-508.

Backman, A.L., 1965, Växtpaleontologiska studier pa Aland: Acta Societas pro Fauna et Flora Fennica, v. 77, p. 1-45.

Bajc, A.F., Warner, B.G. and Morgan, A.V., 1986, Dating and paleontology of a postglacial fossil assemblage from the north shore of Lake Superior, Ontario, Program and Abstracts, 9th Biennial Meeting American Quaternary Association, Champaign, Illinois, p. 70.

Figure 8 *Bark beetle trails on a* Picea *log dated 10,280 ± 100 years B.P. from southern Ontario (Karrow and Warner, 1988).*

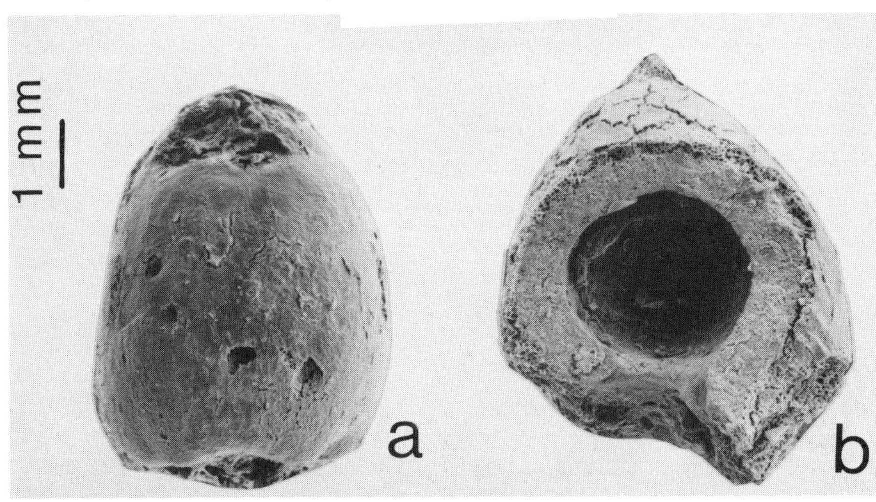

Figure 9 *Fossil* Cladium mariscoides *seeds of mid-Holocene age from a peatland deposit on Manitoulin Island, Ontario (Warner* et al., *1984b).* **(a)** *Intact seed, and* **(b)** *partially eaten seed.*

Baker, R.G., Rhodes, R.S., Schwert, D.P., Ashworth, A.C., Frest, T.J., Hallberg, G.R. and Janssens, J.A., 1986, A full-glacial biota from southeastern Iowa, USA: Journal of Quaternary Science, v. 1, p. 91-107.

Baker, R.G., Sullivan, A.E., Hallberg, G.R. and Horton, D.G., 1989, Vegetational changes in western Illinois during the onset of Late Wisconsinan glaciation: Ecology, v. 70, p. 1363-1376.

Barber, K.E., 1981, Peat Stratigraphy and Climatic Change: A.A. Balkema, Rotterdam, 219 p.

Beals, H.O. and Melhorn, W.N., 1961, An extinct conifer, *Larix* (?) *pleistocenicum*, n. sp. from Yarmouth Interglacial deposits of Indiana: Journal of Paleontology, v. 35, p. 667-672.

Berti, A.A., 1975, Paleobotany of Wisconsinan interstadials, Eastern Great Lakes Region, North America: Quaternary Research, v. 5, p. 591-619.

Birks, H.H., 1973, Modern macrofossil assemblages in lake sediments in Minnesota, *in* Birks, H.J.B. and West, R.G., eds., Quaternary Plant Ecology: Blackwell Scientific Publishers, Oxford, p. 173-189.

Birks, H.H., 1980, Plant macrofossils in Quaternary lake sediments: Ergebnisse der Limnologie, v. 15, p. 1-60.

Birks, H.H., Whiteside, M.C., Stark, D.M. and Bright, R.C., 1976, Recent paleolimnology of three lakes in northwestern Minnesota: Quaternary Research, v. 6, p. 249-272.

Birks, H.J.B., 1976, Late Wisconsinan vegetation history at Wolf Creek, central Minnesota: Ecological Monographs, v. 46, p. 395-429.

Blackwell, W.H., 1982, Fossil "dogwood" from the Pleistocene of central Mississippi: Botanical Gazette, v. 143, p. 395-400.

Brayshaw, T.C., 1985, Pondweeds and burreeds and their relatives of British Columbia: British Columbia Provincial Museum, Occasional Paper 26, 166 p.

Burnham, R.J. and Spicer, R.A., 1986, Forest litter preserved by volcanic activity at El Chichón, Mexico: A potentially accurate record of the pre-eruption vegetation: Palaios, v. 1, p. 158-161.

Collinson, M.E., 1983, Accumulations of fruits and seeds in three small sedimentary environments in southern England and their paleoecological implications: Annals of Botany, v. 52, p. 583-592.

Dawson, J.W., 1857, On the newer Pliocene and Post-Pleistocene deposits in the vicinity of Montreal, with notices of fossils recently discovered in them: Canadian Naturalist and Geologist, v. 2, p. 401-426.

Dawson, J.W., 1866, The evidence of fossil plants as to the climate of the Post-Pliocene period in Canada: The Canadian Naturalist, v. 3, p. 69-76.

Dawson, J.W. and Penhallow, D.P., 1890, On the Pleistocene flora of Canada: Geological Society of America Bulletin, v. 1, p. 311-334.

Delcourt, P.A., Davis, O.K. and Bright, R.C., 1979, Bibliography of taxonomic literature for the identification of fruits, seeds, and vegetative plant fragments: Oak Ridge National Laboratory, Environmental Sciences Division, Publication 1328, 84 p.

Dickson, C.A., 1970, The study of plant macrofossils in British Quaternary deposits, *in* Walker, D. and West, R.G., eds., Studies in the Vegetational History of the British Isles: Cambridge University Press, Cambridge, p. 233-254.

Drake, H. and Burrows, C.J., 1980, The influx of potential plant macrofossils into Lady Lake, north Westland, New Zealand: New Zealand Journal of Botany, v. 18, p. 257-274.

Draycott, W.M., 1948, Pleistocene fossil beetle and vegetational remains in interglacial deposits; A summary report: Southern California Academy of Sciences, Bulletin, v. 47, p. 35-42.

Dunwiddie, P.W., 1987, Macrofossil and pollen representation of coniferous trees in modern sediments from Washington: Ecology, v. 68, p. 1-11.

Ferguson, D.K., 1985, The origin of leaf-assemblages - New light on an old problem: Review of Palaeobotany and Palynology, v.46, p. 117-188.

Gadd, N.R., 1980, Maximum age for a concretion at Green Creek, Ontario: Géographie physique et Quaternaire, v. 34, p. 229-238.

Gagnon, R. and Payette, S., 1981, Fluctuations Holocène de la limite des forêts de mélèzes, Rivière aux Feuilles, Nouveau-Québec: Une analyse macrofossile en milieu tourbeux: Géographie physique et Quaternaire, v. 35, p. 57-72.

Ganong, W.F., 1891, On raised peat-bogs in New Brunswick: Botanical Gazette, v. 16, p. 123-126.

Ganong, W.F., 1897, Upon raised peat-bogs in the Province of New Brunswick: Royal Society of Canada, Transactions, Series 2, v. 3, p. 131-163.

Givens, C.R. and Givens, F.M., 1987, Age and significance of fossil white spruce *(Picea glauca)*, Tunica Hills, Louisiana-Mississippi: Quaternary Research, v. 27, p. 283-296.

Glaser, P.H., 1981, Transport and deposition of leaves and seeds on tundra: A late-glacial analog: Arctic and Alpine Research, v. 13, p. 173-182.

GreatRex, P.A., 1983, Interpretation of macrofossil assemblages from surface sampling of macroscopic plant remains in mire communities: Journal of Ecology, v. 71, p. 773-791.

Grosse-Brauckmann, G., 1986, Analysis of vegetative plant macrofossils, *in* Berglund, B.E., ed., Handbook of Holocene Palaeoecology and Palaeohydrology: J. Wiley and Sons Ltd., Chichester, p. 591-618.

Haynes, C.V., 1985, Mastodon-bearing springs and Late Quaternary geochronology of the Lower Pomme de Terre Valley, Missouri: Geological Society of America, Special Paper 204, 35 p.

Hebda, R.J., Warner, B.G. and Cannings, R.A., 1990, Fossil pollen, plant macrofossils and insects from bushy-tailed woodrat *(Neotoma cinerea)* middens in British Columbia: Géographie physique et Quaternaire, v. 44, in press.

Heer, O.K., 1862, On the fossil flora of Bovey Tracey: Royal Society of London, Philosophical Transactions, v. 152B, p. 1039-1086.

Heusser, C.J., 1973, Age and environment of allochthonous peat clasts from the Bogachiel River Valley, Washington: Geological Society of America, Bulletin, v. 84, p. 797-804.

Hill, R.S., 1981, Consequences of long-distance dispersal of plant macrofossils: New Zealand Journal of Botany, v. 19, p. 241-242.

Hollick, A., 1892, Palaeobotany of the Yellow Gravel at Bridgeton, N.J.: Bulletin of the Torrey Botanical Club, v. 19, p. 330-333.

Holyoak, D.T., 1984, Taphonomy of prospective plant macrofossils in a river catchment on Spitsbergen: New Phytologist, v. 98, p. 405-423.

Hughes, O.L., Harington, C.R., Janssens, J.A., Matthews, J.V., Morlan, R.E., Rutter, N.W. and Schweger, C.E., 1981, Upper Pleistocene stratigraphy, paleoecology, and archeology of the northern Yukon Interior, eastern Beringia, I. Bonnet Plume Basin: Arctic, v. 34, p. 329-365.

Jackson, S.T., 1989, Postglacial vegetational changes along an elevational gradient in the Adirondack Mountains (New York): A study of plant macrofossils: New York State Museum, Bulletin, v. 465, p. 1-29.

Jackson, S.T. and Miller, N.G., 1983, Paleoecology of a fossil plant assemblage from a Pre-Wisconsin till in southern Illinois: American Midland Naturalist, v. 109, p. 120-135.

Jessen, K. and Milthers, V., 1928, Stratigraphical and paleontological studies of interglacial fresh-water deposits in Jutland and Northwest Germany: Danmarks geologiske Undersogelse, Raekke 2, v. 28, p. 1-378.

Karrow, P.F., Anderson, T.W., Clarke, A.H., Delorme, L.D. and Sreenivasa, M.R., 1975, Stratigraphy, paleontology, and age of Lake Algonquin sediments in southwestern Ontario: Quaternary Research, v. 5, p. 49-87.

Karrow, P.F., Harrison, W. and Saunderson, H.C., 1977, Reworked Middle Wisconsinan (?) plant macrofossils from the Brampton Esker, southern Ontario: Canadian Journal of Earth Sciences, v. 14, p. 426-430.

Karrow, P.F. and Warner, B.G., 1988, Ice, lakes, and plants, 13,000 B.P. to 10,000 B.P.: The Erie-Ontario lobe in Ontario, *in* Laub, R.S., Steadman, D.W. and Miller, N.G., eds., Late Pleistocene and Early Holocene Paleoecology and Archeology of the Eastern Great Lakes Region: Buffalo Society of Natural Sciences, Bulletin, v. 33, p. 39-52.

Kerr-Lawson, L.J., 1985, Gastropods and plant macrofossils from the Quaternary Don Formation (Sangamonian Interglacial), Toronto, Ontario, unpublished M.Sc. Thesis, University of Waterloo, 201 p.

Knowlton, F.H., 1896, Report on a collection of fossil plants from Morgantown, West Virginia: American Geologist, v. 18, p. 370-379.

Kokawa, S., 1960, Morphometric reconstruction of the compressed seed remains of *Menyanthes* in Japan: Journal of the Institute of Polytechnics, Osaka City University, v. D-11, p. 79-89.

Lesquereux, L., 1859, On some fossil plants of Recent Formations: American Journal of Science, Second Series, v. 27, p. 359-363.

Lichti-Federovich, S., 1974, *Najas guadalupensis* (Spreng.) Morong. in the Missinaibi Formation, northern Ontario: Geological Survey of Canada, Paper 74-1A, p. 201.

Mangerud, J., Andersen, S.T., Berglund, B.E. and Donner, J.J., 1974, Quaternary stratigraphy of Norden, a proposal for terminology and classification: Boreas, v. 3, p. 109-128.

Mannion, A.M., 1986a, Plant macrofossils and their significance in Quaternary paleoecology. Part I: Introduction: Progress in Physical Geography, v. 10, p. 194-214.

Mannion, A.M., 1986b, Plant macrofossils and their significance in Quaternary paleoecology. Part II: Applications: preglacial, interglacial and interstadial deposits: Progress in Physical Geography, v. 10, p. 364-384.

Mannion, A.M., 1986c, Plant macrofossils and their significance in Quaternary paleoecology. Part III: Applications: late glacial, postglacial and archaeological deposits: Progress in Physical Geography, v. 10, p. 517-546.

Mason, H.L., 1957, A Flora of the Marshes of California: University of California Press, Berkeley, 878 p.

Matthews, J.V., 1975, Incongruence of macrofossil and pollen evidence: A case from the Late Pleistocene of the northern Yukon coast: Geological Survey of Canada, Paper 75-1B, p. 139-146.

Matthews, J.V., 1987a, Late Quaternary plant and insect macrofossils from southern Quebec: International Union for Quaternary Research, 12th International Congress, Ottawa, Canada, Programme with Abstracts, p. 222.

Matthews, J.V., 1987b, Plant macrofossils from the Neogene Beaufort Formation on Banks and Meighen Islands, District of Franklin: Geological Survey of Canada Paper 87-1A, p. 73-87.

Matthews, J.V., 1987c, Potential importance of Arctic and Neogene floras for dating Pleistocene interglacials in northern North America: International Union for Quaternary Research, 12th International Congress, Ottawa, Canada, Programme with Abstracts, p. 222.

Matthews, J.V., Mott, R.J. and Vincent, J.S., 1986, Preglacial and interglacial environments of Banks Island: Pollen and macrofossils from Duck Hawk Bluffs and related sites: Géographie physique et Quaternaire, v. 40, p. 279-298.

Matthews, J.V., Smith, S.L. and Mott, R.J., 1987, Plant macrofossils, pollen, and insect fossils of arctic affinity from Wisconsinan sediments in Chaudiere Valley, southern Quebec: Geological Survey of Canada, Paper 87-1A, p. 165-175.

Mead, J.I., Agenbroad, L.D., Davis, O.K. and Martin, P.S., 1986, Dung of *Mammuthus* in the arid southwest, North America: Quaternary Research, v. 25, p. 121-127.

Meyers, P.A., Leenheer, M.J., Erstfeld, K.M. and Bourbonniere, R.A., 1980, Changes in spruce composition following burial in lake sediments for 10,000 yr: Nature, v. 287, p. 534-536.

Miller, N.G. and Thompson, G.G., 1979, Boreal and western North American plants in the Late Pleistocene of Vermont: Journal of the Arnold Arboretum, v. 60, p. 167-218.

Montgomery, F.H., 1977, Seeds and Fruits of Plants of Eastern Canada and Northeastern United States: University of Toronto Press, Toronto, 232 p.

Mott, R.J., Anderson, T.W. and Matthews, J.V., 1981, Late-Glacial paleoenvironments of sites bordering the Champlain Sea based on pollen and macrofossil evidence, *in* Mahaney, W.C., ed., Quaternary Paleoclimate: GeoAbstracts, Norwich, p. 129-171.

Mott, R.J., Anderson, T.W. and Matthews, J.V., 1982, Pollen and macrofossil study of an interglacial deposit in Nova Scotia: Géographie physique et Quaternaire, v. 36, p. 197-208.

Nathorst, A.G., 1873, Om Skanes nivaförändringar: Geologiska Foreningens i Stockholm Förhandlingar, v. 1, p. 281-194.

Packer, J.G., 1983, Flora of Alberta, 2nd Edition: University of Toronto Press, Toronto, 687 p.

Penhallow, D.P., 1904, Notes on Tertiary plants from Canada and the United States: Royal Society of Canada, Transactions, Section IV, v. 10, p. 57-76.

Porsild, A.E. and Cody, W.J., 1980, Vascular plants of continental Northwest Territories, Canada, National Museums of Canada, Ottawa, 667 p.

Porsild, A.E., Harington, C.R. and Mulligan, G.A., 1967, *Lupinus arcticus* Wats, grown from seeds of Pleistocene age: Science, v. 158, p. 113-114.

Riley, J.L. and MacKay, S.L., 1980, The vegetation and phytogeography of coastal southwestern James Bay: Royal Ontario Museum, Life Science Contributions 124, 81 p.

Ritchie, J.C. and DeVries, B., 1964, Contributions to the Holocene paleoecology of west central Canada. A late glacial deposit from the Missouri Coteau: Canadian Journal of Botany, v. 42, p. 677-692.

Schofield, W.B. and Robinson, H., 1960, Late-glacial and post-glacial plant macrofossils from Gillis Lake, Richmond County, Nova Scotia: American Journal of Science, v. 258, p. 518-523.

Schwert, D.P., Anderson, T.W., Morgan, A., Morgan, A.V. and Karrow, P.F., 1985, Changes in Late Quaternary vegetation and insect communities in southwestern Ontario: Quaternary Research, v. 23, p. 205-226.

Spicer, R.A., 1980, The importance of depositional sorting to the biostratigraphy of plant megafossils, *in* Dilcher, D.L. and Taylor, T.N., eds., Biostratigraphy of Fossil Plants: Dowden, Hutchinson and Ross, Inc., Stroudsburg, p. 171-183.

Spicer, R.A., 1981, The sorting and deposition of allochthonous plant material in a modern environment at Silwood Lake, Silwood Park, Berkshire, England: United States Geological Survey, Professional Paper 1143, p. 1-77.

Spicer, R.A. and Wolfe, J.A., 1987, Plant taphonomy of Late Holocene deposits in Trinity (Clair Engle) Lake, northern California: Paleobiology, v. 13, p. 227-245.

Terasmae, J., 1959, Palaeobotanical study of a buried peat from the Mackenzie River Delta area, Northwest Territories: Canadian Journal of Botany, v. 37, p. 715-717.

Terasmae, J. and Fyles, J.G., 1959, Palaeobotanical study of late-glacial deposits from Vancouver Island, B.C.: Canadian Journal of Botany, v. 37, p. 815-817.

Tralau, H., 1963, The recent and fossil distribution of some boreal and arctic montane plants in Europe: Arkiv för Botanik, v. 5, p. 533-582.

Warner, B.G., 1984a, Commentary on plant indicator taxa in climatic reconstructions: Quaternary Research, v. 22, p. 266-268.

Warner, B.G., 1984b, Late Quaternary paleoecology of eastern Graham Island, Queen Charlotte Islands, British Columbia, Canada, Ph.D. Dissertation, Simon Fraser University, 190 p.

Warner, B.G., 1986, Early work in Quaternary botany in Canada: Geoscience Canada, v. 13, p. 39-44.

Warner, B.G. and Barnett, P.J., 1986, Transport, sorting and reworking of Late Wisconsinan plant macrofossils from Lake Erie, Canada: Boreas, v. 15, p. 323-329.

Warner, B.G. and Chmielewski, J.G., 1987, Biometric analysis of modern and late Pleistocene cones of *Picea* from western Canada: New Phytologist, v. 107, p. 449-457.

Warner, B.G., Clague, J.J. and Mathewes, R.W., 1984a, Geology and paleoecology of a Mid-Wisconsin peat from the Queen Charlotte Islands, British Columbia, Canada: Quaternary Research, v. 21, p. 337-350.

Warner, B.G., Hebda, R.J. and Hann, B.J., 1984b, Postglacial paleoecological history of cedar swamp, Manitoulin Island, Ontario, Canada: Palaeogeography, Palaeoclimatology, Palaeoecology, v. 45, pp. 301-345.

Warner, B.G., Karrow, P.F., Morgan, A.V. and Morgan, A., 1987, Plant and insect fossils from Nipissing sediments along the Goulais River, southeastern Lake Superior: Canadian Journal of Earth Sciences, v. 24, p. 1526-1536.

Warner, B.G., Morgan, A.V. and Karrow, P.F., 1988, A Wisconsinan interstadial Arctic flora and insect fauna from Clarksburg, southwestern Ontario, Canada: Palaeogeography, Palaeoclimatology, Palaeoecology, v. 68, p. 27-47.

Wasylikowa, K., 1986, Analysis of fossil fruits and seeds, *in* Berglund, B.E., ed., Handbook of Holocene Palaeoecology and Palaeohydrology: J. Wiley and Sons Ltd., Chichester, p. 571-590.

Watts, W.A., 1978, Plant macrofossils and Quaternary palaeoecology, *in* Walker, D. and Guppy, J.C., eds., Biology and Quaternary Environments: Australian Academy of Science, Canberra, p. 53-67.

Watts, W.A., 1979, Late Quaternary vegetation of coastal Appalachia and the New Jersey Coastal Plain: Ecological Monographs, v. 49, p. 427-469.

Originally published in
Geoscience Canada Volume 15 Number 2
(June 1988)

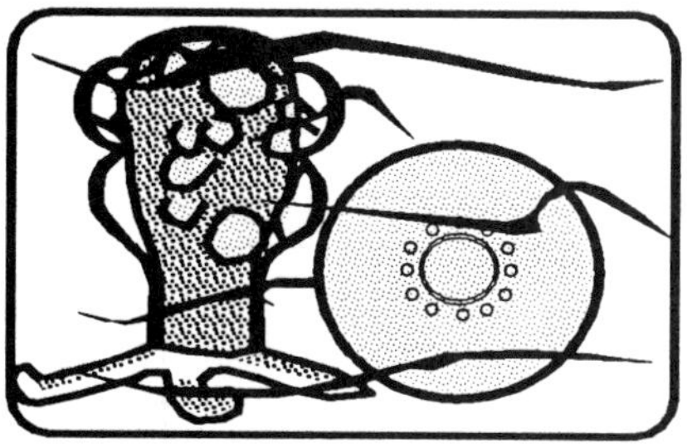

Testate Amoebae (Protozoa)

Barry G. Warner
Department of Earth Sciences
and Quaternary Sciences Institute
University of Waterloo
Waterloo, Ontario N2L 3G1

SUMMARY
Testate amoebae are unicellular animals with a discrete shell enclosing the cytoplasm. The shells are readily preserved, are abundant in freshwater peatland and lake sediments, and can be identified to the species. Fossil testate amoebae were first discovered in the late 1800s in limnic sediments, but shortly thereafter were also found in peat deposits. Though testate amoebae are widely distributed throughout the northern hemisphere, they do not have cosmopolitan habitat preferences. Their most promising role in paleoecology lies in deciphering past hydrological conditions in peatlands. The future of testate amoebae analysis in Canada is bright, particularly in understanding fundamental processes of peatland development, and in peatland management schemes for forestry, and for the agricultural and horticultural industries.

INTRODUCTION
Protozoa literally means "first animals" which is appropriate because most members in the group have the simplest cellular organization of all animals. Following classical classification schemes, Protozoa might be considered a subkingdom within Kingdom Animalia; however, if more modern classification schemes are preferred, Protozoa would be regarded as a subkingdom within Kingdom Protista (Committee on Systematics and Evolution of the Society of Protozoologists, 1980). Most protozoa are uni-cellular and possess at least one well-defined nucleus. The majority range in size from 5 to 250μm (Sleigh, 1973). At present, over 65,000 living and fossil protozoa taxa have been described. Among living forms having importance in paleoecology, there are about 11,300 free-living sarcodines (including approximately 4600 belong to the Foraminiferida) with thousands more species still unnamed (Committee on Systematics and Evolution of the Society of Protozoologists, 1980).

One group of amoeboid protozoa, the testate amoebae, is important in Quaternary ecology. Testate amoebae, also commonly referred to as "rhizopodes", "testaceans", or "thecamoebae", comprise the Subclass Testacealobosia in Superclass Rhizopoda, within the Subphyllum Sarcodina (Committee on Systematics and Evolution of the Society of Protozoologists, 1980). These animals are characterized by a discrete shell or test enclosing the cytoplasm, and by the presence of pseudopodia which protrude through an opening called the pseudostome (Figure 1). This distinctive shell and the shell characters are a major part in the taxonomy of the group. The shells are composed of layers of organic material (tectinous) secreted by the amoebae, or of preformed siliceous or calcareous plates fitted into precise positions, or others may incorporate foreign materials such as mineral grains, fungal hyphae, or diatom frustules into or onto the shell during formation (Sleigh, 1973). After the animal dies, the shells, being resistant to decay in certain environments, may remain in soil or sediment for many years or millenia. Testate amoebae are ubiquitous in freshwater, forest floor litter, damp soil, and mossy habitats. Most genera occupy specialized ecological niches regulated by water, temperature, oxygen, pH and salinity. Estimates of abundance of testate amoebae reach 226,000 per m^2 in oligotrophic-eutrophic lakes (Schönborn, 1962) and 16,000,000 per m^2 in peat (Heal, 1962). The diagnostic taxonomic features of the shell and its durability for fossilization, their abundance, and widespread occurrence in a variety of specific ecological niches, and the relative ease with which testate amoebae may be extracted from fossil deposits makes them a useful group of paleoecological indicators (Figures 2 and 3).

Fossil testate amoebae analysis is in its infancy in North America, with probably less than ten published accounts or passing mention of their occurrence in fossil deposits. Peat deposits, and specifically *Sphagnum* peats, contain abundant and the most diverse fossil faunas with about 40 taxa, although limnic deposits also contain protozoan remains in abundance. This paper reviews field and laboratory procedures for studying Quaternary fossil testate amoebae, the modern ecological data base available for interpreting fossil faunas, the role of testate amoebae in Quaternary ecology, in particular in studies of peatland developmental history, and the promise and prospect of testate amoebae in Quaternary ecology in North America. The paper by Tolonen (1986) gives an up-to-date account on fossil testate amoebae studies in Europe, and another useful introduction to fossil testate amoebae is Beyens (1984).

HISTORICAL DEVELOPMENT
Living Protozoa were first noticed by van Leeuwenhoek in 1674 in Europe. The earliest surveys of Protozoa in North America were not published until Leidy's monograph in 1879. Initially, it was thought that all Protozoa in soil were encysted and originated from lakes and oceans; however, around the turn of the century living Protozoa were found in a wide range of habitats. Thus, it is not surprising that the first reports of fossil testate amoebae were from lake sediments (Lagerheim, 1902; and reviews in: Frey, 1964; Tolonen, 1986). After this period, most studies on fossil faunas concentrated on peatland deposits primarily in Germany (Harnisch, 1927; Grospietsch, 1953; Overbeck, 1975), and later in the Netherlands (Casparie, 1972), Denmark (Aaby, 1976; Aaby and Tauber, 1975), and Finland (Tolonen, 1966, 1968). More recent studies include those of Barber (1981) from Britain and Beyens (1985) from Belgium.

Comparatively little is known about the character and distribution of modern faunas in North America, and even less work has been undertaken on Quaternary faunas. In Canada, Ogden (1960) published the first stratigraphic profiles of testate amoebae from a raised peatland in Nova Scotia. He recorded five taxa, *Amphitrema flavum, A. wrightianum, Arcella artocrea, Assulina seminulum* and *Nebela lageniformis* which comprise typical coastal ombrotrophic peatland faunas. Heusser (1960) noted a similar fauna in surface and fossil peat deposits along the North Pacific coast of North America. One of the most detailed and informative investigations using testate amoebae in paleoecology in North America is the study of Tolonen *et al.* (1985) on raised peatlands in New Brunswick and Maine. They found a dry to moist-loving fauna characteristic of highly humified peat layers, while a moist to wet-loving fauna was associated with less humified *Sphagnum* peat layers. The former faunas were associated with lichenous communities, while the later represented hummock or lawn communities.

Once again, interest has returned to fossil testate amoebae preserved in lake sediments, in both Europe (Schönborn, 1973, 1984; Ruzicka, 1982) and North America (Scott and Medioli, 1983; Ellison and Ogden, 1987). Though lacustrine faunas usually are not as abundant and diverse as peatland faunas, and lakes contain taxonomically poorly known groups, de Vernal *et al.* (1985) found that testate amoebae can be useful in defining relative sea-level changes by identifying marine-freshwater isolation times in coastal lakes. More detailed paleoenvironmental reconstructions must await modern ecological surveys of the important taxa.

The individual silicified plates of disarticulated shells of certain taxa may be recovered from recent lake sediments and identified to genera and sometimes to species (Douglas and Smol, 1987a,b). Again the paleoecological significance of these fossils remains problematical because of the possibility of long-distance transport, reworking, and focussing of the fossil remains in the lake basin.

FIELD SAMPLING AND LABORATORY PROCESSING PROCEDURES

Testate amoebae are recovered best from *Sphagnum* peat deposits, while recovery tends to be poorer in other peat types, limnic sediments, and forest soils. This distribution in part, may relate to soil aeration, temperature, and other conditions which control biological activity which would lead to decomposition of the tests. Recent evidence shows that population turnover from reproduction to decomposition occurs rapidly in well-aerated soil environments (*e.g.*, Lousier, 1984a,b,c, 1985; Lousier and Parkinson,

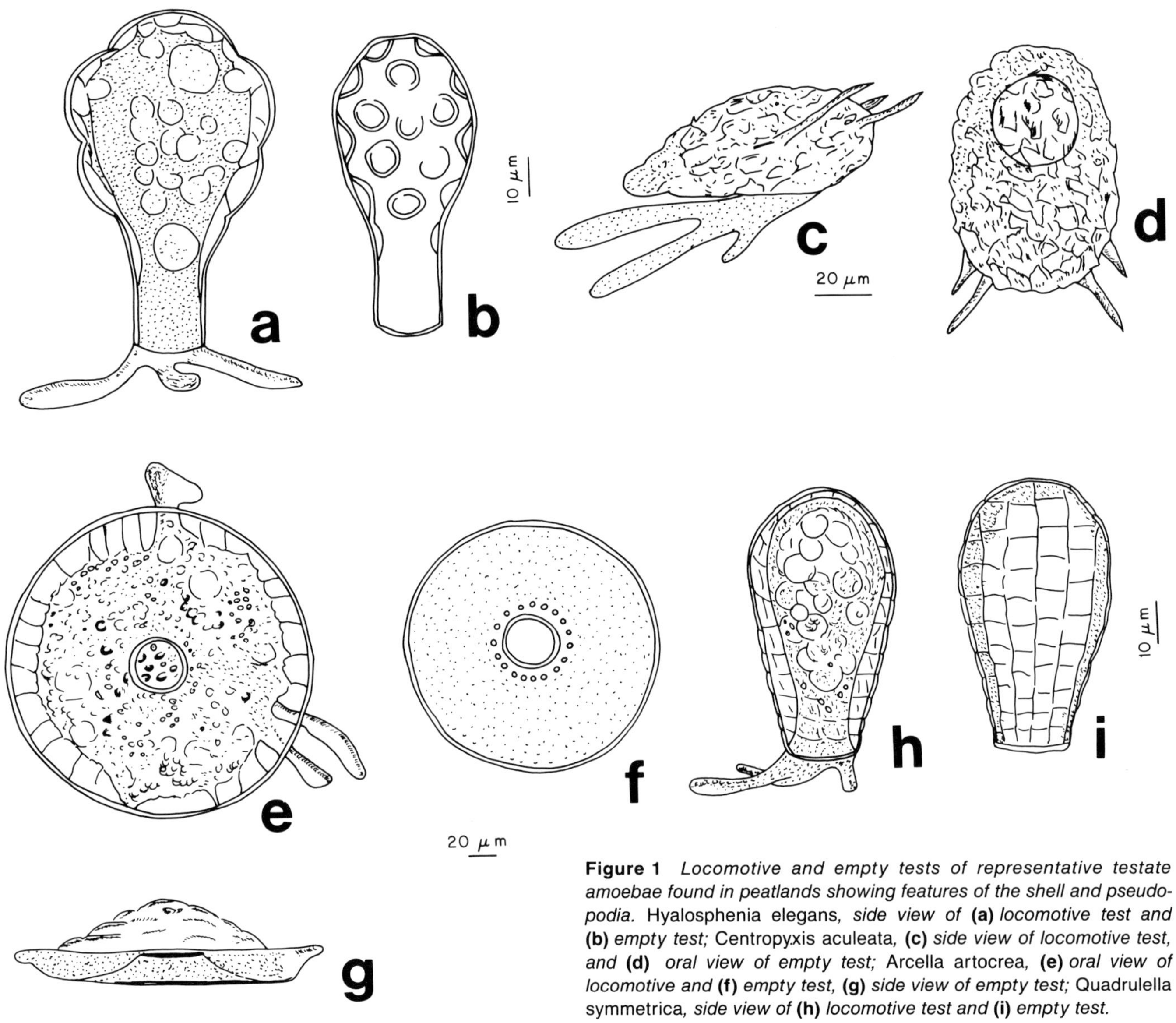

Figure 1 *Locomotive and empty tests of representative testate amoebae found in peatlands showing features of the shell and pseudopodia.* Hyalosphenia elegans, *side view of* **(a)** *locomotive test and* **(b)** *empty test;* Centropyxis aculeata, **(c)** *side view of locomotive test, and* **(d)** *oral view of empty test;* Arcella artocrea, **(e)** *oral view of locomotive and* **(f)** *empty test,* **(g)** *side view of empty test;* Quadrulella symmetrica, *side view of* **(h)** *locomotive test and* **(i)** *empty test.*

1981), whereas in raw humus or *Sphagnum* moss population turnover is much slower (Schönborn, 1986).

Field samples should be stored wet and refrigerated or frozen until processing in the laboratory. Dried samples tend to make the shells brittle and susceptible to fragmentation. Generally, processing procedures follow those of Tolonen (1986) with some modification for more quantitative analyses. A volume of peat or sediment of 1 to 5 cm^3 is measured with a calibrated syringe, weighed, placed in a small beaker, and boiled in water for 5 to 10 minutes to disaggregate the sample. Subsequently, the sample is washed with water on a course screen (*i.e.*, kitchen tea

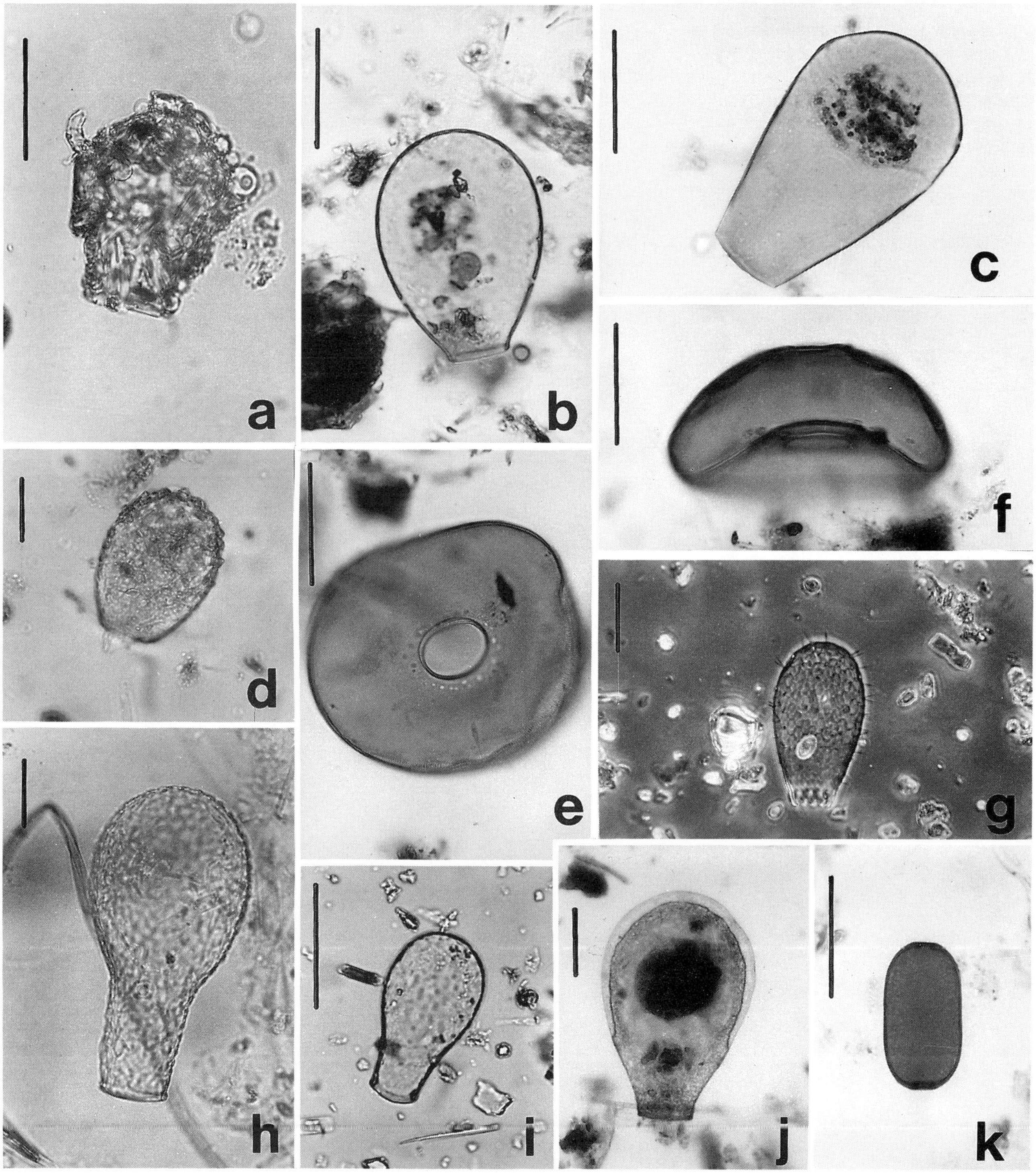

Figure 2 *Selected testate amoebae commonly found in* Sphagnum *peatlands.* **(a)** Difflugia leidyi; **(b)** Hyalosphenia ovalis; **(c)** Hyalosphenia papilio; **(d)** Heleopera petricola; **(e)** Arcella artocrea, *oral view;* **(f)** A. artocrea, *side view;* **(g)** Euglypha strigosa-*type;* **(h)** Nebela lageniformis; **(i)** N. militaris; **(j)** N. carinata; **(k)** Amphitrema flavum. *Scale is 20µm for a,b,c,e,f,i,k, and 44µm for d,g,h,j.*

sieve, mesh diameter about 750μm), and the fine fraction is concentrated in test tubes in a centrifuge. An innoculum of an exotic marker (usually standard pollen tablets manufactured for absolute pollen analysis) is added to each sample following boiling. Preliminary comparative experiments found there to be fewer exotic markers when the samples were innoculated prior to boiling, suggesting that the exotic pollen grains adhered to the *Sphagnum* plants despite thorough washing on the screen. Ratios of the numbers of exotic markers to numbers of shells can be calculated to estimate the numbers of shells per unit volume of sediment using formulae developed for absolute pollen analysis (Birks and Birks, 1980). Finally the concentrate of amoebae shells is stained with safranin-O in glycerin and stored in stoppered vials.

A small drop of the concentrate is smeared on a glass slide, covered with a coverslip, and sealed to the slide with nail varnish. The slide is systematically scanned and all identifiable shells within the field of view on each transect are identified and tabulated. A minimum count of 200 is probably sufficient and no less than 100 is best to achieve a reasonable representation of the number of species in the fossil fauna present (Figure 4).

Unfortunately, there are no standard keys for identification. The best aids are Grospietsch (1958), Loeblich and Tappan (1964), Corbet (1973), Ogden and Hedley (1980), and Lee *et al.* (1985). Monographs on specific genera such as *Arcella* (Deflandre, 1928), *Nebela* (Deflandre, 1936), *Euglypha* (Decloitre, 1962), and *Hyalosphenia* (Grospietsch, 1964) are useful.

DISTRIBUTION AND ECOLOGY

Cosmopolitanism in testate amoebae distributions has been assumed ever since the first studies in the last century (Hoogenraad and De Groot, 1979; and references therein). This view is changing as protozoan surveys expand into previously uncollected areas and the taxonomy of major groups and genera

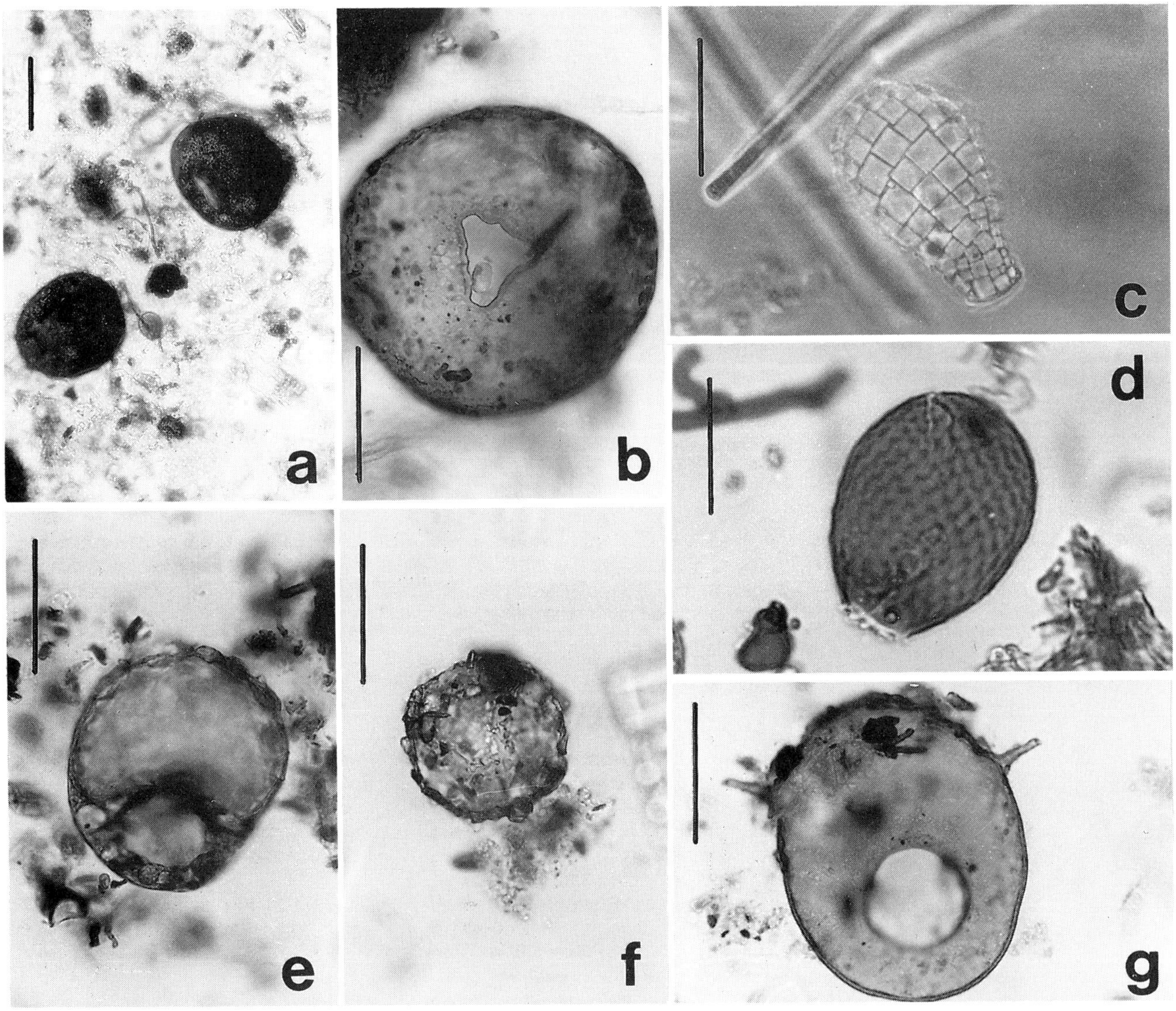

Figure 3 *Selected testate amoebae commonly found in* Sphagnum *peatlands.* **(a)** *Two individuals of* Bullinularia indica; **(b)** Trigonopyxis arcula; **(c)** Quadrulella symmetrica; **(d)** Assulina muscorum; **(e)** Centropyxis cassis-*type;* **(f)** Centropyxis (Cyclopyxis) arcelloides-*type;* **(g)** Centropyxis aculeata-*type. Scale is 20μm for b,c,d,e,f,g, and 44μm for a.*

improves. These factors led Hoogenraad and De Groot (1979) to recognize two distinctive associations in the world: a northern fauna confined to continental masses of Laurasia (North America, Europe, and Asia), and a southern fauna confined to continental masses of Gondwanaland (South America, Africa, Australia, New Zealand, and Antarctica). Central America and Indonesia are transitional zones with species and varieties common to both northern and southern regions. Reasons for these distributional patterns are unclear.

Unfortunately, there are few detailed surveys of testate amoebae in North America for comparison with faunas from other regions. Some important surveys include: Leidy (1879), Wailes (1912, 1913), Fantham and Porter (1945), Hoogenraad and De Groot (1952), and Medioli and Scott (1983). Recent work has revealed some geographical shifts in habitat preferences of certain species. Tolonen (1986) noted differences in habitat preferences of *Difflugiella oviformis* between Finland and England. Beyens *et al.* (1986) traced the habitat changes for *Assulina muscorum* from moss to lichen habitats along a short south-north transect in Greenland. In his study of Ontario bogs, Warner (1987) encountered major differences between sphagnicolous faunas in Canada and Europe. For example, *Hyalosphenia subflava* is characteristic of drained peatlands in Finland (Tolonen, 1986; Tolonen and Warner, in preparation), however this species is common in the virgin *Sphagnum* peatlands of southern Ontario, Canada. *H. subflava* is probably adapted to the xeric conditions brought about artificially by drainage in Finland, and by major natural variations in the seasonal water table position and the late summer xeric conditions in the peatlands of southern Ontario. Though it appears that many species are distributed throughout the northern hemisphere, these examples indicate the dangers of assuming cosmopolitan habitat preferences in paleoecological reconstructions. Clearly, systematic regional surveys of modern faunas are required for interpreting fossil faunas in Canada.

Protozoa depend on water to live because they possess an unprotected cell membrane for feeding. Soil protozoa are confined to the water films around soil particles. Generally, distribution and abundance are correlated with soil organic matter, which for the most part reflects their food sources such as bacteria, fungi, and algae. It is thought that oxygen content and other chemical characteristics of the water have less direct influence. Experiments have shown that testate amoebae are not affected significantly by artificial acidification of substrates (Costan and Planas, 1986). The testate amoebae reach their greatest abundance in soils, in part aided by their ability to survive prolonged dry periods by encysting (Sleigh,

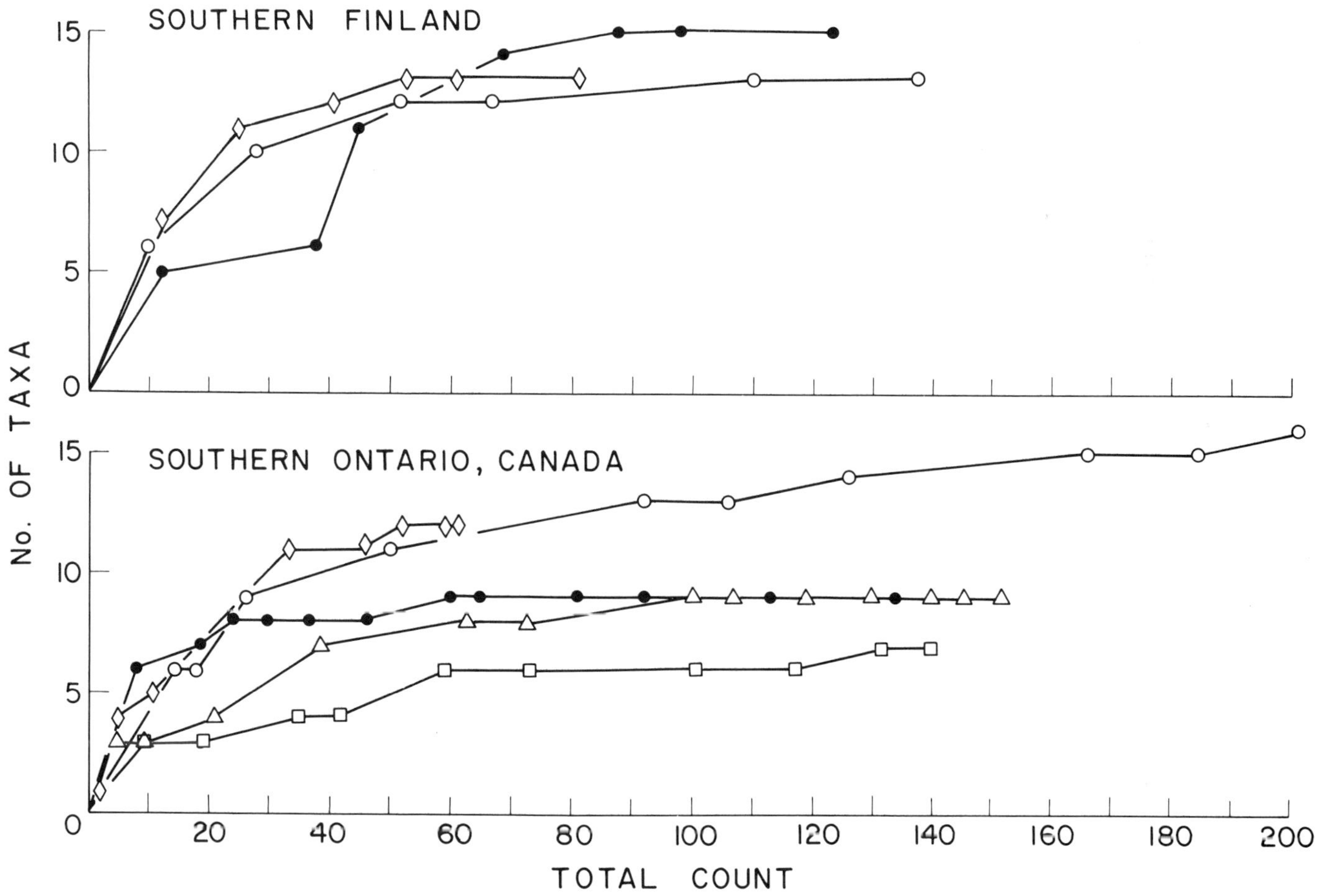

Figure 4 *Graphs showing the number of testate amoebae taxa* versus *total number of testate amoebae counted in samples from* Sphagnum *swards from various habitats throughout southern Ontario, Canada and southern Finland. See Warner (1987) for details on field sampling procedures.*

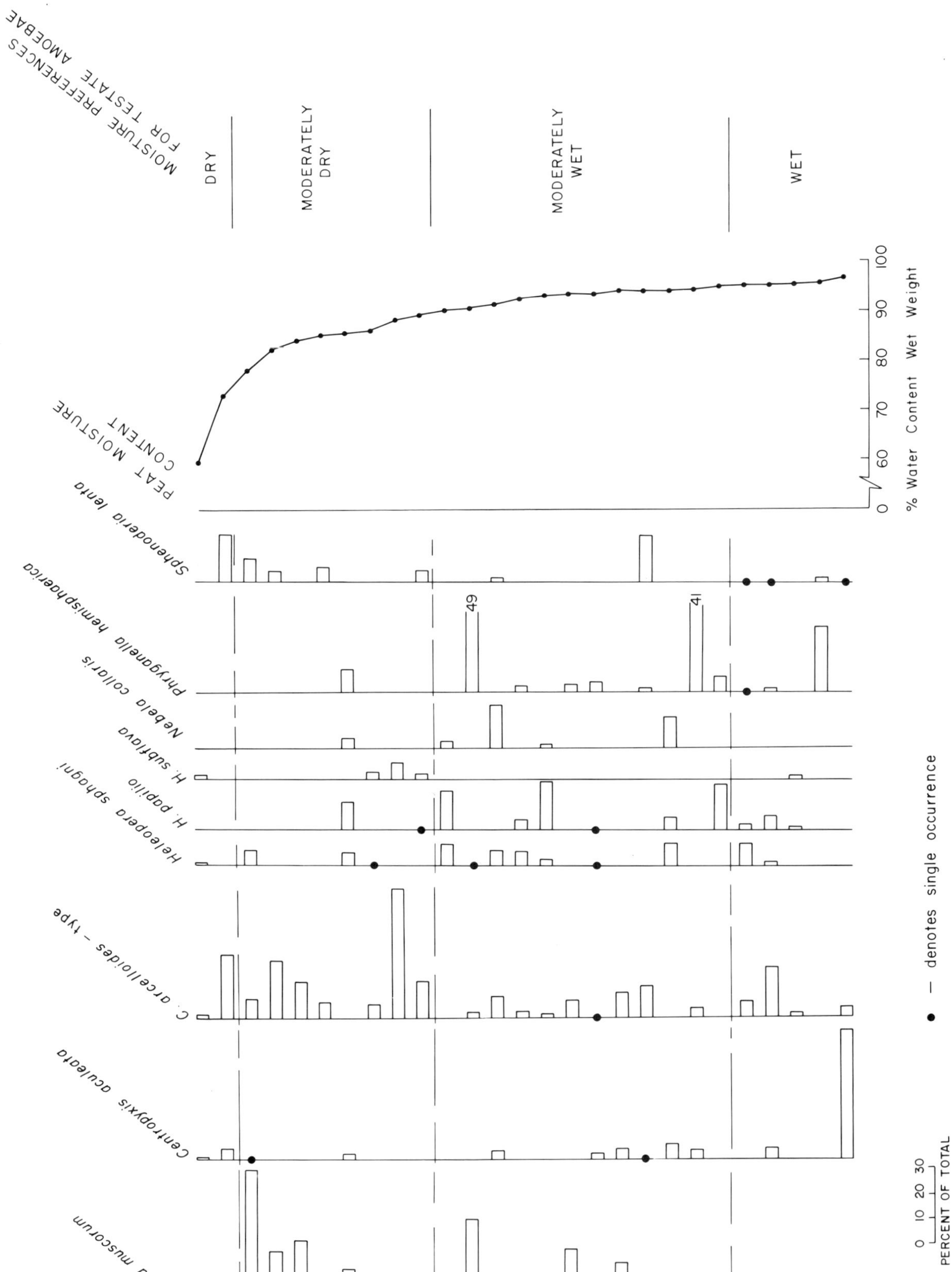

Figure 5 *Summary of important indicator taxa of preferred moisture regimes in* Sphagnum *peatlands, southwestern Ontario.*

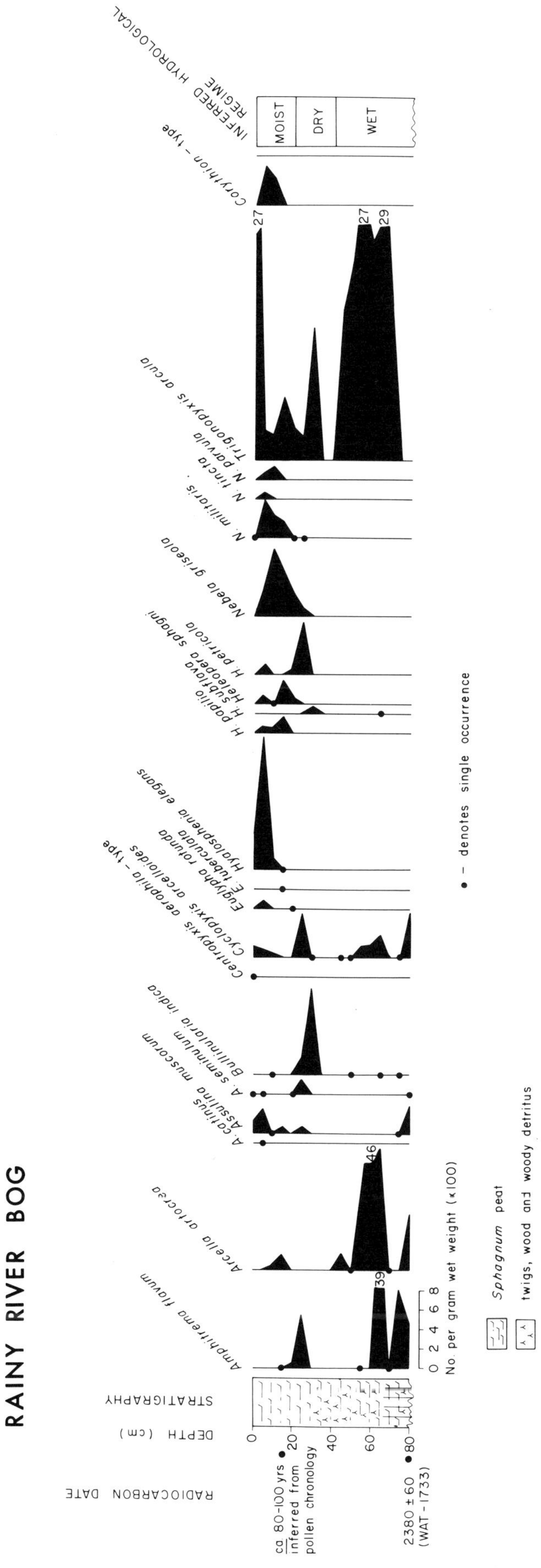

Figure 6 *Fossil testate amoebae profile from the Rainy River Bog, northwestern Ontario.*

1973). Hence, it is these physiological attributes that make testate amoebae useful in paleoecology.

In an elegant study, Meisterfeld (1977) demonstrated the importance of vertical variations of testate amoebae along the *Sphagnum* stem in response to light. Taxa such as *Hyalosphenia papilio* with zoochlorellae and symbiotic algae are restricted to the chlorophyllous segment of the *Sphagnum* stem. Greatest species diversity occurs at the transition from the chlorophyllous to the non-chlorophyllous part of the plant, and highest densities are observed on the brown part of the living stem. It is the communities confined to the brown, non-chlorophyllous segment of the *Sphagnum* sward that are important to paleoecology for they will likely constitute the fossilized assemblage.

The *Sphagnum* bogs of southern Ontario are characterized by great seasonal fluctuations in the moisture regime, and even periodic drying at the peatland surface. The testate amoebae communities are adapted to such dramatic moisture extremes. Three main community types are recognizable based on soil moisture (Warner, 1987). *Assulina muscorum* is indifferent throughout all study sites (Figure 5). *Centropyxis arcelloides*-type, *Hyalosphenia subflava,* and *Sphenoderia lenta* are characteristic of moderately dry sites (78-89% water content), and *Hyalosphenia papilio, Phryganella hemisphaerica (= P. acropodia), Heleopera sphagni,* and *Nebela collaris* are characteristic of moderately wet sites (90-95% water content). Stud-

Table 1 Moisture preferences of selected testate amoebae from bogs in eastern North America. Given is the moisture regime (as percent wet weight) of the habitat where the species was most abundant, followed by the range. (From Tolonen *et al.*, 1985).

Amphitrema flavum	93.2 (87.9–95.1)
Arcella artocrea	93.2 (88.3–93.2)
Assulina muscorum	90.5 (70.8–95.1)
A. seminulum	90.5 (70.8–95.1)
Bullinularia indica	70.8 (70.8–88.3)
Heleopera petricola	90.5 (70.8–95.1)
H. sphagni	90.2 (70.8–95.1)
Hyalosphenia elegans	90.3 (73.0–95.1)
H. ovalis	93.2 (90.2–93.2)
H. papilio	88.3 (88.3–95.1)
H. subflava	73.0 (71.0–93.2)
Nebela militaris	90.5 (70.8–95.1)
Trigonopyxis arcula	73.0 (71.0–88.4)

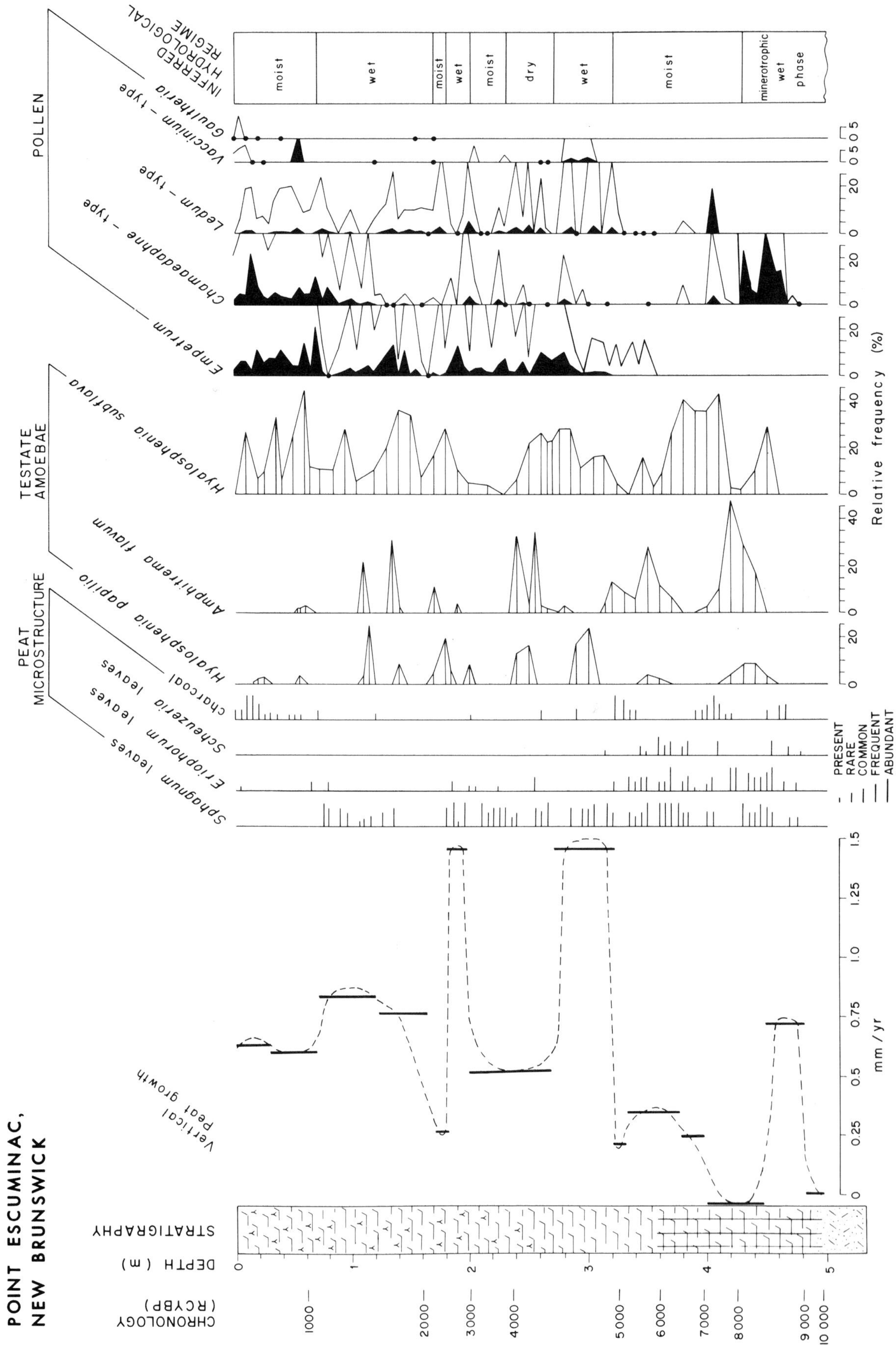

Figure 7 *Summary profile from Point Escuminac, New Brunswick showing vertical peat growth rates, peat microstructure, testate amoebae, and pollen of peatland shrubs. (Adapted from Tolonen et al., 1985, and Warner, unpublished data). The solid curves for the pollen represents the relative frequency, and the open curves represent a 10 × expansion of the relative frequency.*

ies in progress on testate amoebae in *Sphagnum* peatlands in northern Ontario indicate distinct communities associated with different soil moisture regimes. *Trigonopyxis arcula* is abundant in newly burned peatlands which might prove to be a useful indicator of past fires in peatlands (Warner, unpublished data).

Testate amoebae in coastal raised bogs in eastern North America show similar moisture preferences (Table 1). Comparison of faunas from *Sphagnum* bogs in southern Ontario and New Brunswick reveals a marked absence in the former of typical bog pool species such as *Amphitrema* spp., *Hyalosphenia ovalis,* and *Arcella artocrea.* Their absence probably relates to the lack of permanently wet sites throughout the year in the Ontario bogs.

PALEOECOLOGY

From what is known on the ecology of sphagnicolous testate amoebae, their most promising role in paleoecology is in deciphering past hydrological conditions in peatlands (*i.e.*, Beyens and Chardez, 1987). Testate amoebae were used to trace *Sphagnum* bog development in northwestern Ontario (Figure 6). Hygrophilous taxa are characteristic of the early wet minerotrophic *Sphagnum* fen, followed by the dominance of xerophillous species in a dry *Sphagnum* peatland by 2400 years B.P., before development of the modern hydrophillous-xerophillous association in the ombrotrophic peatland. The *Sphagnum* peatland, probably was either in response to a drop in the basin water table, or there was a brief rejuvenation phase as peat accumulation reached the limits of the water table, before onset of ombrotrophy.

In another example, testate amoebae analysis was performed on a raised bog in New Brunswick (Figure 7). *Hyalosphenia papilio*, a hydrophilous species, and *H. subflava*, a xerophilous, species reveal the alternating hydrology between wet and dry conditions. Generally, wet intervals are characterized by rapid peat growth, well-preserved *Sphagnum* leaves, and peaks in *Chamaedaphne*-type pollen, which are followed by dry periods of slow peat growth, poor preservation of *Sphagnum*, and high amounts of charcoal. These observations reveal the impact of fire on bog development. The surface of the peatland became wet after major fires. Thus, the local impact of fire blurrs the regional vegetational and climatic records, and serves to illustrate some difficulty in using testate amoebae analysis and peatlands in general, for regional paleoenvironmental reconstructions.

The future of testate amoebae analysis in Quaternary ecology in Canada is bright, which undoubtedly will contribute to new insights on peatland development and Quaternary history. The degree to which testate amoebae can contribute to paleoecology depends on our understanding of modern faunas in North America. In view of their potential as paleohydrological indicators, testate amoebae will aid in deciphering fundamental processes of bog development in Canada, processes that we already know are different from classical schemes developed in Europe (Warner, unpublished data). The key to the wise use and management of Canada's vast peatland resources lies in a thorough understanding of past developmental trends, for the Canadian peatland resource can be regarded as renewable. Testate amoebae can be an important tool in peatland management schemes in the forestry, agricultural, and horticultural industries. As testate amoebae characterize specific ecological habitats, they may be used to refine peat type classification schemes in the peat by-product industry.

I have made little mention of testate amoebae in paleolimnology, and their role in the paleoecological study of paleosols. Has anyone ever found fossil testate amoebae in coprolites or animal middens, or associated with archeological sites and artifacts?

ACKNOWLEDGEMENTS

I am indebted to Professor K. Tolonen, University of Joensuu for introducing me to testate amoebae analysis. Versions of this manuscript benefited from critical readings by H.J. Kubiw, J.D. Lousier, and C.G. Ogden. N. Bahar drafted the figures. Ongoing research on testate amoebae which contributed to this paper, is supported by the Natural Sciences and Engineering Research Council of Canada.

REFERENCES

Aaby, B., 1976, Cyclic climatic variations in climate over the past 5500 years reflected in raised bogs: Nature, v. 263, p. 281-284.

Aaby, B. and Tauber, H., 1975, Rates of peat formation in relation to degree of humification and local environment as shown by studies of a raised bog in Denmark: Boreas, v. 4, p. 1-17.

Barber, K.E., 1981, Peat stratigraphy and climatic change: A.A. Balkema, Rotterdam, 219 p.

Beyens, L., 1984, A concise survey of testate amoebae analysis: Bulletin de la Société Belge de Géologie, v. 93, p. 261-266.

Beyens, L., 1985, On the Subboreal climate of the Belgian Campine as deduced from diatom and testate amoebae analysis: Review of Palaeobotany and Palynology, v. 46, p. 9-31.

Beyens, L. and Chardez, D., 1987, Evidence from testate amoebae for changes in some local hydrological conditions between c. 5000 B.P. and c. 3800 B.P. on Edgeoya (Svalbard): Polar Research, v. 5, p. 165-169.

Beyens, L., Chardez, D. and De Landtsheer, R., 1986, Testate amoebae populations from moss and lichen habitats in the Arctic: Polar Biology, v. 5, p. 165-173.

Birks, H.J.B. and Birks, H.H., 1980, Quaternary paleoecology: Edward Arnold, London, 289 p.

Casparie, W.A., 1972, Bog development in southeastern Drenthe (The Netherlands): Vegetatio, v. 25, p. 1-271.

Committee on Systematics and Evolution of the Society of Protozoologists, 1980, A newly revised classification of the Protozoa: Journal of Protozoology, v. 27, p. 37-58.

Corbet, S.A., 1973, An illustrated introduction to the testate rhizopods in *Sphagnum*, with special reference to the area around Malham Tarn, Yorkshire: Field Studies, v. 3, p. 801-838.

Costan, G. and Planas, D., 1986, Effects of short-term experimental acidification on a microinvertebrate community: Rhizopoda, Testacea: Canadian Journal of Zoology, v. 64, p. 1224-1230.

de Vernal, A., Lortie, G., Larouche, A., Scott, D.B. and Richard, P.J.H., 1985, Évolution d'un milieu littoral et remontée du niveau relatif de la mer à l'Holocène supérieur au nord de l'île du Cap-Breton, Nouvelle-Écosse: Canadian Journal of Earth Sciences, v. 22, p. 315-323.

Decloitre, L., 1962, Le genre *Euglypha* Dujardin: Archiv für Protistenkunde, v. 106, p. 53-100.

Deflandre, G., 1928, Le genre *Arcella* Ahrenberg: Archiv für Protistenkunde, v. 64, p. 152-167.

Deflandre, G., 1936, Etude monographique sur le genre *Nebela* Leidy (Rhizopoda-Testacea): Annales de Protistologie, v. 5, p. 201-286.

Douglas, M.S.V. and Smol, J.P., 1987a, Siliceous protozoan and chrysophycean microfossils from recent sediments of *Sphagnum* dominated Lake Colden, N.Y., U.S.A.: Verhandlungen Internationale Vereinigung für Theoretische und Angewandte Limnologie, (in press).

Douglas, M.S.V. and Smol, J.P., 1987b, Siliceous protozoan plates in lake sediments: Hydrobiologia, v. 154, p. 13-23.

Ellison, R.L. and Ogden, C.G., 1987, A guide to the study and identification of fossil testate amoebae in Quaternary lake sediments: Internationale Revue der Gesamten Hydrobiologia, v. 72, p. 639-652.

Fantham, H.B. and Porter, A., 1945, The microfauna, especially the Protozoa, found in some Canadian mosses: Zoological Society London, Proceedings, v. 115, p. 97-174.

Frey, D.G., 1964, Remains of animals in Quaternary lake and bog sediments and their interpretation: Ergebnisse der Limnologie, v. 2, p. 1-114.

Grospietsch, T., 1953, Rhizopodenanalytische Untersuchungen an Mooren Ostholsteins: Archiv für Hydrobiologie, v. 47, p. 321-452.

Grospietsch, T., 1958, Wechseltierchen (Rhizopoden): Kosmos, Stuttgart, 86 p.

Grospietsch, T., 1964, Monographische Studie der Gattung *Hyalosphenia* Stern: Hydrobiologia, v. 26, p. 211-241.

Harnisch, O., 1927, Eige Daten sur rezenten und fossilen testaceen Rhizopodenfauna der Sphagnen: Archiv für Hydrobiologie, v. 18, p. 245-360.

Heal, O.W., 1962, The abundance of microdistribution of testate amoebae (Rhizopoda: Testacea) in *Sphagnum*: Oikos, v. 13, p. 35-47.

Heusser, C.J., 1960, Late Pleistocene environments of North Pacific North America: American Geographical Society, New York, 308 p.

Hoogenraad, H.R. and De Groot, A.A., 1952, Thekamöbe moosrhizopoden aus Nordamerika: Archiv für Hydrobiologie, v. 47, p. 229-262.

Hoogenraad, H.R. and De Groot, A.A., 1979, Die Geographische Verbreitung der Süsswasser-Rhizopoden: Hydrobiological Bulletin, v. 13, p. 152-171.

Lagerheim, G., 1902, Om lämningar af rhizopoder, heiliozer och tintinnider i Sveriges och Finlands lakustrina kvartäraflagringar: Geologiska Föreningar Forhandlingar, v. 23, p. 469-520.

Lee, J.J., Hutner, S.H. and Bovee, E.C. eds., 1985, An illustrated guide to the Protozoa: Society of Protozoologists, St. Lawrence, Kansas, 629 p.

Leidy, J., 1879, Freshwater rhizopods of North America: Report of the United States Geological Survey, Monograph 12, Washington, 324 p.

Loeblich, A.R. and Tappan, H., 1964, Protista 2, Sarcodina, *in* Moore, R.C., ed., Treatise on Invertebrate Paleontology, vol. 1: Geological Society of America and University of Kansas Press, p. C16-C54.

Lousier, J.D., 1984a, Population dynamics and production studies of species of Nebelidae (Testacea, Rhizopoda) in an aspen woodland soil: Acta Protozoologica, v. 23, p. 145-159.

Lousier, J.D., 1984b, Population dynamics and production studies of *Phryganella acropodia* and *Difflugia oviformis* (Testacea, Rhizopoda, Protozoa) in an aspen woodland soil: Pedobiologia, v. 26, p. 331-347.

Lousier, J.D., 1984c, Population dynamics and production studies of species of Euglyphidae (Testacea, Rhizopoda, Protozoa) in an aspen woodland soil: Pedobiologia, v. 26, p. 309-330.

Lousier, J.D., 1985, Population dynamics and production studies of species of Centropyxidae (Testacea, Rhizopoda) in an aspen woodland soil: Archiv für Protistenkunde, v. 130, p. 165-178.

Lousier, J.D. and Parkinson, D., 1981, The disappearance of the empty tests of litter- and soil-testate amoebae (Testacea, Rhizopoda, Protozoa): Archiv für Protistenkunde, v. 124, p. 312-336.

Medioli, F.S. and Scott, D.B., 1983, Holocene Arcellacea (Thecamoebians) from eastern Canada: Cushman Foundation for Foraminiferal Research, Special Publication 21, 63 p.

Meisterfeld, R., 1977, Die horizontale und verteilung der Testaceen (Rhizopoden, Testacea) in *Sphagnum*: Archiv für Hydrobiologie, v. 79, p. 319-356.

Ogden, C.G. and Hedley, R.H., 1980, An atlas of freshwater testate amoebae: British Museum (Natural History), Oxford University Press, London, 222 p.

Ogden, J.G., 1960, Recurrence surfaces and pollen stratigraphy of a postglacial raised bog, Kings County, Nova Scotia: American Journal of Science, v. 258, p. 341-353.

Overbeck, F., 1975, Botanisch-Geologische Moorkunde: K. Wachhooltz Verlag, Neumünster, 719 p.

Ruzicka, E., 1982, Die subfossilen Testaceen des Krottensees (Salzburg, sterreich): Limnologica (Berlin), v. 14, p. 49-88.

Schönborn, W., 1962, Die Ökologie der Testaceen im oligotrophen See, dargestellt am Beispiel des Grossen Stechlinsees: Limnologica (Berlin), v. 1, p. 111-182.

Schönborn, W., 1973, Paläolimnologische Studien an Testaceen aus Bohrkernen des Latnjajaure (Abisko-Gebiet; Schwedisch-Lappland): Hydrobiologia, v. 42, p. 63-75.

Schönborn, W., 1984, Studies on remains of Testacea in cores of the Great Woryty Lake (NE-Poland): Limnologica, v. 16, p. 185-190.

Schönborn, W., 1986, Population dynamics and production biology of testate amoebae (Rhizopoda, Testacea) in raw humus of two coniferous forest soils: Archiv für Protistenkunde, v. 132, p. 325-342.

Scott, D.B. and Medioli, F.S., 1983, Agglutinated rhizopods in Lake Erie: Modern distribution and stratigraphic implications: Journal of Paleontology, v. 57, p. 809-820.

Sleigh, M., 1973, The biology of the Protozoa: Edward Arnold, London, 315 p.

Tolonen, K., 1966, Stratigraphic and rhizopod analyses on an old raised bog, Varrassuo, in Hollola, South Finland: Annales Botanici Fennici, v. 3, p. 147-166.

Tolonen, K., 1968, Zur Entwicklung der Binnenfinnland-Hochmoore: Annales Botanici Fennici, v. 5, p. 17-33.

Tolonen, K., 1986, Rhizopod analysis, *in* Berglund, B.E., ed., Handbook of Holocene Paleoecology and Palaeohydrology: John Wiley and Sons Ltd., Chichester, p. 645-666.

Tolonen, K., Huttunen, P. and Junger, H., 1985, Regeneration of two coastal raised bogs in eastern North America: Stratigraphy, radiocarbon dates and rhizopod analysis from sea cliffs: Annales Academiae Scientiarum Fennicae, Series A, v. 139, p. 1-51.

Wailes, G.H., 1912, Freshwater Rhizopoda and Heliozoa from the States of New York, New Jersey, and Georgia, U.S.A.: Zoological Journal of the Linnean Society, v. 32, p. 201-218.

Wailes, G.H., 1913, Freshwater Rhizopoda from North and South America: Zoological Journal of the Linnean Society, v. 32, p. 122-161.

Warner, B.G., 1987, Abundance and diversity of testate amoebae (Rhizopoda, Testacea) in *Sphagnum* peatlands in southwestern Ontario, Canada: Archiv für Protistenkunde, v. 133, p. 173-189.

Originally published in
Geoscience Canada Volume 15 Number 4
(December 1988)

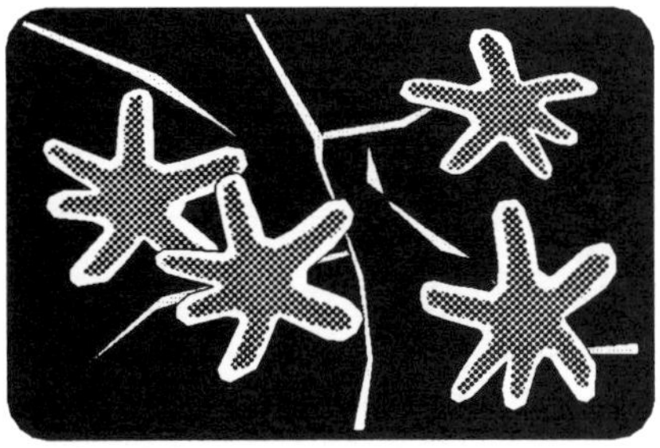

Freshwater Sponges

Frederick W. Harrison
Department of Biology
Western Carolina University
Cullowhee, North Carolina 28723

SUMMARY

In recent years, there has been an increased awareness of the value of fossilized remains of freshwater sponges in Quaternary paleoecological reconstructions. Because the taxonomy and ecology of extant forms have been clarified, and because a well-defined fossil record exists that exhibits continuity with extant species, the stage is set for widespread utilization of freshwater sponge remains in Quaternary environmental reconstructions. This paper presents an introduction to the freshwater sponges, techniques for their utilization in Quaternary paleoecological studies, and summarizes some of the most recent work performed in North America.

INTRODUCTION

Despite the astonishment of many individuals, sponges do exist in most freshwater habitats and in fact, have existed for at least 100 million years. We are familiar with marine sponges. Their often brilliant hues attract attention and even if most sponges purchased from the local merchant consist of a non-natural plastic, many of us still remember the day when a "bath sponge" was non-synthetic. But what of these freshwater sponges? What do they look like (Figures 1 and 2), where can one find them, are they *truly* sponges, and how are they important in Quaternary ecology?

FRESHWATER SPONGE BIOLOGY: AN OVERVIEW

Freshwater sponges are members of the great poriferan class Demospongiae, a taxonomic grouping that contains most species of marine sponges. The majority of freshwater sponges are members of the family Spongillidae, a widely distributed group characterized by a siliceous spicular skeleton and by asexual reproductive structures, gemmules. Fortunately for the paleoecologist, both these diagnostic sponge parts are found as fossils in lacustrine and peat deposits.

As with all sponges, the physiology of freshwater forms is organized around a system of canals that courses through the body of the animal (Figure 3). Water is driven through the canals by clusters of flagellated cells, choanocytes; thus, the animal is in intimate contact with the aquatic environment. Because of this degree of dependency, the sponges are delicate monitors of water quality, degenerating or dying in the presence of adverse conditions. But under favourable circumstances, the sponges utilize their aquatic environment, filtering their bacterial food from it, assimilating dissolved oxygen, and releasing waste products into the excurrent flow of water in their canal systems.

In terms of most criteria, the freshwater sponges are organized and exist upon the same ancient organizational and physiological patterns as most marine sponges. They differ from most marine sponges in their mode of reproduction. While they do reproduce sexually, freshwater sponges also survive the

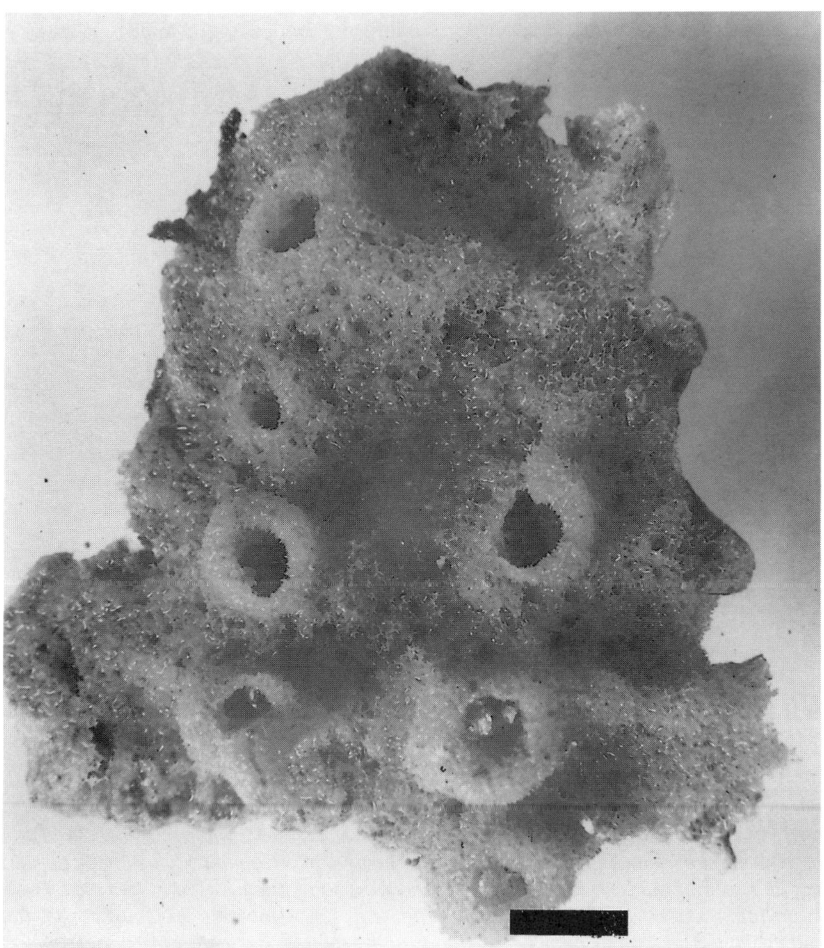

Figure 1 *Modern freshwater sponge,* Eunapius carteri. *The large openings are oscules. Scale bar = 1 cm.*

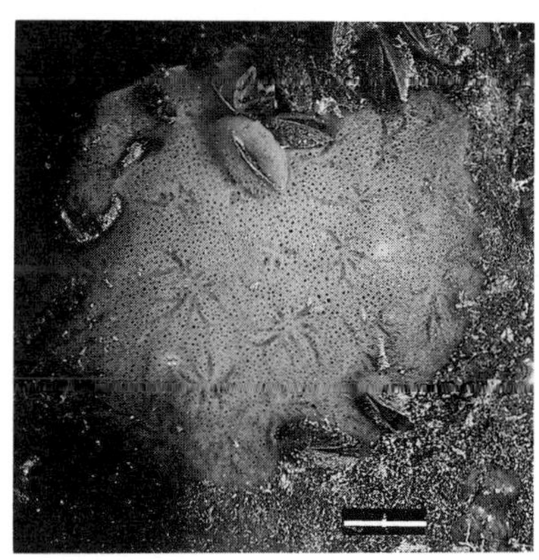

Figure 2 *Modern freshwater sponge,* Eunapius fragilis. *Scale bar equals 1 cm.*

harsh conditions of northern winters or the warm waters of southern summers by entering an asexual phase of resistant gemmules.

The habit of gemmule formation apparently evolved as a response to unstable environmental conditions. This is seen in a few marine species, particularly those found in harsh, fluctuating intertidal environments.

Seasonal environmental changes trigger the release of endogenous stimuli, perhaps cyclic nucleotides, in the body of the sponge. The ameboid cells of the interior of the sponge migrate and mass in spherical aggregates. As gemmule development proceeds, cells at the aggregate periphery assume a secretory function, depositing a shell of spongin, a form of collagen. Cells in the aggregate interior accumulate yolk in lens-shaped platelets which crowd the cytoplasm of these reproductive cells. However, the gemmules with their protective shells of spongin survive and await the coming of more favourable conditions. With the return of warmer, or cooler, water temperatures, a burst of germinative activity sends the yolk cells of the gemmule migrating out over its surface, spreading upon the substrate, and differentiating into a young sponge. There the sponge grows, forming an encrusting green, tan, or darkly coloured mass on emergent vegetation, upon floating logs, or in the folds and ridges of submerged stones.

The preceding paragraphs offer a brief overview of limited aspects of freshwater sponge biology. The biology of sponges has been reviewed in detail in recent publications. Those seeking further information about these animals should begin with the excellent book by Simpson (1984), *The Cell Biology of Sponges*.

Eventually, siliceous spicules comprising the skeleton of the sponge, and the hard, resistant shells of gemmules become incorporated in the sediment. Although spicules and gemmules which are easily recognizable and diagnostic are preserved in the fossil record, they have not played a significant role in paleoecological reconstructions. Their use has been overlooked by paleoecologists due to a dearth of knowledge of the ecology of extant freshwater sponge species and long-standing confusion about the taxonomy of the group. Recent advances in both areas of study finally allow the utilization of fossilized sponge remains in paleoecology.

FRESHWATER SPONGES IN PALEOECOLOGY: THE HISTORICAL PERSPECTIVE

For freshwater sponges to be useful in paleoecological studies, it is necesary for three factors to exist. First, it is necessary to have a well-defined fossil record that exhibits continuity with extant species. Second, an orderly systematic framework must be available so that the non-specialist on sponges may identify his materials. Finally, the environmental preferences, including physicochemical parameters, of species must be understood. Great effort has been made within the past two decades to meet these three conditions, thus setting the stage for widespread use of this group by paleoecologists.

Much of our understanding of freshwater sponge systematics dates from the monumental study by A.A. Racek (Penney and Racek, 1968). Racek, in

Figure 3 *Diagram of the anatomy of a freshwater sponge. "Knobbed" arrows indicate incurrent flow of water. Arrows without "knobs" indicate excurrent flow. A superior epithelium consisting of a layer of fusiform external pinacocytes and an inner layer of endopinacocytes (black cells) is pierced by pores (**p**). Environmental water passes through the surface epithelium, into a sub-epithelial space (atrium, **at**), and into the loosely organized mesohyl (**m**). Water is both drawn into and expelled from choanocyte chambers (**c**) by the beating of choanocyte (collar cell) flagella. The excurrent system terminates in oscular chimneys (**o**). Water exits the sponge by oscules, openings at the top of the chimney. Ameboid cells (**a**) migrate through the mesohyl. Skeletal elements, spicules (**sp**), and the basopinacoderm (**b**) attachment surface respectively, are bound into fasicles or attach through the matrix material, spongin (**s**).*

completing the study only just initiated before the death of Professor Penney, brought order out of the chaos of spongillid systematics by recognizing global evolutionary patterns among spongillid species. He demonstrated the artificial nature of the two spongillid subfamilies Spongillinae and Meyeninae, erecting a new genus, *Radiospongilla*, for those sponges showing morphological similarities to the two subfamilies. Although lacking fossil evidence, Racek proposed that the two artificial subfamilies had diverged from a common ancestral stock, a stock that had also given rise to the genus *Radiospongilla*. The fossil spongillid, *Paleospongilla chubutensis*, recovered from Cretaceous strata, exhibits spicular characters combining those found in the two subfamilies, thus suggesting the derivation of both subfamilies and appearing to be the most direct ancestral form of the genus *Radiospongilla* (Racek and Harrison, 1975). With the clarification of spongillid systematics, recent taxonomic treatments and keys (Pennak, 1978) have followed the Penney and Racek (1968) nomenclature. Therefore, as spicules and gemmules (Figures 4 to 8), the same elements that are preserved as fossils, are diagnostic criteria for fossil sponge identification, specialists may identify sponges both extant and fossil with the same level of confidence. Recently, increasing work, particularly on North American species has allowed for a fuller understanding of sponge ecology. The ecology of individual species of North American freshwater sponges is presented in a series of papers by Harrison (1974, 1977, 1979), Harrison *et al.* (1977), Harrison and Harrison (1977, 1979), Jewell (1935, 1939), Old (1932), and Poirrier (1969, 1974, 1977).

Consequently, the three necessary criteria have been met. A fossil record with direct relationships to extant forms dates from Cretaceous time. Freshwater sponge species exist within a well-defined systematic framework and, important for limnologic analysis, the presence of particular species may be useful in characterizing and defining local aquatic environments.

Therefore, building upon earlier European studies (Frey, 1964) and upon pioneering studies by Racek (1966, 1970, 1974) of deposits from Europe, Jordan, and Central America, there has been increasing interest in freshwater deposits particularly in North America (Hall and Hermann, 1980; Harrison *et al.*, 1979; Turner, 1985; Harrison and Warner, 1986).

FIELD SAMPLING AND LABORATORY EXTRACTION TECHNIQUES

In the field, lake and peatland cores can be collected with standard coring equipment. For spicule analysis, subsamples are boiled in 1:1 mixture of concentrated nitric and sulphuric acids until a dry white dust remains. Following washing, this residue is retained for microscopic examination. In addition, specific volumes (usually 20 cm^3) of fresh sediment are washed on a 250 μm mesh sieve, and the remaining residue is examined for macroscopic gemmules under a dissecting microscope.

Identification of freshwater sponge remains relies upon analysis of spicules and gemmules. There are three spicule types in freshwater sponges: megascleres, gemmoscleres, and microscleres. The largest, megascleres, are skeletal spicules which form a supportive framework for the delicate sponge tissues. Megascleres are of limited taxonomic value although they are the predominant spicule type in the sponge. Gemmoscleres which are components of the gemmule coat form the armature of gemmules. Their morphology and placement in the gemmule coat are of considerable diagnostic value in taxonomy. Microscleres, the small "dermal"

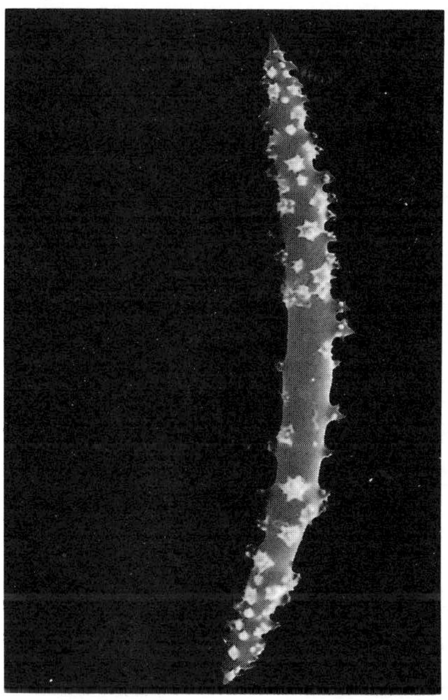

Figure 4 *Scanning electron micrograph of a microsclere of* Spongilla lacustris. *Although this spicule type is not found in all freshwater sponge species, when present, the microscleres are of considerable diagnostic value. Magnification* 2400 ×.

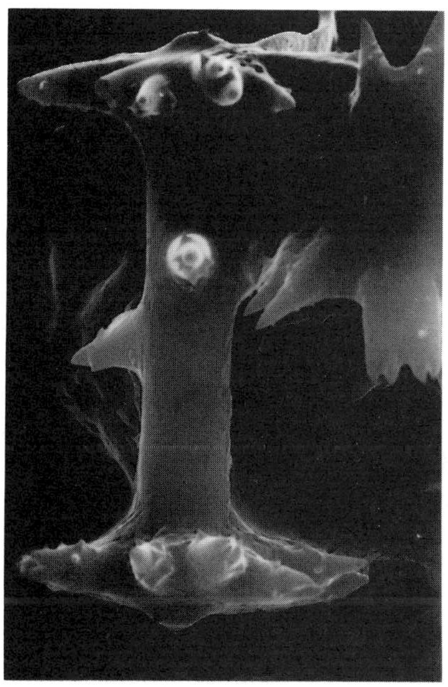

Figure 5 *Scanning electron micrograph of a birotulate gemmosclere of* Ephydatia fluviatilis. *The rotule of another spicule is seen in the upper right of the figure. Magnification* 2000 ×.

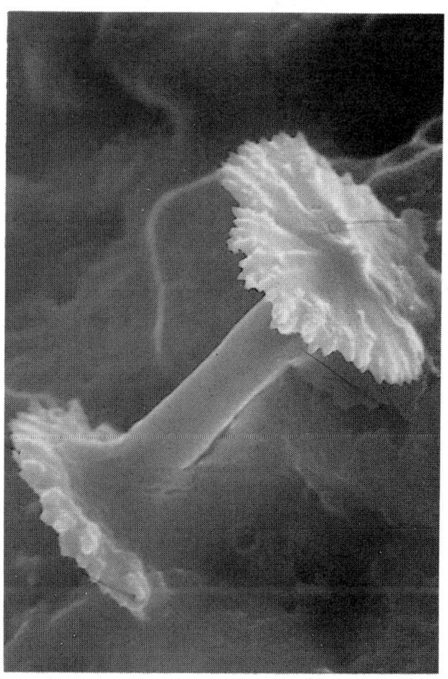

Figure 6 *Scanning electron micrograph of a gemmosclere of* Anheteromeyenia ryderi. *Compare rotule morphology with that of the* E. fluviatilis *gemmosclere seen in Figure 5. This spicule is partly embedded in the spongin coat of the gemmule. Magnification* 1800 ×.

spicules, are not found in all species. However, their presence or absence and their morphology are of considerable value in taxonomic determinations. A taxonomic key to North American freshwater sponges is presented in Pennak (1978). The original Penney and Racek (1968) monograph is valuable because of the clarity of illustrations. The scanning electron microscope promises to be of exceptional value in taxonomic analysis, but at present, has been rarely used (Harrison, 1981).

QUATERNARY ENVIRONMENTAL RECONSTRUCTIONS

Although there have been few studies utilizing freshwater sponge remains, the most recent studies involve examination of lake sediments from Canada and from Kentucky, USA. Analyses of fossil spicules and whole gemmules of freshwater sponges were performed on early Holocene sediments collected from Serendipity Bog, a bog lake on eastern Graham Island, Queen Charlotte Islands, British Columbia (Harrison and Warner, 1986). Their study established the essential criterion of corroboration of data obtained from analyses of freshwater sponge remains with data derived independently from other fossilized components, *e.g.*, pollen and plant macrofossils.

Examination of the spicular components in limnic sediments from the Serendipity Bog site revealed a flourishing freshwater sponge population between about 9400 and 8500 years B.P., composed of *Spongilla lacustris* and a member of the genus *Anheteromeyenia*. Of the two species, apparently, *Anheteromeyenia* was the community dominant throughout this period. Two peaks of fossil spicules, one at 3.15-3.10 m and another at 2.95-2.90 m depth, indicate that there were two major periods exceptionally favourable to the growth of sponges. Of the species represented in the Serendipity core, members of the genus *Anheteromeyenia* generally are more specific in their environmental requirements than is the cosmopolitan *S. lacustris*.

Sponges of the genus *Anheteromeyenia* prefer waters of moderate to low alkalinities (known range, 2-80 $mgCaCO_3 \cdot L^{-1}$; Harrison, 1974, 1977) that have high specific conductances (known range, 37-750 μmhos; Old, 1932; Jewell, 1939; Poirrier, 1969; Harrison, 1974, 1977). They prefer waters that are slightly acidic, but will grow well in circum-neutral to slightly basic habitats (known range, pH 4.2-8.5; Harrison, 1974, 1977).

Extant *S. lacustris* exhibits broader habitat tolerances, although Poirrier (1969) notes that the species is common in acidic water low in bicarbonate; that it exists through a wide range of pH; and that it shows a slight tolerance to basic waters of higher specific conductances, silicon concentrations, and colour (*i.e.*, dissolved organics) (known ranges, total hardness as $CaCO_3$, 28.8- 250.0 $mg \cdot L^{-1}$; pH 5.3-9.0; specific conductance, 9.4-470 μmhos; silicon 0-20.5 ppm; colour 0-202; Harrison, 1974). Neither of these species is capable of withstanding high levels of turbidity, since suspended silt and clay particles tend to clog the canal system of the sponge's body.

Spicule development in *S. lacustris* and in other freshwater sponges is related directly to silicon levels (Jewell, 1939). Conspicuously abnormal (*i.e.*, underdeveloped) skeletal elements from *S. lacustris* were not observed in the fossil sequence, which suggested moderate to high levels of silicon in the water, an interpretation proffered in consideration of the quartzo-feldspathic sand deposits that lie beneath the Serendipity basin.

Available ecological information confirms that *S. lacustris* prefers habitats with high light intensity. Modern colonies of *S. lacustris* which were attached to small crystalline rocks, associated with *Isoetes echinospora* on the shore of Hippa Lake, west of Graham Island were observed in water about 15 cm deep. Spicules of *S. lacustris* appeared to be more abundant near the base of the Serendipity core, and progressively declined upward. Paleobotanical data indi-

Figure 7 *Scanning electron micrography of a gemmule of* Anheteromeyenia ryderi. *Close examination of the figure will reveal the two gemmosclere forms characteristic of the species. The shorter, "flat-head" rotule form is also seen in Figure 6. Magnification* 440 ×.

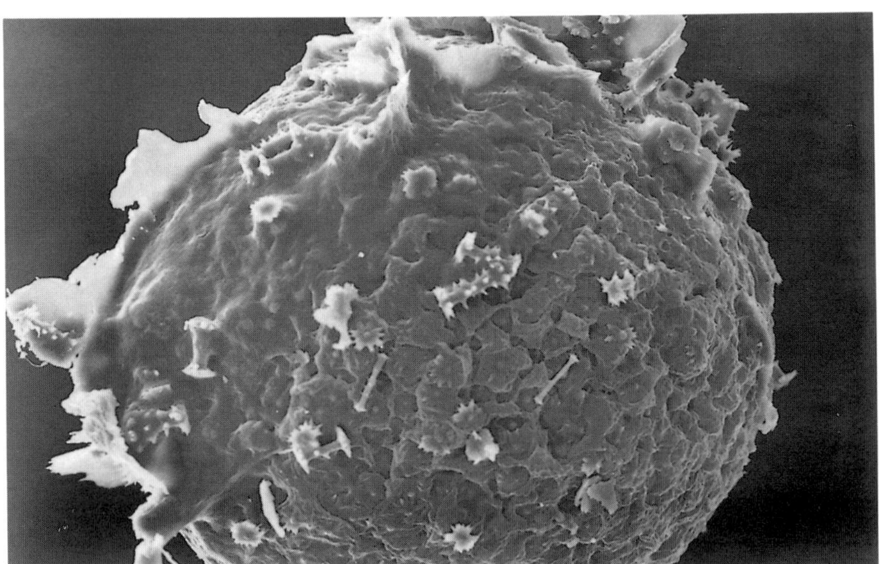

Figure 8 *Scanning electron micrograph of a gemmule of* Ephydatia fluviatilis. *Gemmules of this species exhibit only one class of gemmosclere (Figure 5) as opposed to* A. ryderi *(Figures 6 and 7). Magnification* 180 ×.

cated the presence of submerged and floating-leaved aquatic macrophytes (Warner, 1984), which may suggest higher light intensities during earlier phases of development of the Serendipity basin. Subsequent macrophytic growth and progression to a fen community might have increased shading, and hence may have contributed to a reduction in the population of *S. lacustris*.

Members of the *Anheteromeyenia* group express variable responses to light. Perhaps the fossil remains of *Anheteromeyenia* were in fact those of *A. ryderi*, a species known to prefer shaded habitats. This factor may explain its better representation in the fossil sequence, inasmuch as it would be able to survive the shading effect of associated prolific macrophytic growth.

The presence of fossil materials from the submerged macrophyte *Potamogeton filiformis* corroborated limnological tolerances indicated by the freshwater sponge fauna. Like *S. lacustris* and *Anheteromeyenia* sp., *P. filiformis* occurs in circum-neutral to basic waters with pH usually above 5.9, and is indifferent to alkalinity of the water and specific conductance. However, *P. filiformis* almost always is confined to mineral substrates and to water from between 0.5 and 2.0 m deep (Warner, 1984).

Therefore, both the sponge fauna and the submerged aquatic flora as reconstructed from fossil evidence reflect that the early Serendipity basin was shallow. This basin contained standing or slow-moving waters that exhibited circum-neutral to slightly basic pH with moderate to low carbonate alkalinity and high specific conductance, silicon, and dissolved organic content.

The site of a second study utilizing fossil remains of freshwater sponges is Jackson Pond, a sink-hole lake located in the north-central region of Kentucky. The pond lies in the northern portion of the Interior Low Plateaus and within 190 km of the full-glacial ice margin (Wilkins, 1985; Delcourt *et al.*, 1986). Jackson Pond sediments have been radiocarbon-dated from 20,000 years B.P. to the present (Wilkins, 1985).

Pollen records from Jackson Pond (Wilkins, 1985) indicate that the full-glacial (20,000 to 17,000 years B.P.) and late-glacial (17,000 to 11,500 years B.P.) intervals in central Kentucky were periods of cool, wet climate with high sedimentation rates in the pond, indicative of sheetwash erosion from the nearby uplands. These high sedimentation rates were associated with conditions of high levels of soil moisture (Delcourt, 1985; Wilkins, 1985). During the full-glacial interval, the region was occupied by closed boreal forest dominated by spruce with jack pine as a subdominant. After 17,000 years B.P., a period of open boreal woodlands characterized by both tree populations, but with an increase in herbaceous plants, persisted until the Holocene transition.

This transition may mark a shift to a more moderate, warm, wet environment. The open boreal woodland was replaced by an open deciduous woodland dominated by *Quercus* and *Ostrya/Carpinus*.

The mid-Holocene interval (7500 to 3500 years B.P.) was a warmer, drier period with low soil moisture. The pollen record indicates an upland mosaic of vegetation with open woodland and prairie patches. The upland deciduous forest was characterized by tree populations of oak, hickory, and chestnut. On mesic slopes surrounding Jackson Pond, forests contained black ash, willow, sweetgum and blackgum.

During late Holocene time (3500 years B.P. to the present), the climate was warm and moist. Willow populations decreased near Jackson Pond. Open grasslands succeeded at the expense of upland woodlands; grasses increased to comprise up to 20% of the upland pollen assemblage. Fossil pollen of *Petalostemon* occurred throughout this interval, indicating the areal expansion of prairie (Wilkins, 1985).

Freshwater sponge spicule analysis was performed on the complete sediment profile from 20,000 years B.P. to the present. The analyses were performed at about 1000-year intervals. During the full-glacial and late-glacial intervals (20,000 to 11,500 years B.P.), the dominant freshwater sponge species in Jackson Pond was *Heteromeyenia latitenta*. Populations of this sponge co-existed with cool-temperate populations of aquatic plants such as naiads and pondweeds (such as *Najas gracillima, N. flexilis*, and *Potamogeton spirillus* found together today in the Great Lakes and New England regions). Populations of this cold-water sponge declined in Jackson Pond in the early Holocene and disappeared during the warm, dry mid-Holocene interval. No *H. latitenta* spicules were found after 6000 years B.P. The known distributional range of extant populations of this sponge is limited to the northeastern United States. This is consistent with the relict boreal populations of spruce, persisting until about 6000 years ago in the cool microclimate around Jackson Pond.

A population of the sponge, *Ephydatia fluviatilis* co-existed with *H. latitenta* and has persisted to the present. This species is favoured by water conditions of high conductivity, $CaCO_3$ hardness, and alkalinity. It can withstand silt-laden, turbid waters. Macrophyte abundance of oospores of marl-forming charophytes confirm conditions of $CaCO_3$ hardness and alkalinity, particularly between 20,000 and 7000 years B.P. Between 8000 and 7000 years B.P., a third sponge species, *Anheteromeyenia ryderi*, colonized Jackson Pond. This coincided in time with the fossil evidence for the establishment of upland populations of warm-temperate trees such as chestnut and the colonization of lower slopes and wetlands at the margin of Jackson Pond by warm-temperate species of sweetgum, willow, black gum, buttonbush, Virginia willow, and alder. By 7000 years B.P., the bottomland species of willow, buttonbush, Virginia willow and alder formed extensive thickets in the shallow water of the peripheral littoral zone. This complex of swamp forest and shrub thicket probably served as a vegetative screen to filter out mineral sediment eroding from the uplands, thus reducing turbidity levels of water in Jackson Pond. In the middle Holocene, the marginal thicket may have provided an organic substrate for the direct colonization by *A. ryderi*, a sponge species inhibited by high turbidity. Between 7500 and 3500 years ago, the climate was warmer and drier, water tables were lower, and siltation levels probably decreased. Correspondingly, the turbidity-inhibited populations of *A. ryderi* flourished as the turbidity-tolerant populations of *E. fluviatilis* decreased significantly. Around 3500 years B.P., with a return to the warm and moist climate of the late Holocene, the water level rose in the basin, substantially reducing the area occupied by swamp forest, and reducing the effectiveness of swamp/thicket barrier in filtering out mineral sediment carried from the uplands. *Ephydatia fluviatilis* once again became the dominant species in Jackson Pond.

The Serendipity and Jackson Pond studies and current studies involving analysis of sediments from northern Ontario indicate that fossilized remains of freshwater sponge spicules and gemmules are well-preserved in most Quaternary wetland and aquatic deposits. Because spicule and gemmule morphologies are species-specific, the most commonly utilized criteria of neosystematics are available to the paleoecologist.

ACKNOWLEDGEMENTS

All scanning electron micrographs are used through the courtesy of Dr. Michael Poirrier of the University of New Orleans. I wish to express my thanks to Mrs. Shirley Weeks for typing the original manuscript. Drs. Hazel and Paul Delcourt, University of Tennesse, Knoxville provided access to their Jackson Pond core and have been an unfailing resource.

REFERENCES

Delcourt, P.A., 1985, The influence of late-Quaternary climatic and vegetational change on paleohydrology in unglaciated eastern North America: Ecologia Mediterranea, v. 11, p. 17-26.

Delcourt, H.R., Delcourt, P.A., Wilkins, G.R. and Smith, E.M., Jr., 1986, Vegetational history of the Cedar Glades regions of Tennessee, Kentucky, and Missouri during the past 30,000 years: Association of Southeastern Biologists, Bulletin, v. 33, p. 128-137.

Frey, D.G., 1964, Remains of animals in Quaternary lake and bog sediments and their interpretation: Ergebnisse der Limnologie, v. 2, p. 1-114.

Hall, B.V. and Hermann, S.J., 1980, Paleolimnology of three species of freshwater sponges (Porifera: Spongillidae) from a sediment core of a Colorado semidrainage mountain lake: American Microscopical Society, Transactions, v. 99, p. 93-100.

Harrison, F.W., 1974, Sponges (Porifera: Spongillidae), *in* Hart, C.W. and Fuller, J.L.H., eds., Pollution Ecology of Freshwater Invertebrates: Academic Press, New York, p. 29-66.

Harrison, F.W., 1977, The taxonomic and ecological status of the environmentallly restricted spongillid species of North America. III. *Corvomeyenia carolinensis* Harrison 1971: Hydrobiologia, v. 56, p. 187-190.

Harrison, F.W., 1979, The taxonomic and ecological status of the environmentally restricted spongillid species of North America. V. *Ephydatia subtilis* Weltner and *Stratospongilla penneyi* sp. nov.: Hydrobiologia, v. 65, p. 99-105.

Harrison, F.W., 1981, Scanning electron microscopy of taxonomic diagnostic criteria of the freshwater sponge, *Heteromeyenia tubisperma* (Potts 1881) (Porifera: Spongillidae): Hydrobiologia, v. 77, p. 257-259.

Harrison, F.W., Gleason, P.J. and Stone, P.A., 1979, Paleolimnology of Lake Okeechobee, Florida: an analysis utilizing spicular components of freshwater sponges (Porifera: Spongillidae): Notulae Naturae of the Academy of Natural Sciences of Philadelphia, v. 454, p. 1-6.

Harrison, F.W. and Harrison, M.B., 1977, The taxonomic and ecological status of the environmentally restricted spongillid species of North America. II. *Anheteromeyenia biceps* (Lindenschmidt, 1950): Hydrobiologia, v. 55, p. 167-169.

Harrison, F.W. and Harrison, M.B., 1979, The taxonomic and ecological status of the environmentally restricted spongillid species of North America. IV. *Spongilla heterosclerifera* Smith, 1918: Hydrobiologia, v. 62, p. 107-111.

Harrison, F.W., Johnston, L., Stansell, K.B. and McAndrew, W., 1977, The taxonomic and ecological status of the environmentally restricted spongillid species of North America. I. *Spongilla sponginosa* Penney, 1957: Hydrobiologia, v. 53, p. 99-202.

Harrison, F.W. and Warner, B.G., 1986, Fossil freshwater sponges (Porifera: Spongillidae) from western Canada: An overlooked group of Quaternary paleoecological indicators: American Microscopical Society, Transactions, v. 105, p. 110-120.

Jewell, M.E., 1935, An ecological study of the freshwater sponges of northeastern Wisconsin: Ecological Monographs, v. 5, p. 463-504.

Jewell, M.E., 1939, An ecological study of the freshwater sponges of northeastern Wisconsin, II. The influence of calcium: Ecology, v. 20, p. 11-28.

Old, M.C., 1932, Environmental selection of the freshwater sponges (Spongillidae) of Michigan: American Microscopical Society, Transactions, v. 51, p. 129-136.

Pennak, R.W., 1978, Freshwater Invertebrates of the United States: John Wiley and Sons, New York, 803 p.

Penney, J.T. and Racek, A.A., 1968, Comprehensive revision of a worldwide collection of freshwater sponges (Porifera: Spongillidae): Bulletin of the United States National Museum, v. 272, p. 1-184.

Poirrier, M.A., 1969, Louisiana freshwater sponges: ecology, taxonomy, and distribution, unpublished Ph.D. Thesis, Louisiana State University, 172 p.

Poirrier, M.A., 1974, Ecomorphic variation in gemmoscleres of *Ephydatia fluviatilis* Linnaeus (Porifera: Spongillidae) with comments upon its systematics and ecology: Hydrobiologia, v. 44, p. 337-347.

Poirrier, M.A., 1977, Systematic and ecological status of *Anheteromeyenia ryderi* (Porifera: Spongillidae) in Louisiana: American Microscopical Society, Transactions, v. 96, p. 62-67.

Racek, A.A., 1966, Spicular remains of freshwater sponges: Memoirs of the Connecticut Academy of Arts and Sciences, v. 17, p. 78-83.

Racek, A.A., 1970, The Porifera, *in* Hutchinson, G.E., ed., Ianula: An Account of the History and Development of the Lago di Monterosi, Latium, Italy: American Philosophical Society, Transactions, v. 60, p. 143-149.

Racek, A.A., 1974, The waters of Merom: A study of Lake Huleh. IV. Spicular remains of freshwater sponges (Porifera): Archiv für Hydrobiologie, v. 74, p. 137-158.

Racek, A.A. and Harrison, F.W., 1975, The systematic and phylogenetic position of *Paleospongilla chubutensis* (Porifera: Spongillidae): Linnean Society of New South Wales, Proceedings, v. 99, p. 157-165.

Simpson, T.L., 1984, The Cell Biology of Sponges: Springer-Verlag, New York, 661 p.

Turner, J., 1985, Sponge gemmules from lake sediments in the Puget Lowland, Washington: Quaternary Research, v. 24, p. 240-243.

Warner, B.G., 1984, Late Quaternary paleoecology of eastern Graham Island, Queen Charlotte Islands, British Columbia, Canada, unpublished Ph.D. Thesis, Simon Fraser University, British Columbia, 190 p.

Wilkins, G.R., 1985, Late-Quaternary vegetational history at Jackson Pond, Larue County, Kentucky, unpublished M.Sc. Thesis, University of Tennessee, Knoxville, 172 p.

Originally published in
Geoscience Canada Volume 15 Number 3
(September 1988)

Cladocera

B.J. Hann
Department of Zoology
University of Manitoba
Winnipeg, Manitoba R3T 2N2

SUMMARY

Cladocera are among the best represented groups of aquatic invertebrates that leave fossils in lake sediments. The skeletal fragments of Cladocera are abundant, can be identified to species in most cases, and represent taxa from a variety of ecological niches in a lake. Unlike more conventional paleoecological indicators such as pollen or beetles, Cladocera are not useful directly in paleoclimatic reconstructions. Instead, Cladocera microfossil assemblages are best used to reconstruct past lacustrine environments, to investigate the effects of human disturbance (eutrophication, acidification, contaminants), and to assess long-term population and community changes. Improved understanding of the ecological tolerances of living Cladocera, increased comparisons with analogue communities, and application of multivariate analytical methods will expand our interpretive capabilities for both modern and fossil assemblages. Once we understand how and why cladoceran communities responded to environmental events in the past, we may be better able to predict the direction and magnitude of future change.

INTRODUCTION

The Cladocera, commonly known as water fleas, are a major component of the microcrustacean fauna in freshwater lakes and ponds. They occur in both the offshore (pelagic) regions, where the families Daphniidae and Bosminidae predominate, as well as in the shallow littoral zone where the diverse members of the family Chydoridae are most abundant. Among the chydorids, some species are found among submersed vegetation, whereas others more commonly live on sand, mud, or rocky substrata. In contrast to lakes, the microfauna of bogs are primarily macrothricids.

Cladocerans must moult their exoskeletons periodically in order to permit growth in size. This process occurs several times during the lifetime of each individual. The exoskeletons are chitin, a remarkable chemically inert material which preserves details of shell sculpturing, setae (or their attachment sites), as well as other features critical to species identification. Most fragments can be identified to species or at least can be assigned to a species group. These exoskeletons accumulate and become incorporated in the sediments. All exoskeletons (except the very earliest ones) produced by an individual appear to have an equal chance of being preserved.

Cladoceran exoskeletons are not preserved intact, but instead disarticulate into their component parts: head-shield, carapace, postabdomen, claws, antennules, mandibles, ephippia (Figures 1 and 2). The majority of skeletal remains consist of members of the cladoceran families Bosminidae and Chydoridae. Other planktonic Cladocera are poorly preserved and are encountered rarely as fossils in the sediments except as post-abdominal claws, ephippia, and mandibles. However, in an extremely detailed study of Cladocera remains in surficial sediments of several Finnish lakes, Cotten (1985) identified a greatly expanded repertoire of fragments from other families.

This paper demonstrates the utility of Cladocera microfossil remains in the context of Canadian or North American research progress. Specifically, the intent is (a) to review briefly the historical development of the field in North America, (b) to describe the nature and occurrence of microfossils in lakes, and the methods of preparation and identification of remains, (c) to suggest how cladoceran fossils can provide a basis for inferring or reconstructing paleolimnological or paleoecological events, for expanding our biogeographical database, and for theoretical studies in community structure, development, and ecology, and lastly, (d) to provide suggestions for additional work in the area of cladoceran microfossil research to extend their usefulness in Quaternary paleoecology.

HISTORICAL DEVELOPMENT

Examination of Quaternary fossil Cladocera began in the late 19th century in Europe, but not until the 1900s in North America. The majority of early literature records (see Frey, 1964) simply report the occurrence of a few Cladocera taxa with no attempt to reconstruct the entire cladoceran community or interpret the data in an ecological context. Deevey (1942) carried out the first quantitative study of animal microfossils in North America. The abundantly represented *Bosmina* exoskeletons proved useful in analyzing the developmental history of Linsley Pond (Deevey, 1942; Vallentyne and Swabey, 1955) and Rogers Lake (Deevey, 1964, 1969). These studies stimulated a number of fundamental investigations of accumulation of remains and sedimentation rate, and internal transport of fossils within a basin (Livingstone, 1957; Livingstone *et al.*, 1958).

A major break-through in cladoceran microfossil studies came with Frey's (1959) study of the headpores on the headshields of the chydorid Cladocera; the well-preserved headshields could now be identified to species. Goulden and Frey (1963) demonstrated that the 2 subgenera of the Bosminidae could be distinguished on the basis of the location of the headpores on the headshields. As the chydorid and bosminid remains are typically the most abundant in the sediments, this new diagnostic tool greatly improved the resolution of microfossil studies. Since those landmark studies, cladoceran microfossil evidence has been used to examine past ecological development of lakes in many parts of the world: temperate zone (Goulden, 1964; Megard, 1964; Harmsworth, 1968; DeCosta, 1968; Hofmann, 1978, 1986), tropical regions (Goulden, 1966a,b; Megard, 1967), and the Arctic (Bradbury and Whiteside, 1980).

One of the first systematic studies of fossil Cladocera in Canada was by Sreenivasa (1973). More recent investigations include: Warner *et al.* (1984), Hann and Karrow (1984), Smol and Boucherle (1985), Boucherle *et al.* (1986), and Hann and Warner (1987).

FIELD SAMPLING

The abundance and species composition of the cladoceran community vary greatly both seasonally and in different habitats within a lake. Fortunately, the remains of species from different microhabitats in the littoral zone, occurring in all seasons, appear to be integrated (to varying degrees) with remains of planktonic taxa in the deep-water sedimentary record (Frey, 1960; Mueller, 1964; Boucherle, 1982) as a consequence of redeposition of littoral remains offshore. Thus, the entire chydorid fauna of the lake will be represented in the sediments, and percentage species composition of the community can be determined by analyzing the microfossils in a sediment core (Goulden, 1969a).

The extent of integration of littoral and pelagic remains is a function of distance from shore and water depth (Mueller, 1964; Boucherle, 1982). Microfossil remains of littoral species are well represented in the littoral sediments, constituting a rapidly declining proportion of total cladoceran fossils with increasing distance from shore. In contrast, remains of planktonic species are rare in shallow-water sediments, and increase in abundance offshore. In some lakes, they appear to form a "concentration zone" at the top of the hypolimnion (Mueller, 1964), perhaps a function of lake morphometry. Nevertheless, this distributional feature must be taken into consideration in lakes where water levels have fluctuated, and where the concentration zone may periodically be represented in the core. Although the concentration of remains in the surficial sediment does vary across a lake basin, the percentage species composition of cladoceran remains is close to uniform in the deep-water sediments (Boucherle, 1982).

The design of strategies for sampling (both spatial and stratigraphic) must consider the mechanisms that affect sediment (and microfossil) redistribution as governed by basin morphometry. These include: changes in lake level, thermal stratification, sediment focussing, bioturbation, diagenesis, and resuspension (Hakanson, 1977; Davis *et al.*, 1984; Hilton, 1985; Dearing, 1986). In most instances, these difficulties have

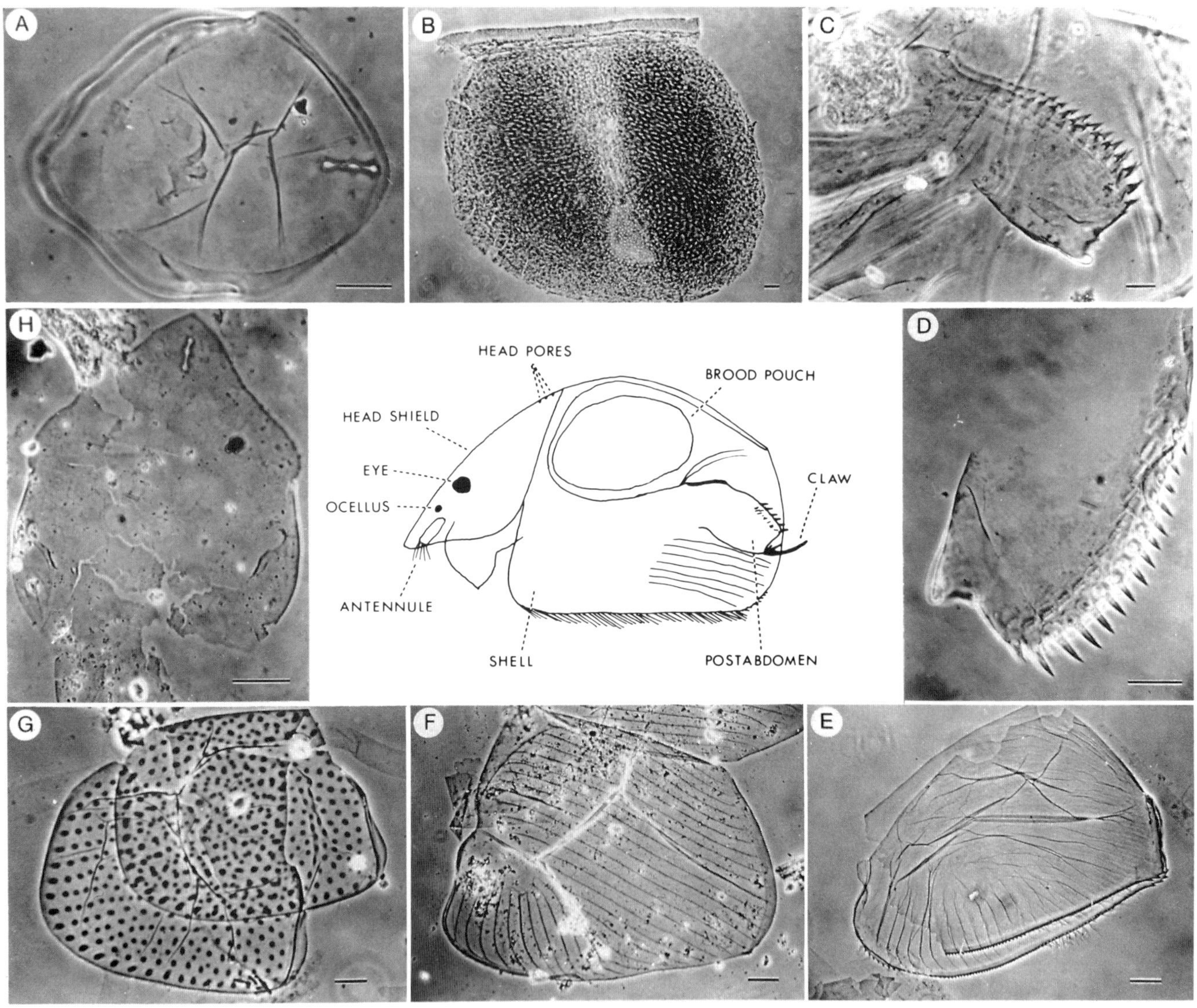

Figure 1 *Selected fossil remains of Cladocera from the Don and Scarborough Formations at Toronto, Ontario. All scale bars represent* 50 μm. *See Hann and Karrow (1984) for details.* **(a)** Alona circumfimbriata, *headshield;* **(b)** Daphnia *sp., ephippium;* **(c)** Alona quadrangularis, *postabdomen;* **(d)** Alona quadrangularis, *postabdomen;* **(e)** Pleuroxus denticulatus, *carapace;* **(f)** Acroperus *cf.* harpae, *carapace;* **(g)** Alona rustica, *carapace;* **(h)** Alona quadrangularis, *headshield.*

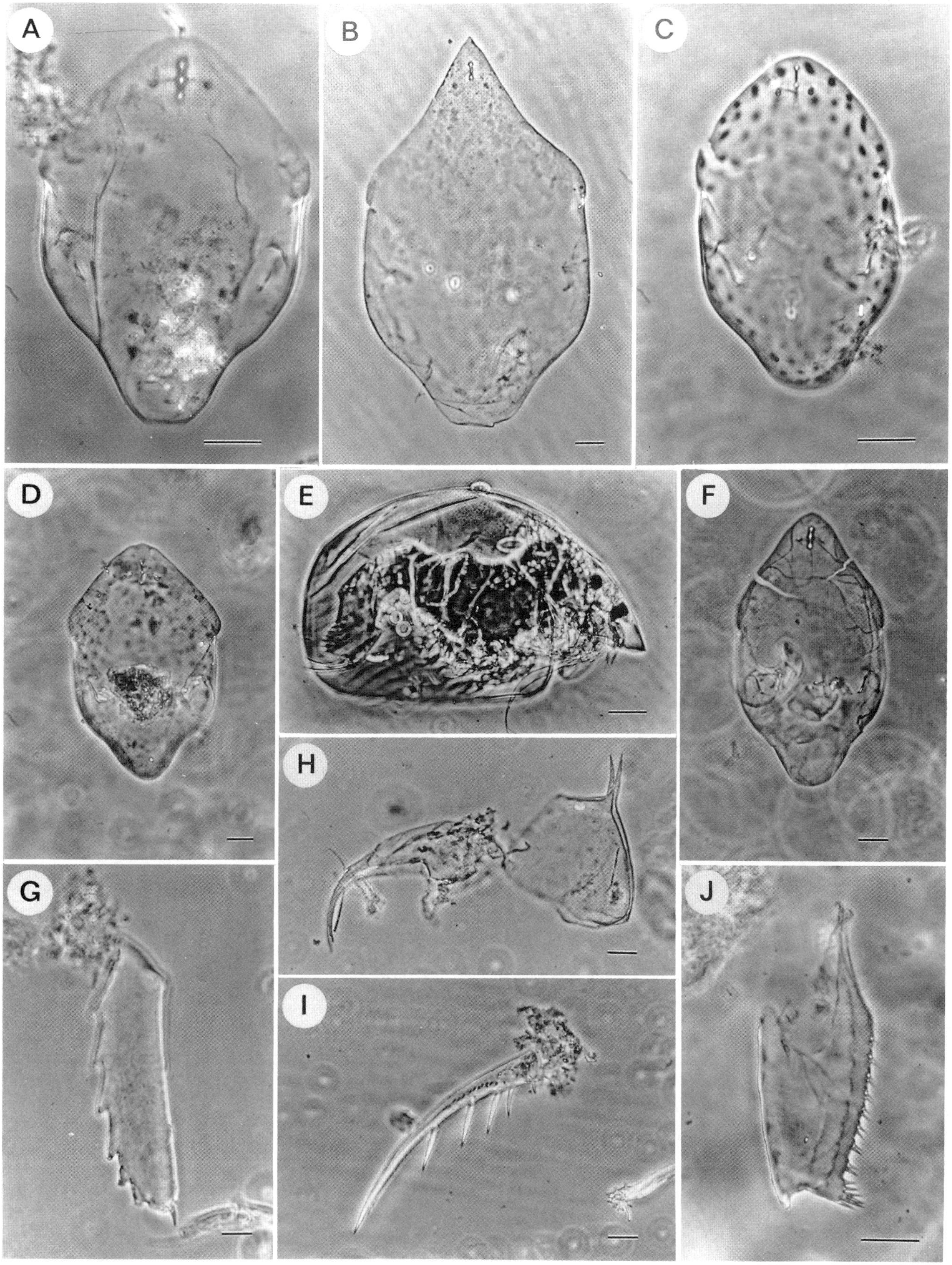

Figure 2 *Fossil remains of Cladocera from Manitouwadge, Ontario. All scale bars represent* 50 μm. **(a)** Alona cirumfimbriata, *headshield;* **(b)** Alona affinis, *headshield;* **(c)** Alona rustica, *headshield;* **(d)** Alona costata, *headshield;* **(e)** *Alona lapdicola, whole animal;* **(f)** Alona quadrangularis, *headshield,* **(g)** Sida crystallina, *antennal segment;* **(h)** Bosmina longirostris, *headshield and carapace;* **(i)** Sida crystallina, *postabdominal claw;* **(b)** Pleuroxus denticulatus, *postabdomen.*

been ignored, and one central reference core (assumed to represent the integrated sedimentary record of the lake) has been obtained from the deepest part of a lake. This approach may be adequate in broad-scale regional comparisons of lakes. However, where the goal of the research is a more detailed understanding of the events within the catchment and waterbody, transects of cores or even a stratified or systematic design of core samples will be required to represent adequately the depositional environments. Methods for examining multiple-core correlation within time-stratigraphic units are available (Birks, 1986; Dearing, 1986), but have not been applied to cladoceran microfossil studies.

Coring of sediments, especially annually-laminated sediments requires care to retrieve the stratigraphic sequences intact and with minimal distortion. Various types of freeze-corers have been designed for recent sediments (Huttunen and Meriläinen, 1978; Wright, 1980; Renberg, 1981). Gravity or piston corers are used to obtain longer sediment cores in lakes (Aaby and Digerfeldt, 1986).

LABORATORY TECHNIQUES AND IDENTIFICATION

Concentrations of cladoceran microfossils in lake sediment may vary from less than 100 to several hundred thousand per cm^3. To estimate the number of exuviae per cm^3 of fresh sediment for each species present at each level, different amounts of sediment must be processed according to the microfossil densities. The volumes assessed by preliminary determinations may range from as little as 1 cm^3 in gyttja to 10-20 cm^3 per stratigraphic level in mineral sediments.

Unnecessarily harsh chemical treatment is to be avoided as it may increase fragmentation or degradation of remains and make identification difficult. Frey (1986b) has summarized various treatments needed for different types of sediments. The main steps include: gentle heating and stirring of the sediment in 10% KOH, followed by careful sieving through a 37 μm mesh sieve to retain the smallest exoskeletal pieces. The residue retained on the screen is transferred to a vial and topped up to a known volume, with a few drops of formalin or alcohol added as a preservative.

To prepare quantitative slides, volumetric aliquots of the residue (typically 0.05 mL) are transferred to microscope slides using a precision pipette, with glycerin jelly as the mounting medium, lightly stained with lignin pink.

Identification of cladoceran microfossils is greatly assisted by a reference collection of slides of whole animals and moulted exoskeletons from the region under study. Literature sources useful for identification of North American material include Brooks (1959) and Pennak (1978). It is essential to supplement these sources with original literature descriptions of the many taxa not included therein or inadequately illustrated (Hann, 1981, 1982; Hann and Chengalath, 1981; Chengalath and Hann, 1981a,b; Frey, 1980, 1987). In addition, it has been clearly demonstrated that several chydorid species do not have cosmopolitan distributions, as previously assumed, and this pattern of more restricted distributions of species is undoubtedly prevalent (Frey, 1980; Michael and Frey, 1983, 1984). Hence, accurate identifications of taxa may often require

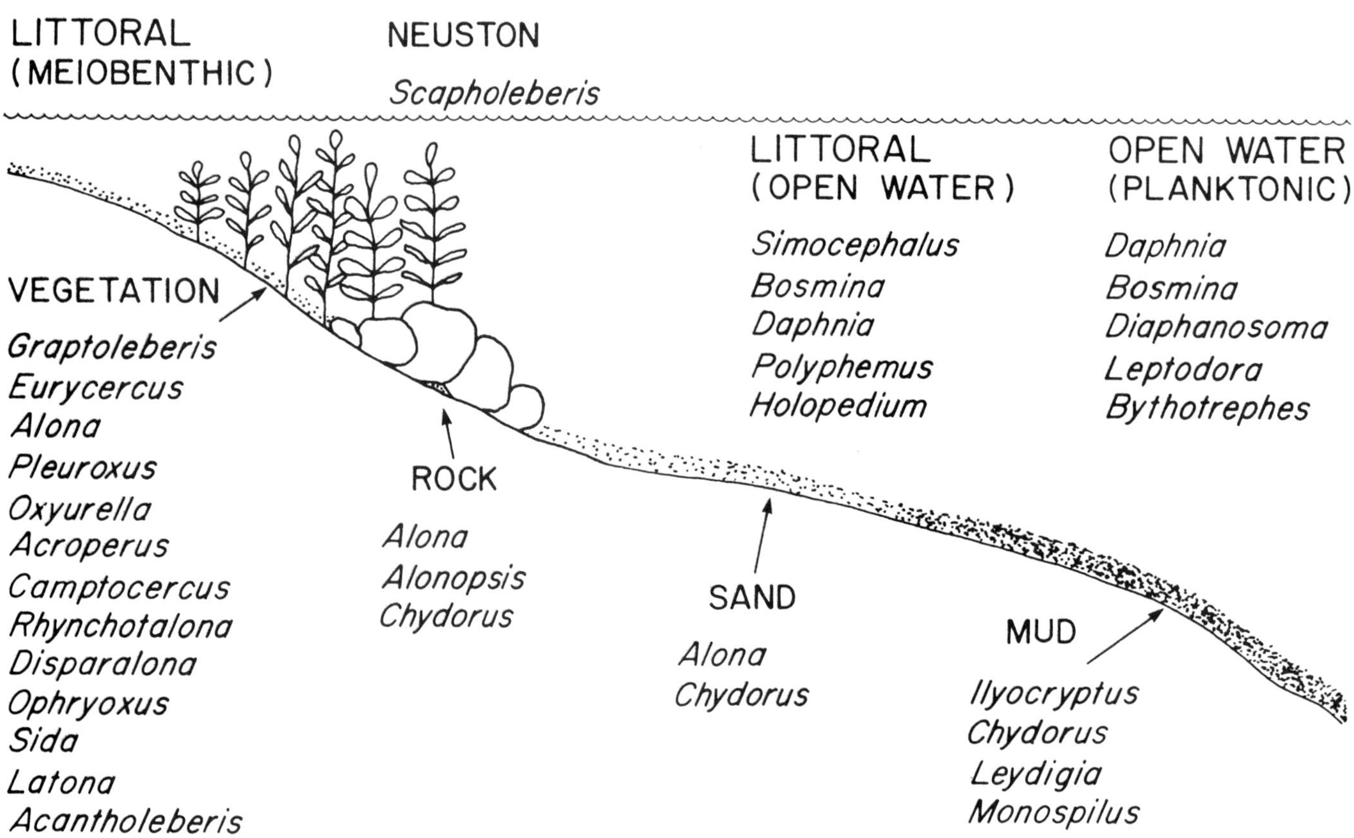

Figure 3 *Ecological niches occupied by Cladocera include: the pelagic zone with planktonic genera, the shallow water littoral zone with planktonic genera, and the littoral zone with meiobenthic genera segregated into several subhabits: vegetation, rock, sand, mud, and the neuston.*

use of primary species descriptions, and verification by taxonomic experts, familiar with the local fauna.

Numbers of each exoskeletal fragment (headshield, carapace, postabdomen, claw) are tabulated for each species. Of the several methods available for assessing population size for each taxon (see Frey, 1986b), a common choice is to use the largest category count as the estimate of species abundance at each level for each species. These numbers can be presented as (1) percentage that each species constitutes of the total at each level, (2) concentration (numbers per gram of dry sediment) or (3) accumulation rate (remains per cm^3 × sedimentation rate (cm/yr)).

The most common graphical representation of data is a relative frequency diagram for the most abundant taxa (Figure 3), and a plot of percentage littoral *versus* pelagic taxa throughout the stratigraphic record. Species diversity, typically estimated by the Shannon-Wiener diversity index, and equitability (Lloyd and Ghelardi, 1964) are also calculated for each level. If adequate dating of the sediment sequence is available, presentation of abundance diagrams constructed using all three computation methods listed above provides the most comprehensive basis for analysis. Each method has advantages and disadvantages in interpretation of the lacustrine sedimentary record (Brugam and Speziale, 1983; Binford, 1986). Renewed attention must be focussed on critical examination of statistical and analytical methods applied to fossil Cladocera communities (see Whiteside and Swindoll, 1988).

USES AND APPLICATIONS IN PALEOECOLOGY

A wide range of problems can be addressed using cladoceran microfossil assemblages: (a) paleolimnological and paleoecological reconstructions, (b) comparative regional limnology, (c) theoretical community ecology, and (d) biogeography (Frey, 1986a).

Cladoceran microfossil evidence provides a primary tool for the reconstruction of past lacustrine environments, from interglacial periods (Frey, 1962; Hann and Karrow, 1984) through to historical time (Kerfoot, 1974; Birks *et al.*, 1976; Brugam, 1978). In the only two studies of cladoceran microfossils in interglacial sediments, the Eemian in Europe (Frey, 1962) and the Sangamonian in North America (Hann and Karrow, 1984), no morphological change in any cladoceran species was detected between interglacial representatives and modern counterparts. Similarly, there was stability in species composition of the assemblage and the relative abundance of taxa was comparable to that in modern communities. Given this evidence of morphological and ecological uniformity in the cladoceran community, at least since the last interglaciation, modern ecological data can be validly used for paleoecological reconstructions within this time period.

Cladocerans inhabit diverse habitats, ranging from the littoral to the limnetic zone, and among vegetation, on rocks, sand, and other substrata in the littoral zone (Figure 4). By distinguishing the remains of littoral taxa from those that

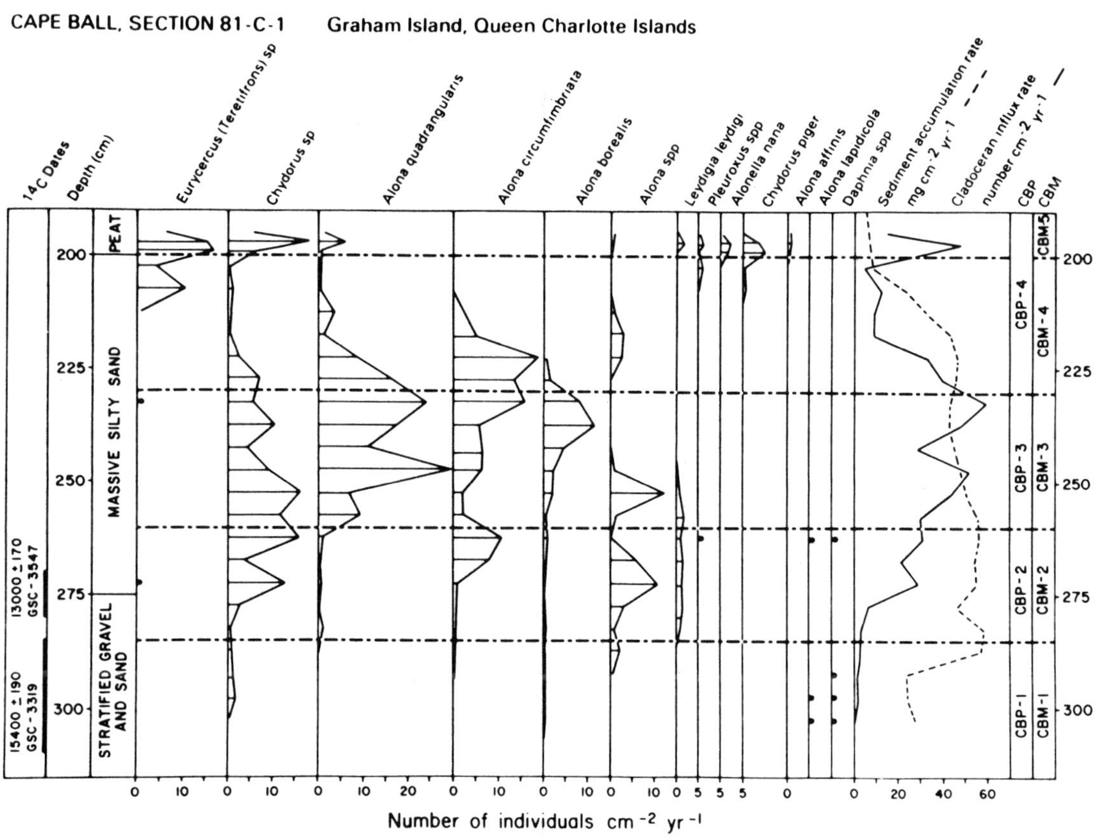

Figure 4 *Representative microfossil Cladocera diagram illustrating species accumulation rates, bulk sediment accumulation rates, and local pollen and macrofossil zones provided for comparison. Only those taxa comprising at least 5% of the total cladoceran number in one or more levels are included in the diagram. (After Hann and Warner, 1987).*

inhabited the limnetic zone of a lake, a planktonic/littoral (P:L) ratio can be determined. This reflects, in part, the relative volumes of the lake that have contributed remains to the sediments, and can be used to assess fluctuations in lake levels (Mueller, 1964; Alhonen, 1970) and, perhaps of greater significance, changes in littoral area providing habitat suitable for development of the littoral fauna. Chydorid Cladocera can be subdivided into vegetation- and substrate-dwelling groups (Quade, 1969; Whiteside, 1974; Whiteside *et al.*, 1978). The majority of chydorid species which have been carefully investigated are quite habitat-specific; for example, *Alona borealis* is typically found only in littoral habitats with sandy substrate, with little vegetation, and in nutrient-poor waters (Chengalath and Hann, 1981a). Therefore, the composition of the cladoceran assemblage recovered from the sediments can reveal a great deal about the lacustrine environment that existed in the past.

Do cladoceran microfossil assemblages reflect regional conditions in the watershed or local limnological conditions? Unlike the pollen record in lakes which is usually an amalgam from sources within the local catchment as well as from long distances, the Cladocera assemblage represents the product of the immediate aquatic environment, which greatly buffers the direct impact of climatic change. If cores can be obtained from a number of lakes in a region, it may be possible to separate the regional influences from the strictly local ones. Goulden (1964) and Harmsworth (1968) studied two small lakes in the English Lake District and found similar patterns of cladoceran relative abundance and distribution in time, suggesting parallel developmental histories. However, planktonic Cladocera predominated at EsthwaiteWater, while littoral forms were dominant at Blelham Tarn, probably a consequence of differing lake size and littoral zone area (Harmsworth, 1968). The overall similarity of the littoral chydorid community argues for regionally similar limnological parameters and composition of the macrophyte community, governed in turn by regional climatic impact on the watershed.

The abrupt decline in hemlock in eastern North America 4800 years ago (Webb, 1982; Allison *et al.*, 1986) provides another example of a broad scale environmental factor impinging upon the biota of lakes. Boucherle *et al.* (1986) studied the effects of the loss of hemlock from the watersheds of three lakes in southern Ontario. Although each lake experienced a change in trophic status, the response of the cladoceran community was unique in each lake, the differences probably related to lake size and the ratio of littoral to pelagic zone volume.

Thus, for two very different regional environmental factors, climate and hemlock decline, the evidence suggests that a hierarchical approach to interpreting the influence of environmental factors, as recommended by Hann and Warner (1987) and Warner and Hann (1987), will provide the most parsimonious result. This involves taking into consideration regional, local, and microhabitat characteristics, in turn, as contributing to the overall variation in community structure and composition.

Inference of paleotemperatures from cladoceran assemblages has drawn on studies relating present latitudinal distributions of Cladocera with climatic regimes (DeCosta, 1964; Harmsworth, 1968), classifying species as arctic, north or south temperate, or eurytopic. However, because of taxonomic difficulties (*e.g.*, north-south sibling species), geographic affinities are subject to considerable misrepresentation and must be approached with caution. Many of the more common taxa, as currently defined, have broad distributions and, thus, are not useful in climatic interpretations.

Species diversity as well as species composition change during development of a community through the sedimentary sequence. To interpret these patterns, comparison with modern analogue communities is a useful approach. As Whiteside (1970) pioneered application of the technique to chydorid assemblages, his study will be described in some detail. Whiteside (1970) examined assemblages of chydorid remains from surficial mud samples derived from 77 Danish lakes for which extensive limnological data were available. The lakes could be classified subjectively into 3 types (clear-water lakes, pond and bog lakes, and culturally disturbed clear water lakes), the groupings subsequently validated using multiple discriminant analysis of the limnological parameters. Analysis of correlations between individual chydorid species abundances in the surficial assemblages and limnological parameters revealed strong negative correlations between many species and alkalinity, pH, conductivity, and transparency.

More useful than relationships for individual species are those characterizing species assemblages. Multiple discriminant analysis (MDA) of average relative abundances (percentages) of microfossil chydorid species in each lake type provided a measure of the importance of species in defining a lake type by species. Generally, there was a good relationship between species groups and lake types, lending support to the hypothesis that chydorid assemblages can be used to reconstruct limnological conditions in paleoecological studies.

Whiteside (1970) next tested his modern analogue data from surficial chydorid assemblages against assemblages from cores of two Danish lakes, and attempted to infer past lake conditions. In one instance, Esrom Sø, use of the modern analogue communities as delimited by MDA was useful because the major factors affecting relative abundances in the past had been included in the original construction of the analogues. In Grane Langsø, in contrast, the analysis was less successful because the impact of climate during the developmental history of the lake was considerable, a factor which had not been incorporated into the initial survey data as the lakes were all within close geographic proximity. Hence, "the more accurate an investigator pinpoints the important factors (abiotic and biotic) affecting the distribution and abundance of organisms (and uses these factors to generate the discriminant function), the more accurate will be the description of that organism's niche" (Whiteside, 1970, p. 114), and consequently, the stronger the basis for interpretation of paleoassemblages (Whiteside and Swindoll, 1988).

Synerholm (1979) analyzed microfossil chydorid assemblages from surficial sediments in 32 lakes in Minnesota and North Dakota, on a transect from acidic, low conductivity waters in coniferous forest to circum-neutral pH waters in deciduous forest to alkaline waters with high dissolved ion concentration on the prairies. Each lake type had some characteristic chydorid species, although several species were generalists, occurring in all lake types. The strong negative correlation between chydorid species diversity and specific conductance found in this study supports the relationship detected by Whiteside (1970) between chydorid abundance and diversity and oligotrophic-type lakes of low productivity. Similarly, Whiteside and Harmsworth (1967) found a positive relationship between diversity and water transparency in 20 lakes in Denmark and in 14 lakes in northern Indiana, and an inverse relationship between species diversity and primary productivity.

Crisman (1980) examined subtropical chydorid assemblages in 52 lakes in Florida. The lakes were separated into three groups on the basis of four limnological parameters (phosphorus, conductivity, alkalinity, and Secchi disk transparency). Microfossil chydorid assemblages extracted from surficial sediment samples from the lakes were expressed as percentage representation by each species in each lake. Principal components analysis (R-mode) was used to examine the associations among individual chydorid species, and to define the limnological parameters controlling the groupings. The chydorid fauna was separated into soft-water and hard-water assemblages. Although both Synerholm (1979) and Crisman (1980) established the correlations between chydorid species and limnological parameters, neither attempted to use their analogues to interpret and reconstruct past lake development or examine chydorid community changes in their region.

Changes in species diversity have often been attributed to change in trophic status of the waterbody. Increased productivity of algae is typically associated with reduced transparency and decreased diversity of chydorid Cladocera. Other explanations must be considered, however, as emphasized by Whiteside (1983). For instance, increased overall productivity of the lake may be largely attributable to higher littoral zone productivity, primarily macrophytes. Thus, the available habitat for chydorids should increase, as would species diversity. In those few studies where remains of both cladocerans and aquatic macrophytes have been analyzed (Warner *et al.*, 1984; Hann and Warner, 1987), correlations of species diversity and overall concentration (remains/cm^3) or accumulation rates for the two groups lend support to this hypothesis.

The interaction of climate, morphoedaphic factors, and lake ontogeny with cladoceran community composition is complex. Only through incorporation of a variety of complementary indicator groups into the matrix, along with the Cladocera microfossils, can we hope to tease apart major controlling factors from significant interactions.

The species replacement of *Bosmina longispina* by *B. longirostris* is frequently considered indicative of eutrophication of the lake (Crisman and Whitehead, 1978; Hofmann, 1978). However, a change in dominant predators, for example, from planktivorous fish to invertebrate predators, could lead to the same replacement of a large bosminid by a smaller one. Kerfoot (1974, 1981) has demonstrated that a shift in major predators concurrent with nutrient enrichment could explain the changes in *Bosmina* morphology during the postglacial history of Frains Lake, Michigan. The planktonic Cladocera community in Lake Harriet, Minnesota also responded primarily to change in predators rather than to trophic change (Brugam and Speziale, 1983). Changes in size structure and cladoceran community composition in Peter-Paul Lake, Michigan have been attributed to different predation intensity in two halves of the lake subject to contrasting experimental manipulation (Kitchell and Kitchell, 1980). Nilssen (1978) has summarized the major influence of vertebrate and invertebrate predators on the zooplankton community as recorded in the microfossil remains.

The impact of human disturbance can also be read in the recent microfossil record. The acidification history of lakes has been studied using short cores from lakes in sensitive areas and examining the response of the cladoceran assemblages (Paterson, 1985). The modern analogue approach was employed to examine the relationship between pH and subfossil chydorid assemblages in surficial sediment samples from 37 lakes and a core from each of 3 lakes in Maine (Brakke *et al.*, 1984). A positive relationship was observed between species richness, diversity, and pH. Using a cluster analysis technique, lakes with low pH tended to cluster together on the basis of their cladoceran community, as did high- and low-altitude lakes. In the cores, chydorid assemblages were analyzed and separation of clusters suggested levels at which pH change was likely to have occurred.

Cotten (1985) examined surficial sediment samples for cladoceran remains from 46 Finnish lakes and recovered 71 taxa belonging to 9 families. The relationship between lake parameters and cladoceran assemblages, assessed using a multivariate ordination technique, detrended correspondence analysis, revealed 5 lake types with distinctive cladoceran communities. Nutrient conditions, pH, and transparency were the chief lake discriminators. Transfer functions were developed to permit reconstruction of past lake conditions based on cladoceran remains recovered from lake sediments.

The chydorid Cladocera have been described as a taxocene, displaying taxonomic and ecological homogeneity (Deevey, 1969). Theoretical community ecology has gained from several pioneering studies of temporal variation in community structure of the chydorid fossil taxocene. Goulden (1966a,b, 1969a,b,c) defined three phases of development of the community: initially, diversity increases as species immigrate into a new habitat, then species numbers stabilize and equitability, or evenness, increases. Finally, species number, or richness, may increase slightly as additional rare species colonize the habitat. At this point in development, the chydorid community appears to fit the Type I or broken-stick model of MacArthur (1957), or any one of several other related distributions (Binford *et al.*, 1983). A taxocene will approach this condition gradually, as demonstrated for a variety of organisms (Tsukada, 1967; Goulden, 1969a; Deevey, 1969) in lakes around the world. Perturbations in species diversity or equitability may signal climatic change (Goulden, 1966a), catastrophic natural disturbances such as nearby volcanic eruptions (Tsukada, 1967), fire (Baas and Boucherle, 1984), eutrophication (Bradbury and Megard, 1972; Boucherle and Zullig, 1983; Brugam and Speziale, 1983; Hofmann, 1986), agriculture (Goulden, 1966b), or other human disturbances. As a consequence of disturbance, one species tends to become super-abundant and the others rare. When the disturbance ceases, the community readjusts and, depending upon the severity of the perturbation, may or may not return to the initial structure of the community. Examination of the effects of stress on the cladoceran community and subsequent recovery, as preserved and integrated in the sediments, may provide an excellent environmental monitoring tool that is only beginning to be refined.

Cladoceran remains have been used less extensively than plants or beetles to address biogeographic problems. In part, this stems from the long-held presumption of cosmopolitan distributions of most Cladocera. As the endemism of each continental faunule is recognized and documented, the utility of the Cladocera in testing vicariance theory or in examining colonization history is greatly enhanced.

Investigation of colonization patterns and timing of species invasions, governed by glacial advances and retreats, might be accomplished using cladoceran remains as suggested by Frey (1986a). In a study of cladoceran remains from a site on the Queen Charlotte Islands, British Columbia (Hann and Warner, 1987), a relatively high proportion of remains were recovered which represented unknown species, especially in the oldest sediments. Attempts are on-going to determine whether these unknown taxa were colonists from a refugium in Alaska or from the southern coastal regions, both areas for which the modern Cladocera are poorly studied.

Remains in the sediments may be used to define past ranges of well-delimited species. In particular, the presumed southward migration of northern taxa into southern United States at the time of the last glaciations may be detectable, or conversely, the northward expansion of ranges of southern taxa (Frey, 1986a). However, the occurrence of north-south sibling species pairs in several chydorid species (Hann, 1982; Shan and Frey, 1983), whose remains cannot be differentiated, complicates this historical sedimentary approach to colonization history.

In any paleoecological reconstruction, one must deal with differential preservation and distortions that may bias the interpretation. Deevey (1964) examined the fossilization process for selected zooplankton, finding that *Bosmina* exuviae fossilize well, in contrast to those of copepods which decompose rapidly.

Typically, the *Bosmina* headshield: carapace ratio is roughly equal to one, but can become distorted in zones of secondary deposition where breakage generally increases (Vallentyne and Swabey, 1955). No studies have systematically examined headshield: carapace ratios in chydorid remains, ephippia:parthenogenetic carapace ratios, nor other instances of marked distortion in representation of exoskeletal components. However, Deevey (1955) reported changes in frequency of ephippia of the overwhelmingly dominant *Pleuroxus hastirostris* in Pyramid Valley, New Zealand, perhaps in response to environmental stress. Goulden (1966a) concluded that changing ratios of carapaces of parthenogenetic females: ephippial females (of all Cladocera considered together) reflected response to fluctuations in water level.

Evidence has accrued suggesting that exuviae from early instars of bosminids and chydorids do not fossilize as well as those from later instars (Kerfoot, 1974; Culver *et al.*, 1981; Frey and Hann, 1985). Methods which take into account this selective preservation of exuviae are being developed (Hann, unpublished) prior to an attempt to estimate biomass of various components of the taxocene. Then it may be possible to evaluate past productivity of cladoceran species, and to correlate productivity of different trophic levels, as first suggested by Frey (1960).

Changes in species richness, diversity and the planktonic/littoral ratio in the sedimentary record are frequently used to assist paleoenvironmental reconstructions. In sediment stratigraphies where rates of sedimentation are relatively constant, diversity measures are representative of community structure. However, in recent sediment sequences particularly, patterns in diversity indices can be severely distorted by variable sedimentation rates (Smol, 1981). Although parallel responses in lakes to climate and regional geology can be useful in deriving empirical relationships (Duarte *et al.*, 1986), differences among lake histories can be indicative of unique events in a watershed or in the lake itself, *e.g.*, meromixis. As meromictic lakes often have annually laminated sediments, they are especially useful in paleoecological studies (Culver *et al.*, 1981; Smol and Boucherle, 1985).

Though Cladocera are useful in approaching a diversity of major problems in Quaternary ecology, nevertheless, there are several limitations that require refinement in their application to Quaternary problems: (1) There is a serious lack of ecological data for modern species, particularly relating to tolerances to physical, chemical and biotic parameters. Until such information becomes available, our ability to extrapolate from modern conditions of existence of a species into past analogous occurrence of the species is severely limited. (2) There do not appear to be well-defined "indicator" species in the Cladocera, although this may change as we recognize more and more species with clearly defined ranges, a radical reversal of the previously popular idea of globally distributed species (Frey, 1982, 1986a). Discrepancies in the literature regarding tolerance to various environmental factors for a given species may be coincidentally resolved. (3) Too little is known about distribution of cladoceran species generally, so that if a species occurs in small numbers in the fossil record, we cannot determine whether it is a rare species that is nonetheless well-established in the region, or if it is a broadly distributed common species that is infrequent at this locality because it is at the edge of its range. Therefore, ranges need to be defined in terms of relative abundance as well as occurrence (presence/absence).

FUTURE DIRECTIONS

To more fully realize the potential of Cladocera microfossil communities in Quaternary paleoecology, new approaches until now applied primarily to pollen analysis (Prentice, 1986) are necessary, particularly in the area of data analysis and interpretation. Whiteside's (1970) use of multiple discriminant analysis proved useful in defining axes upon which predetermined types of lakes could be differentiated. Chydorid assemblages could then be segregated into parallel groups. Cluster analysis, in attempting to define the groups themselves, has been used to distinguish pH-defined clusters of chydorid assemblages (Brakke *et al.*, 1984; Cotten, 1985; Paterson, 1985). Boucherle (1982) used a stratigraphically constrained clustering algorithm to delimit cladoceran zones, akin to the standard pollen zonation. This method provides an objective, quantitative approach to assessing change in the microfossil assemblage and merits further attention.

The use of modern analogue communities for comparison with microfossil assemblages must be expanded to encompass a broader range of habitat types. For a given region, if a catalogue of analogue communities are described from surficial sediment samples, then transfer functions developed using multiple discriminant analysis or detrended correspondence analysis (Gauch, 1982), can be used on down-core microfossil assemblages to make inferences regarding past environmental conditions (Cotten, 1985).

The absence of clear-cut indicator species of Cladocera has focussed attention on evaluating changes in the entire community. However, individual taxa of Cladocera may prove to be useful as pollution indicators. Deformities have been documented in a variety of chydorid species; in fact, some species have been described based on single deformed specimens. Morphological deformities in chironomids have been shown to have great potential as a screening technique for detection and assessment of contaminants in aquatic ecosystems (Warwick, 1985). The chydorids, too, may act as sophisticated environmental sensors, perhaps complementary to the chironomids, as the sediment-associated chydorids could be affected predominantly by sediment-bound contaminants, in contrast to species associated with the macrophytes or entirely free-swimming that would be influenced more strongly by contaminants in the water column. If particular contaminants could be shown to cause specific types of deformities or species-specific responses, the historical incidence of environmental contamination could be traced using the chydorid microfossil record.

Finally, preliminary attempts to examine the population structure and survivorship of certain species of Cladocera using microfossils have been successful (Frey and Hann, 1985). Detailed studies of such death assemblages using a matrix approach (Green, 1979) may permit estimation of life history parameters in populations over long periods of time.

Ultimately, the potential of cladoceran microfossils lies in providing a long-term record of response to environmental conditions, both stable and changing. The qualitative and quantitative responses to disturbance, whether physical or biotic, can be assessed and predictive models developed and tested. Understanding the processes that operated in the past will be the key to predicting the impact of present-day forces impinging on cladoceran populations in aquatic ecosystems. The clearest comprehension of the processes will come through a synthetic approach, integrating data derived from a wide variety of plant and animal fossil remains, and necessitating a co-ordinated multi-disciplinary approach.

ACKNOWLEDGEMENTS

Part of this work was supported in part by Natural Sciences and Engineering Research Council of Canada grants. I thank B.G. Warner, M.W. Binford, and M.C. Whiteside for helpful criticism which greatly improved the clarity of the manuscript.

REFERENCES

Aaby, B., and Digerfeldt, G., 1986, Sampling techniques for lakes and bogs, *in* Berglund, B.E., ed., Handbook of Holocene Palaeoecology and Palaeohydrology: John Wiley and Sons, New York, p. 181-194.

Alhonen, P., 1970, On the significance of the planktonic/littoral ratio in the cladoceran stratigraphy of lake sediments: Commentationes Biologiae (Helsinki), v. 35, p. 1-9.

Allison, T.D., Moeller, R.E., and Davis, M.B., 1986, Pollen in laminated sediments provides evidence for a Mid-Holocene forest pathogen outbreak: Ecology, v. 67, p. 1101-1105.

Baas, D.S., and Boucherle, M.M., 1984, Changes in Cladocera community structure and fossil pigment stratigraphy as compared to the fire history of Elk Lake, Clearwater Co., MN, USA: Internationale Vereinigung für Limnologie, Verhandlungen, v. 22, p. 1366-1371.

Binford, M.W., 1986, Ecological correlates of net accumulation rates of Cladocera remains in lake sediments: Hydrobiologia, v. 143, p. 123-128.

Binford, M.W., Deevey, E.S., and Crisman, T.L., 1983, Palaeolimnology: an historical perspective on lacustrine ecosystems: Annual Review of Ecology and Systematics, v. 14, p. 255-286.

Birks, H.J.B., 1986, Numerical zonation, comparison and correlation of Quaternary pollen-stratigraphical data, *in* Berglund, B.E., ed., Handbook of Holocene Palaeoecology and Palaeohydrology: John Wiley and Sons, New York, p. 743-774.

Birks, H.H., Whiteside, M.C., Stark, D.M. and Bright, R.C., 1976, Recent palaeolimnology of three lakes in north-western Minnesota: Quaternary Research, v. 6, p. 249-272.

Boucherle, M.M., 1982, An ecological history of Elk Lake, Clearwater Co., Minnesota, based on Cladocera remains, Ph.D. Thesis, Indiana University.

Boucherle, M.M. and Züllig, H., 1983, Cladoceran remains as evidence of change in trophic state in three Swiss lakes: Hydrobiologia, v. 103, p. 141-146.

Boucherle, M.M., Smol, J.P., Oliver, T.C., Brown, S.R. and McNeely, R., 1986, Limnologic consequences of the decline in hemlock 4800 years ago in three Southern Ontario lakes: Hydrobiologia, v. 143, p. 217-225.

Bradbury, J.P. and Megard, R.O., 1972, Stratigraphic record of pollution in Shagawa Lake, northeastern Minnesota: Geological Society of America, Bulletin, v. 85, p. 2639-2648.

Bradbury, J.P. and Whiteside, M.C., 1980, Palaeolimnology of two lakes in the Klutlan Glacier region, Yukon Territory, Canada: Quaternary Research, v. 14, p. 149-168.

Brakke, D.F., Davis, R.B. and Kenlan, K.H., 1984, Acidification and changes over time in the chydorid Cladocera assemblage of New England lakes, *in* Hendrey, G.R., ed., Early Biotic Responses to Advancing Lake Acidification: Butterworth Publishers, Boston, p. 85-104.

Brooks, J.L., 1959, Cladocera, *in* Edmondson, W.T., ed., Freshwater Biology, Second Edition: John Wiley, New York.

Brugam, R.B., 1978, Human disturbance and the historical development of Linsley Pond: Ecology, v. 59, p. 19-36.

Brugam, R.B. and Speziale, B.J., 1983, Human disturbance and the paleolimnological record of change in the zooplankton community of Lake Harriet, Minnesota: Ecology, v. 64, p. 578-591.

Chengalath, R. and Hann, B.J., 1981a, Two new species of *Alona* (Chydoridae, Cladocera) from western Canada: Canadian Journal of Zoology, v. 59, p. 377-389.

Chengalath, R. and Hann, B.J., 1981b, A new species of *Chydorus* (Cladocera: Chydoridae) from Ontario, Canada: American Microscopical Society, Transactions, v. 100, p. 229-238.

Cotten, C.A., 1985, Cladoceran assemblages related to lake conditions in eastern Finland, PhD thesis, Indiana University, Bloomington.

Crisman, T.L., 1980, Chydorid cladoceran assemblages from subtropical Florida, *in* Kerfoot, W.C., ed., Evolution and Ecology of Zooplankton Communities: University Press of New England, Hanover, p. 657-668.

Crisman, T.L. and Whitehead, D.R., 1978, Palaeolimnological studies on small New England (U.S.A.) ponds. II. Cladoceran community responses to trophic oscillations: Polskie Archiwum Hydrobiologie, v. 25, p. 75-86.

Culver, D.A., Vaga, R.M., Munch, C.S. and Harris, S.M., 1981, Paleoecology of Hall Lake, Washington: a history of meromixis and disturbance: Ecology, v. 62, p. 848-863.

Davis, M.B., Moeller, R.E. and Ford, J., 1984, Sediment focusing and pollen influx, *in* Haworth, E.Y. and Lund, J.W., eds., Lake Sediments and Environmental History: Leicester University, p. 261-293.

Dearing, J.A., 1986, Core correlation and total sediment influx, *in* Berglund, B.E., ed., Handbook of Holocene Palaeoecology and Palaeohydrology: John Wiley and Sons, New York, p. 247-270.

DeCosta, J.J., 1964, Latitudinal distribution of chydorid Cladocera in the Mississippi Valley, based on their remains in surficial lake sediments: Investigations of Indiana Lakes and Streams, v. 6, p. 65-101.

DeCosta, J.J., 1968, The history of the chydorid (Cladocera) community of a small lake in the Wind River Mountains, Wyoming, U.S.A.: Archiv für Hydrobiologie, v. 64, p. 400-425.

Deevey, E.S., 1942, Studies on Connecticut lake sediments. III. The biostratonomy of Linsley Pond: American Journal of Science, v. 240, p. 233-264, 313-324.

Deevey, E.S., 1955, Paleolimnology of the Upper Swamp deposit, Pyramid Valley: Records of the Canterbury Museum, v. 6, p. 291-344.

Deevey, E.S., 1964, Preliminary account of fossilization of zooplankton in Rogers Lake: Internationale Vereinigung für Limnologie, Verhandlungen, v. 15, p. 981-992.

Deevey, E.S., 1969, Specific diversity in fossil assemblages, *in* Diversity and Stability in Ecological Systems: Brookhaven Symposia in Biology No. 22, Brookhaven National Laboratory, Upton, New York, p. 224-241.

Duarte, C.M., Kalff, J. and Peters, R.H., 1986, Patterns in biomass and cover of aquatic macrophytes in lakes: Canadian Journal of Fisheries and Aquatic Science, v. 43, p. 1900-1908.

Frey, D.G., 1959, The taxonomic and phylogenetic significance of the head pores of the Chydoridae (Cladocera): Internationale Revue der gesamten Hydrobiologie, v. 44, p. 27-50.

Frey, D.G., 1960, The ecological significance of cladoceran remains in lake sediments: Ecology, v. 41, p. 684-699.

Frey, D.G., 1962, Cladocera from the Eemian Interglacial of Denmark: Journal of Palaeontology, v. 36, p. 1133-1154.

Frey, D.G., 1964, Remains of animals in Quaternary lake and bog sediments and their interpretation: Archiv für Hydrobiologie, Supplementband, Ergebnisse der Limnologie, v. 2, p. 1-116.

Frey, D.G., 1980, On the plurality of *Chydorus sphaericus* (O.F. Müller) (Cladocera, Chydoridae), and designation of a neotype from Sjaelsø, Denmark: Hydrobiologia, v. 69, p. 83-123.

Frey, D.G., 1982, Questions concerning cosmopolitanism in Cladocera: Archiv für Hydrobiologie, v. 93, p. 484-502.

Frey, D.G., 1986a, The non-cosmopolitanism of chydorid Cladocera: implications for biogeography and evolution, *in* Gore, R.H. and Heck, K.L., eds., Crustacean Issues, v. 4, Crustacean Biogeography: A.A. Balkema, Rotterdam, p. 237-256.

Frey, D.G., 1986b, Cladocera analysis, *in* Berglund, B.E., ed., Handbook of Holocene Palaeoecology and Palaeohydrology: John Wiley and Sons, New York, p. 667-692.

Frey, D.G., 1987, The North American *Chydorus faviformis* (Cladocera, Chydoridae) and the honeycombed taxa of other continents: Royal Society of London, Philosophical Transactions, v. B315, p. 353-402.

Frey, D.G. and Hann, B.J., 1985, Growth in Cladocera, *in* Wenner, A.M., ed., Crustacean Issues, v. 2, Crustacean Growth: A.A. Balkema, Rotterdam, p. 315-335.

Gauch, H.G., Jr., 1982, Multivariate Analysis in Community Ecology: Cambridge University Press, Cambridge.

Goulden, C.E., 1964, The history of the cladoceran fauna of Esthwaite Water (England) and its limnological significance: Archiv für Hydrobiologie, v. 60, p. 1-52.

Goulden, C.E., 1966a, La Aguada de Santa Ana Vieja: an interpretative study of the cladoceran microfossils: Archiv für Hydrobiologie, v. 62, p. 373-404.

Goulden, C.E., 1966b, The animal microfossils, *in* The History of Laguna de Petenxil, a small lake in northern Guatemala: Connecticut Academy of Arts and Science, Memoirs, v. 17, p. 1-126.

Goulden, C.E., 1969a, Developmental phases of the biocoenosis: National Academy of Science, Proceedings, v. 62, p. 1066-1073.

Goulden, C.E., 1969b, Temporal changes in diversity, *in* Diversity and Stability in Ecological Systems: Brookhaven Symposia in Biology No. 22, Brookhaven National Laboratory, Upton, New York, p. 96-102.

Goulden, C.E., 1969c, Interpretative studies of cladoceran microfossils in lake sediments: Internationale Vereinigung für Limnologie, Mitteilungen, v. 17, p. 43-55.

Goulden, C.E. and Frey, D.G., 1963, The occurrence and significance of lateral head pores in the genus *Bosmina* (Cladocera): Internationale Revue der gesamten Hydrobiologie, v. 48, p. 513-522.

Green, R.H., 1979, Matrix population models applied to living populations and death assemblages: American Journal of Science, v. 279, p. 481.

Hakanson, L., 1977, The influence of wind, fetch and water depth on the distribution of sediments in Lake Vanern, Sweden: Canadian Journal of Earth Science, v. 14, p. 397-412.

Hann, B.J., 1981, Occurrence and distribution of littoral Chydoridae (Crustacea, Cladocera) in Ontario, Canada and taxonomic notes on some species: Canadian Journal of Zoology, v. 59, p. 1465-1474.

Hann, B.J., 1982, Two new species of *Eurycercus* (*Bullatifrons*) from eastern North America (Chydoridae, Cladocera). Taxonomy, ontogeny, and biology: Internationale Revue der gesamten Hydrobiologie, v. 67, p. 585-610.

Hann, B.J. and Chengalath, R., 1981, Redescription of *Alonella pulchella* Herrick, 1884 (Chydoridae, Cladocera), and a description of the male: Crustaceana, v. 41, p. 249-262.

Hann, B.J. and Karrow, P.F., 1984, Pleistocene paleoecology of the Don and Scarborough Formations, Toronto, Canada, based on cladoceran microfossils at the Don Valley Brickyard: Boreas, v. 13, p. 377-391.

Hann, B.J. and Warner, B.G., 1987, Late Quaternary Cladocera from coastal British Columbia, Canada: a record of climatic or limnologic change?: Archiv für Hydrobiologie, v. 110, p. 161-177.

Harmsworth, R.V., 1968, The developmental history of Blelham Tarn (England) as shown by animal microfossils, with special reference to the Cladocera: Ecological Monographs, v. 38, p. 223-241.

Hilton, J., 1985, A conceptual framework for predicting the occurrence of sediment focusing and sediment redistribution in small lakes: Limnology and Oceanography, v. 30, p. 1131-1143.

Hofmann, W., 1978, Analysis of animal microfossils from the Grösser Segeberger See (F.R.G.): Archiv für Hydrobiologie, v. 82, p. 316-346.

Hofmann, W., 1986, Developmental history of the Grösser Plöner See and the Schöhsee (north Germany): cladoceran analysis, with special reference to eutrophication: Archiv für Hydrobiologie, Supplementband, v. 74, p. 259-287.

Huttunen, P. and Meriläinen, J., 1978, New freezing device providing large, unmixed sediment samples from lakes: Annales Botanici Fennici, v. 15, p. 128-130.

Kerfoot, W.C., 1974, Net accumulation rates and the history of cladoceran communities: Ecology, v. 55, p. 51-61.

Kerfoot, W.C., 1981, Long-term replacement cycles in cladoceran communities: a history of predation: Ecology, v. 62, p. 216-233.

Kitchell, J.A. and Kitchell, J.F., 1980, Size-selective predation, light transmission, and oxygen stratificaton: evidence from the recent sediments of manipulated lakes: Limnology and Oceanography, v. 25, p. 389-402.

Livingstone, D.A., 1957, On the sigmoid growth phase in the history of Linsley Pond: American Journal of Science, v. 255, p. 364-373.

Livingstone, D.A., Bryan, K., Jr. and Leahy, R.G., 1958, Effects of an arctic environment on the origin and development of freshwater lakes: Limnology and Oceanography, v. 3, p. 192-214.

Lloyd, M. and Ghelardi, R.J., 1964, A table for calculating the "equitability" component of species diversity: Journal of Animal Ecology, v. 33, p. 217-225.

MacArthur, R.H., 1957, On the relative abundance of bird species: National Academy of Science, Proceedings, v. 43, p. 293-295.

Megard, R.O., 1964, Biostratigraphic history of Dead Man Lake, Chuska Mountains, New Mexico: Ecology, v. 45, p. 529-546.

Megard, R.O., 1967, Late-Quaternary Cladocera of Lake Zeribar, western Iran: Ecology, v. 48, p. 179-189.

Michael, R.G. and Frey, D.G., 1983, Assumed amphi-Atlantic distribution of *Oxyurella tenuicaudis* (Cladocera, Chydoridae) denied by a new species from North America: Hydrobiologia, v. 106, p. 3-35.

Michael, R.G. and Frey, D.G., 1984, Separation of *Disparalona leei* (Chien, 1970) in North America from *D. rostrata* (Koch, 1841) in Europe (Cladocera, Chydoridae): Hydrobiologia, v. 114, p. 81-108.

Mueller, W.P., 1964, The distribution of cladoceran remains in surficial sediments from three northern Indiana lakes: Investigations of Indiana Lakes and Streams, v. 6, p. 1-63.

Nilssen, J.P., 1978, Selective vertebrate and invertebrate predation some paleolimnological implications: Polskie Archiwum Hydrobiologie, v. 25, p. 307-320.

Paterson, M.J., 1985, Paleolimnological reconstruction of cladoceran response to presumed acidification of four lakes in the Adirondack Mountains (New York), MA Thesis, Indiana University, Bloomington.

Pennak, R.W., 1978, Freshwater Invertebrates of the United States, Second Edition: John Wiley and Sons, New York.

Prentice, I.C., 1986, Multivariate methods for data analysis, *in* Berglund, B.E., ed., Handbook of Holocene Palaeoecology and Palaeohydrology: John Wiley and Sons, New York, p. 775-797.

Quade, H.W., 1969, Cladoceran faunas associated with aquatic macrophytes in some lakes in northwestern Minnesota: Ecology, v. 50, p. 170-179.

Renberg, I., 1981, Improved methods for sampling, photographing and varve-counting of varved lake sediments: Boreas, v. 10, p. 255-258.

Shan, R.K. and Frey, D.G., 1983, *Pleuroxus denticulatus* and *P. procurvus* (Cladocera, Chydoridae) in North America: distribution, experimental hybridization, and the possibility of natural hybridization: Canadian Journal of Zoology, v. 61, p. 1605-1617.

Smol, J.P., 1981, Problems associated with the use of "species diversity" in paleolimnological studies: Quaternary Research, v. 15, p. 209-212.

Smol, J.P. and Boucherle, M.M., 1985, Postglacial changes in algal and cladoceran assemblages in Little Round Lake, Ontario: Archiv für Hydrobiologie, v. 103, p. 25-49.

Sreenivasa, B.A., 1973, Paleoecological studies of Sunfish Lake and its environs, PhD Thesis, University of Waterloo, Waterloo.

Synerholm, C.C., 1979, The chydorid Cladocera from surface lake sediments in Minnesota and North Dakota: Archiv für Hydrobiologie, v. 86, p. 137-151.

Tsukada, M., 1967, Fossil Cladocera in Lake Nojiri and ecological order: Quaternary Research (Tokyo), v. 6, p. 101-110. [In Japanese]

Vallentyne, J.R. and Swabey, Y.S., 1955, A reinvestigation of the history of Lower Linsley Pond, Connecticut: American Journal of Science, v. 253, p. 313-340.

Warner, B.G. and Hann, B.J., 1987, Aquatic invertebrates as paleoclimatic indicators?: Quaternary Research, v. 28, p. 427-430.

Warner, B.G., Hebda, R.J. and Hann, B.J., 1984, Postglacial paleoecological history of a cedar swamp, Manitoulin Island, Ontario, Canada: Palaeogeography, Palaeoclimatology, Palaeoecology, v. 45, p. 301-345.

Warwick, W.F., 1985, Morphological abnormalities in Chironomidae (Diptera) larvae as measures of toxic stress in freshwater ecosystems: indexing antennal deformities in *Chironomus* Meigen: Canadian Journal of Fisheries and Aquatic Sciences, v. 42, p. 1881-1914.

Webb, T., 1982, Temporal resolution in Holocene pollen data: Third North American Paleontological Convention, Proceedings, v. 2, p. 569-571.

Whiteside, M.C., 1970, Danish chydorid Cladocera: modern ecology and core studies: Ecological Monographs, v. 40, p. 79-118.

Whiteside, M.C., 1974, Chydorid (Cladocera) ecology: seasonal patterns and abundance of populations in Elk Lake, Minnesota: Ecology, v. 55, p. 538-550.

Whiteside, M.C., 1983, The mythical concept of eutrophication: Hydrobiologia, v. 103, p. 107-111.

Whiteside, M.C. and Harmsworth, R.V., 1967, Species diversity in chydorid (Cladocera) communities: Ecology, v. 48, p. 664-667.

Whiteside, M.C. and Swindoll, M.R., 1988, Guidelines and limitations to cladoceran paleoecological interpretations: Palaeogeography, Palaeoclimatology, Palaeoecology, v. 62, p. 405-412.

Whiteside, M.C., Williams, J.B. and White, C.P., 1978, Seasonal abundance and pattern of chydorid Cladocera in mud and vegetative habitats: Ecology, v. 59, p. 1177-1188.

Wright, H.E., 1980, Cores of soft lake sediments: Boreas, v. 9, p. 107-114.

Originally published in
Geoscience Canada Volume 16 Number 1
(March 1989)

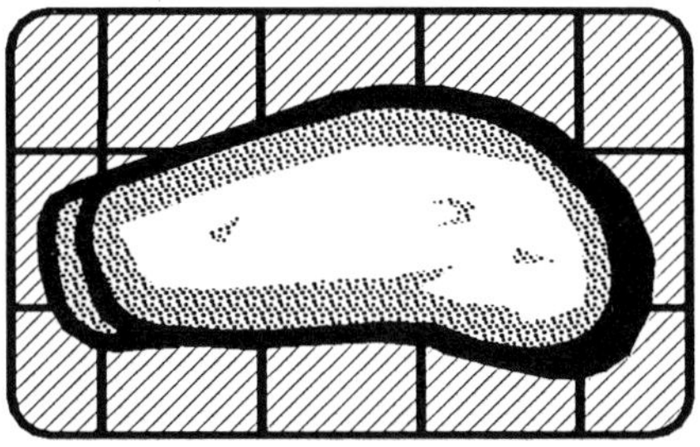

Freshwater Ostracodes

L. Denis Delorme
National Water Research Institute
Canada Centre for Inland Waters
Burlington, Ontario L7R 4A6
and *Quaternary Sciences Institute*
University of Waterloo
Waterloo, Ontario N2L 3G1

SUMMARY

Ostracodes that have lived in lakes, ponds, and streams are recovered as fossil shells from lacustrine and some fluvial sediments. Knowing some details of the physical and chemical habitat in which ostracode species live (autecology) allows one to decipher past habitats from fossil bearing sediments. Based on changes in ostracode faunal associations, changes in the physical and chemical habitats, can be interpreted. Climate, the driving force behind most changes in lakes and ponds, can also be delineated. The history of ostracode paleoenvironmental research as it pertains to Canada and to some extent North America is the subject of this review paper.

INTRODUCTION

The word ostracode is derived from the Greek οστρακφδηϛ (ostracodes) which means testaceous, from *ostracon* — shell of a testacean (Moore, 1961). Most British and some American writers use the spelling "ostracod", while the *Treatise on Invertebrate Paleontology* (Q)Arthropoda 3 (Moore, 1961) prepared by a group of international scientists have adopted the spelling "ostracode".

Ostracodes have two calcitic valves, hinged dorsally to form a carapace. The shells are readily preserved as fossils in well-buffered sediments of oceans, lakes, ponds, and streams. The age range is from upper Cambrian to Recent. In terms of systematics, ostracodes belong to phylum Arthropoda, class Crustacea, and subclass Ostracoda (Moore, 1961). Most of the freshwater ostracodes of North America can be found in the suborder Podocopina (Table 1).

HISTORICAL DEVELOPMENT

Ostracodes were first formally identified by the Danish naturalist O.F. Müller (1776). These were freshwater forms. The first freshwater ostracodes identified from Canada were reported on by Sars (1915) from samples collected from Georgian Bay by E.M. Walker; and samples from Ottawa, Medicine Hat (Alberta), Old Man River (Alberta), Port au Port (Dolphin and Union straight NWT) collected by J.B. Tyrrell, F. Johansen, and A.G. Huntsman (Sars, 1926). The first checklist of Quaternary fossil and modern ostracodes for Canada was given by Delorme (1964). The only detailed systematic work on Canadian freshwater ostracodes was carried out by Delorme (1967, 1970a,b,c,d, 1971a).

In the United States, workers of note are Hoff (1942), Kesling (1951a,b, 1952, 1954), Furtos (1933), Staplin (1963a,b), and others. Compilation of ostracodes in text books have been given by Pennak (1953) and Tressler (1959). Zoologists dealing with benthic invertebrates frequently encounter ostracodes. Most are simply reported in the literature as Crustacea or Ostracoda or both.

Quaternary freshwater species are not used in biostratigraphy. The length of time needed for the evolutionary process to take place has not been long enough to define chronostratigraphic units. In lake basins, zones may be defined by changes in the ostracode fauna brought about by changes in the habitat. Interglacial ostracodes and a paleoenvironmental interpretation were reported on by Klassen *et al.* (1967). Pleistocene ostracodes were described by Delorme (1968) from the Old Crow Basin, Yukon Territory.

An important step in the use of freshwater ostracodes as indicators of past environmental conditions was the collection of autecological data starting in 1962 (Delorme, 1965) and continuing until 1976. This included the collection of modern ostracodes and related physical, chemical, and climatic data, from over 6700 sites. The sites were located in the accessible areas of Alberta, Saskatchewan, Manitoba, Ontario, and along the McKenzie Valley transportation corridor of the Northwest and Yukon Territories. The sites, other than those in the Territories and British Columbia, were collected from within an eight mile by eight mile grid. Each site was sampled once except for some large lakes where a number of samples were taken at various depths along a transect. It was not the intent to study the indigenous ostracode fauna of a given lake, but rather to study as many different types of physical and chemical ostracode habitats as possible. The chemical parameters, for which data were collected, were Ca, Mg, Na, K, SO_4, Cl, HCO_3, CO_3, OH^-, pH, CO_2, O_2. Physical parameters recorded were depth of water, bottom and surface water temperature, air temperature one metre above the water surface, and distance to the nearest point of the shoreline. As well, the presence and/or absence of aquatic vegetation was noted, as was the terrestrial vegetational association. Climatic data from the three closest meteorological stations to the site were obtained from the Monthly Records published by the Canadian Atmospheric Service. The modern climatic values, used in the paleo-interpretive model, are based on five-year averaged data, which included the year the sample was collected plus the two preceding and two following years. These data form the basis for ostracode paleoenvironmental interpretations for Canada and the north central part of the United States. A prelimin-

Table 1 Classification scheme for freshwater and brackish water ostracodes (Moore, 1961; the treatise is in revision, other classifications are available, see Delorme, 1990a).

Subclass OSTRACODA
Order PODOCOPIDA
Suborder PODOCOPINA
- Superfamily CYPRIDACEA
  - Family CYPRIDIDAE
    - Subfamily CYPRIDINAE
    - Subfamily CYPRIDOPSINAE
    - Subfamily HERPETOCYPRIDINAE
  - Family CYCLOCYPRIDIDAE
  - Family CANDONIDAE
  - Family ILYOCYPRIDIDAE
    - Subfamily ILYOCYPRIDINAE
  - Family NOTODROMADIDAE
- Superfamily DARWINULACEA
  - Family DARWINULIDAE
- Superfamily CYTHERACEA
  - Family CYTHERIDEIDAE
    - Subfamily NEOCYTHERIDEIDINAE
  - Family ENTOCYTHERIDAE
    - Subfamily ENTOCYTHERINAE
  - Family LIMNOCYTHERIDAE
  - Family LOXOCONCHIDAE

ary interpretive model was developed for use with fossil assemblages from Holocene exposures and cores (Cvancara *et al.*, 1971; Delorme, 1971b,c) culminating in Delorme *et al.* (1977). Since then, efforts have concentrated on improving the model so the predictive capabilities of interpreted climatic parameters are acceptable.

PRESERVATION AND MORPHOLOGY OF OSTRACODE SHELLS

Preservation of the calcareous shell dictates the enclosing sediments must be well buffered. Some ostracode species can live in water with a pH of less than 8.3. Above this level, the carbonate shell is stable, but below this value the carbonate will dissolve as the epidermal lining of the shell disintegrates after the animal dies. Calcareous shells are only rarely preserved in sediments, such as bogs and marshes, where the pH is below 8.3. Ostracodes are best preserved in marl deposits.

North American ostracode shells range in size from 0.4 to 5.0 mm (Figure 1) but most fall in the range of 0.6 to 2.0 mm (Delorme, 1969). A microscope is required to study the shape, size, and ornamentation of ostracode shells. The shells are commonly rectangular to subelliptical in outline and may be compressed to inflated in the dorsal view. Many of the shells are smooth (Figures 1a-g) on the exterior surface with some pore canals interrupting the surface. Other shells may have a reticulate surface to the more extreme scabrous surface (Figure 1h). These forms may also have protuberances on the surface, variously called alae or pustules, As well, there may be depressions called sulci. In many species, the interior of the valve is rimmed by a shelf of calcite called the inner lamella. Muscle scar patterns are distinctive at the generic and higher taxonomic levels. The shells, generally referred to as the hard parts, are what is preserved and are taxonomically significant to the species level. If the two valves remain intact after the organism dies, the interior void may contain the more resistant chitinous appendages such as claws and Zenker's organs. This preservation of appendages is very rare.

With reference to shell morphology, dimorphism is fairly common. In most species, male shells are generally larger and more robust than the female shells. Some species have no males. Further problems arise with some species where the left valve is not a mirror image of the right valve. Therefore, the ostracode taxonomist must be able to recognise four distinct valve types as belonging to one species. Juveniles are not identified to species level and might not show dimorphism.

ECOLOGY

In order to collect fossil ostracodes, it is helpful to know in what environments their modern counterparts live. Ostracodes are benthic or bottom-dwelling organisms of lakes, ponds, and streams. Some species have limited swimming power which allows them to swim between plants. The true benthic forms crawl over the substrate with their

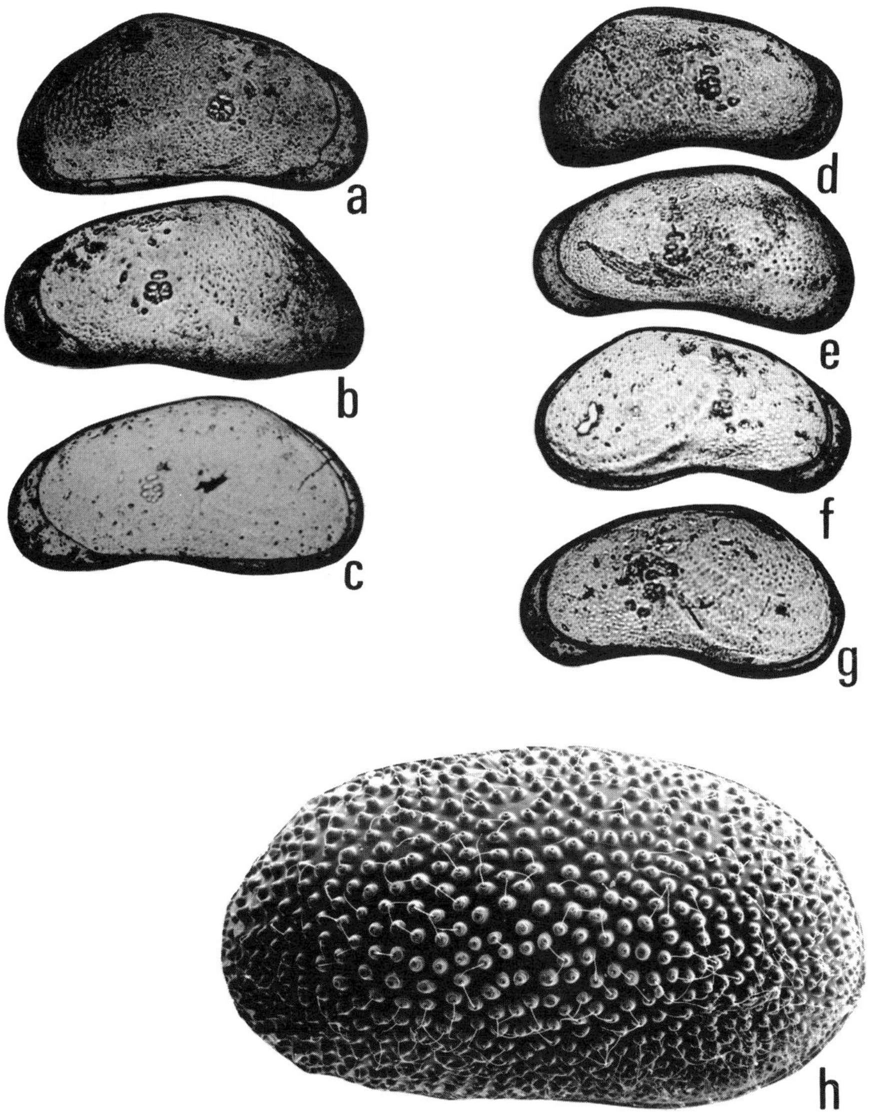

Figure 1 *Exterior views of freshwater ostracodes.*
a-c. Candona acuminata *(Fischer), 1854; 40 ×*
- **(a)** *female right valve*
- **(b)** *female left valve*
- **(c)** *male left valve*

d-g. Candona acuta *Hoff, 1942; 40 ×*
- **(d)** *female right valve*
- **(e)** *female left valve*
- **(f)** *male right valve*
- **(g)** *male left valve*

h. Cypricercus horridus *Sars, 1926; 100 × SEM photograph of the left valve.*

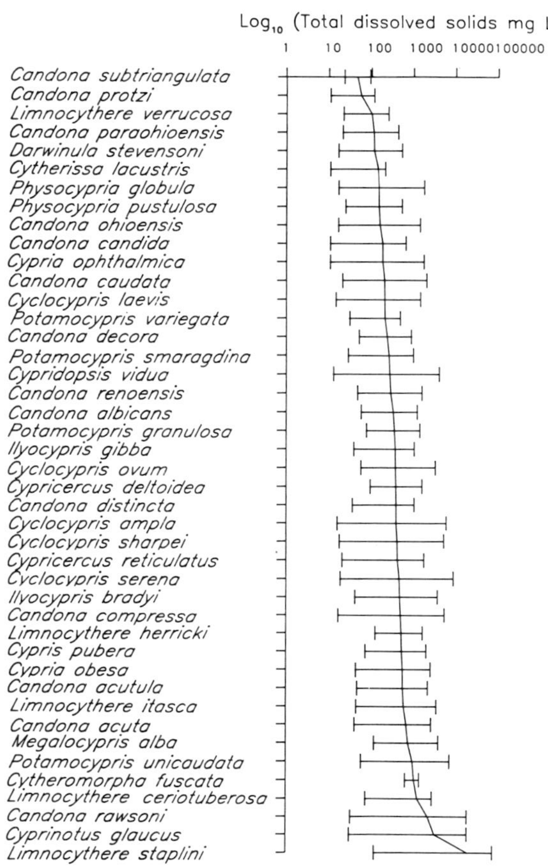

Figure 2 *The total dissolved solids range for some freshwater ostracodes. The solid line represents the mean, while the bars represent the minimum and maximum values.*

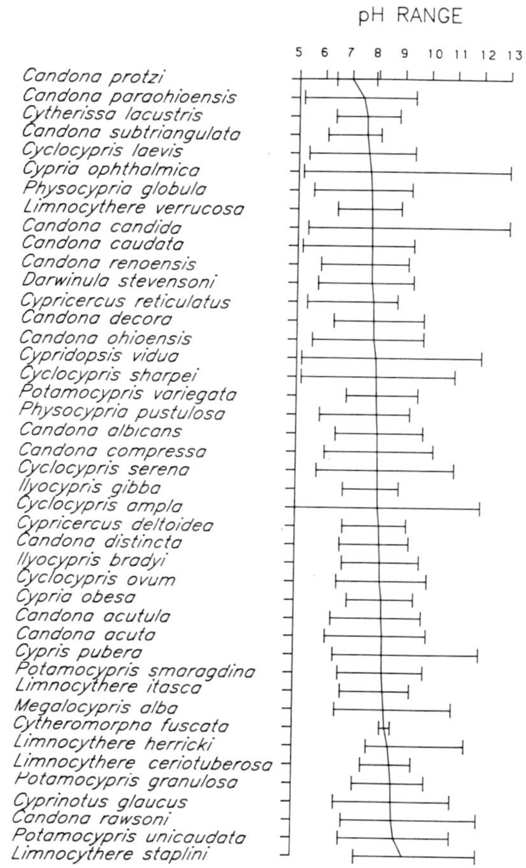

Figure 3 *The pH range for some freshwater ostracodes. The solid line represents the mean, while the bars represent the minimum and maximum values.*

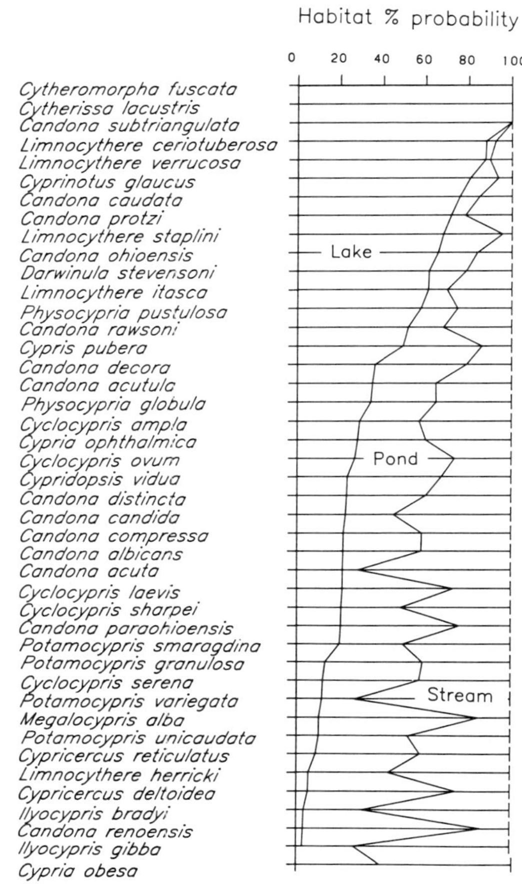

Figure 4 *The % probability of finding freshwater ostracodes in a lake, pond, or stream.*

shells slightly agape. Other groups, in particular the limnocytherids, crawl between the interstices just below and at the sediment-water interface. Ostracodes generally do not live in high energy environments except where there may be fine sand to coarse silt sediments. As detritus feeders, this group of animals prefers an organic rich, well defined, clay to fine silt textured sediment at the sediment-water interface.

The presence and/or absence of ostracodes in a habitat may be controlled by salinity (total dissolved solids), solute composition, dissolved oxygen, pH, depth of water, and availability of food. In general species diversity (the number of species in an assemblage) decreases with an increase in the thickness of the water column and/or the salinity. It also decreases with a reduction of food supply, or when the pH falls below 7 or increases above 10. Figure 2 shows the range of salinity that can be tolerated by a few species and Figure 3 shows the range of pH for the same species. Figure 4 illustrates the range of physical habitat as a percent probability. As all of the habitats (lakes, ponds, streams) were not sampled equally, the number of each habitat was taken to a common base. If A_{i-n} is the sum of sites for each habitat collected and B_{i-n} is the count of the number of samples species *X* occurs in for each of the habitats, then $B_i/A_i = C$. Using the sum of these values $(C_1 + C_2 + C_3 = D)$ to arrive at a common base, the percentage of species *X* for a particular habitat would be $(C_1/D) \times 100$.

FIELD AND LABORATORY TECHNIQUES

Ostracode shells are difficult to recognize with the naked eye in exposures or cores. When wet, the shells are translucent and about the same size as the sediment particles. There are, however, some rules of thumb that can be used to recognize lacustrine sediments which may contain ostracode shells. Sediments containing molluscan gastropods and pelecyopods (*i.e., Pisidium* or *Sphaerium*) will normally contain ostracode shells. Calcareous gyttja or marl will always be rich in ostracode remains. They will be present in most carbonate-buffered pond or littoral lake deposits. Organic laminae may have abundant shells (Cvancara *et al.*, 1971). Coarse-grained sediments should be avoided. Varves are generally devoid of ostracode shells. Diamicton may occasionally contain ostracode shells (Westgate *et al.*, 1987).

Standard procedures should be followed when collecting lacustrine samples with coring devices (Delorme, 1990b). When extruding cores, the sample interval will be a function of the purpose of the study. One centimetre intervals are standard in ostracode work. It may well be that not all the samples will be used, however, the samples are then available if more stratigraphic or paleoenvironmental detail is required from a segment of the core. Sample volume should not be less than 50 mL.

Ostracode shells may be separated from the sediment matrix in two ways. First, one may gently wash the sediments through a bank of sieves (#20 mesh, and #60 mesh). The #60 mesh will retain the majority of ostracode shells although the larger cyprids will be retained on the #20 mesh. The residue should then be oven or freeze-dried. The major disadvantage of this method is the loss of sediment and other microfossils. The second method, which is preferred, is by freeze-drying the sediment. In this procedure, the slice of the core is placed in a vial, frozen, and then placed into an evacuation chamber under a vacuum. After about 48 hours, depending on the size of the freeze dryer, the ice will have been sublimated and disaggregated sediment is left. With all of the sediment retained, analyses for fossil pollen and diatoms (Delorme *et al.*, 1986), size analyses (Duncan and LaHaie, 1979), specific gravity (Holloway and Delorme, 1987), some geochemical work (Wong *et al.*, 1984), ^{210}Pb (Turner and Delorme, 1988) and ^{14}C dating can be carried out on the same sample.

The adult shells are identified and counted using a stereoscopic microscope. Because of the large number of adult shells usually encountered in a sediment sample, only five to ten valves of each species are picked and mounted onto micropaleontological slides. Juvenile shells are not counted. One of the oculars should be equipped with a micrometer so that measurements of length and height of shells can be readily made to aid in species identification. Magnifications up to 750× are adequate for general taxonomic work. No special preparation of shells is needed. Scanning electron microscopy is useful for detailed work on sculpturing, muscle scar patterns, and internal structures of the valves. As well, soft part morphology for modern specimens may be studied through scanning electron microscopy.

QUATERNARY PALEOENVIRONMENTAL MODELS

Lacustrine sediment cores have been used for many years to tell us something about lakes and ponds. Paleo-interpretive models are constantly being improved in order to enhance their predictive capabilities and hence paleoenvironmental reconstruction. The major task is to identify significant changes that have occurred in the associated habitats of organisms and hence in the lake.

It is necessary that fossils be recovered from the sediments and that autecological data be available for each species. There then remains the task of taking the qualitative and quantitative autecological information on the species and making some sense out of it.

The modern geographical range of a species in many instances is very restrictive. Species distribution crosses temperature and precipitation boundaries as well as land vegetation zones. The water chemistry of the species habitat will also vary from place to place within its range. In a slice of a sediment core from a pond or lake, there are generally several fossil species represented. When the modern physical and chemical characteristics of the habitats and the geographic distributions of each species are combined into one data set, then an interpretation of what the conditions were like when the sediments were deposited and the fossil organisms lived is possible. Paleoenvironmental parameters have been interpreted based on a specified interpretive model from fossil freshwater ostracodes and their associated autecological data by Delorme, (1965, 1971b,c, 1972, 1982) and Delorme *et al.* (1977), Delorme and Zoltai (1984).

If one accepts the assumption that correlations do exist between climatic parameters and chemical and biological species, then the next step is to interpret paleoclimate from autecological data referenced by fossil ostracodes. A review of the model used to interpret past climatic and chemical data will be given.

First, one must take the theoretical faunal assemblage (TFA, Delorme *et al.*, 1977) from a core slice and determine the corresponding net geographical distribution for the modern species counterpart (Delorme, unpublished data). The gross geographic distribution contains all those modern ostracode sample sites where the species, identified by the TFA, occur; while the net geographical distribution is restricted to the common area where members of the TFA occur. This model includes the autecological data base for those modern samples of the TFA species within the net geographical distribution. These normalized (log_{10}) modern physical, chemical, and climatic data define the attributes for the habitats of the fossil ostracodes. The modern species tolerance limit for each parameter is represented by mean, minimum, and maximum values (Delorme *et al.*, 1977). The *preliminary-interpreted* mean value, *e.g.*, temperature, is the mean value of the parameter weighted using the reciprocal of the variance and the number of the shells in the core sample. Simply, this says if the variance of annual temperature is small for a given species of an assemblage, it will greatly influence the interpreted annual temperature for the sample. Similarly, if the species is represented by a large number of shells in the sample, then the interpreted value will likewise be influenced. For chemical and physical parameters, other than climatic ones, the preliminary-interpreted values are the *final-interpreted* parameters.

The number of years taken to deposit the sediment thickness of the sample may be longer or shorter than a year. The preliminary-interpreted value of the parameter is then an integrated value of maybe a year's duration or more. To compare the interpreted climatic parameter to the observed parameter it may be necessary to integrate several years of observed data, represented in a slice of a core, into one year to produce *revised-observed* data.

The next step in the analysis is to determine a relationship between the theoretical faunal assemblage (TFA) and the twelve monthly climatic values. *Species dependent files* are produced from preliminary-interpreted data from several cores and revised-observed data from nearby meteorological stations. The upper parts of the cores used correspond to the length of time represented by the climatic record of the closest meteorological station. The uncompacted mid-depth of the core corresponding to the first year that climatic data is available, at the nearest meteorological station, dictates the number of samples or lines of data available per core. Data for those ostracodes not represented in the upper portion of the core being analyzed may be obtained from the upper portion of cores in other geographic areas were the fossil species are present.

The relationship between preliminary-interpreted and revised-observed climatic data is expressed as an equation. Transfer functions are established by applying stepwise multiple regression analyses to both revised-observed (independent variable) and preliminary-interpreted derived climatic data (dependent variable) in the species dependant file. The regression analyses produce 12 monthly equations for every species combination found in the TFA for both temperature and precipitation. Preliminary examples of final-interpreted mean annual temperatures (Figure 5) and total annual precipitation (Figure 6) have been provided by Delorme (unpublished data) for the Clearwater Lake, Saskatchewan, core. Wherever that species composition is encountered in a core slice, the linear equation uses the preliminary-interpreted data to produce a *final interpreted* value of the climatic parameter. The selection of the appropriate equation is based on the climatic parameter selected, correlation coefficient, and the species composi-

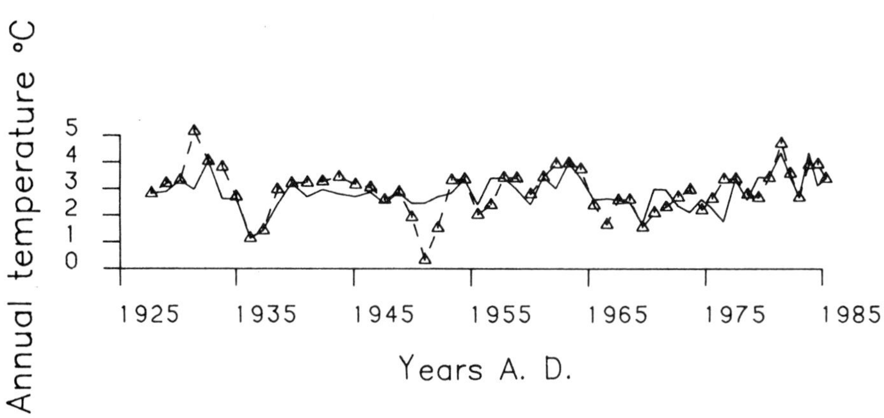

Figure 5 *Mean annual temperature profile based on revised-observed (dashed line with triangle) and final-interpreted (solid line) values for the last 60 years of sediment deposition from Clearwater Lake, Saskatchewan, core. (After Delorme, unpublished data).*

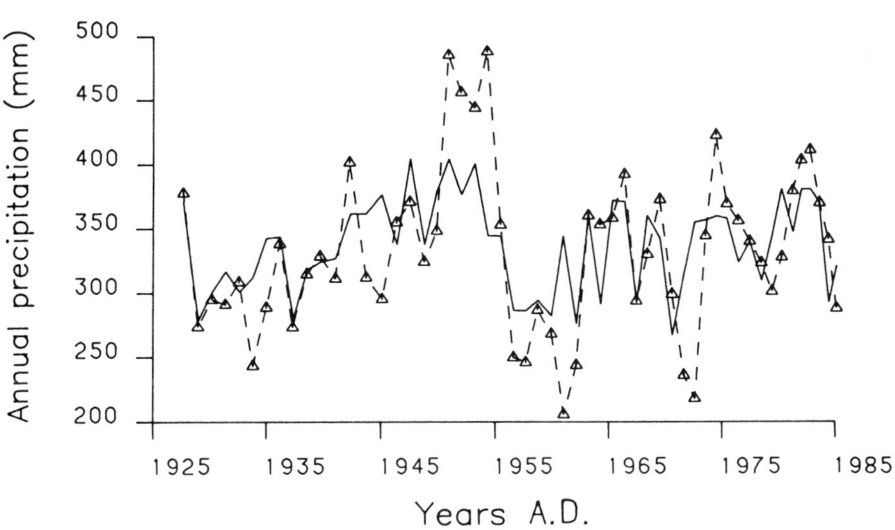

Figure 6 *Total annual precipitation profile based on revised-observed (dashed line with triangle) and final-interpreted (solid line) values for the last 60 years of sediment deposition from Clearwater Lake core. (After Delorme, unpublished data).*

tion. The selection is based on the highest correlation coefficient derived from the stepwise multiple regression of the species dependant files.

The predictive capability of the final-interpreted value (y) when compared to the revised-observed value (x) is given by the linear expression $y = a + bx$. When the y-intercept $a = 0$ and the slope $b = 1$ then $y = x$ and the relationship is perfect. As the values for the y-intercept approach zero and b approaches one, in the preliminary examples used by Delorme (unpublished data), 100% of the final-interpreted 12 monthly temperature and precipitation values are acceptable when a is within the range of ± one standard error of a. One hundred (100) percent of the final-interpreted 12 monthly temperature values and 83% of the precipitation values are acceptable when b is within the range of ± one standard error of b.

PALEOENVIRONMENTAL RECONSTRUCTION

The sedimentary record may read like a book, providing the researcher with an insight of what has happened throughout the pond or lake basin's history. In the glaciated regions of Canada, this record started when the glaciers melted from over or near the pond or lake exposing the landscape as we know it today. Ostracodes can tell us a great deal about this record by providing an interpretation of the physical and chemical habitat in which they lived. Several parameters which characterize the habitats will be discussed and examples given where possible.

Habitat.
Some ostracode species have a high probability (Figure 4) of being found only in a lake (*Candona subtriangulata, Cytherissa lacustris*; Delorme, 1978, 1982). Other species are found predominantly in ponds *(Candona renoensis, Cypricerus deltoidea)* while others are mainly confined to moving waters, such as streams and springs *(P. variegata, Ilyocypris gibba, Cypria obesa, Ilyocypris bradyi)*. This means when these indicator species in combination with other ostracode species are recovered, the physical habitat can be interpreted for that core slice. Based on the fossil ostracodes from the sequence of sediment slices the physical habitat is reconstructed for the total length of the core. In addition, the depth of water above the sediment-water interface at the collection site can be interpreted as can the distance to the nearest point of the shore. As successional development progresses, one should be able to see the physical habitat getting smaller by virtue of sediment infilling which reduces the water volume and depth of water (Delorme, 1971a). An increase in the volume of water would indicate a change in climate, sill level or erosion (Delorme, 1982).

Salinity.
The optimal salinity or total dissolved solids for the habitat of Canadian freshwater ostracodes is between 100 and 1000 $mg \cdot L^{-1}$, although the maximum can go up to 18800 $mg \cdot L^{-1}$ (*Limnocythere staplini*, Figure 3). In closed basins, salinity values can get very high, while flow-through systems have a more constant flushing rate and a lower salinity (Delorme, 1971a,b). Implicit in the interpretation of the habitat's salinity is the ability of the model to interpret average concentrations of the major ions for each core slice.

Oxygen/anoxia.
A number of ostracode species have become notably absent from the central basin of Lake Erie (Benson and MacDonald, 1963; Delorme, 1978). It has been shown that both the life cycle and the requirement for dissolved oxygen have played a role in the survival of *Candona subtriangulata* and *C. caudata* (Delorme, 1982) in Lake Erie. *C. subtriangulata* has a life cycle of over one year and also has a minimum requirement for dissolved oxygen of 5.6 $mg \cdot L^{-1}$. When the central basin of Lake Erie becomes anoxic during the summer, oxygen levels go below 5.6 $mg \cdot L^{-1}$ and the species does not reach sexual maturity. Over a few years, this species' egg pool becomes depleted and the species becomes locally extinct. Another species, *C. caudata*, has a life cycle of three to four weeks and a minimum dissolved oxygen requirement of 2.3 $mg \cdot L^{-1}$. Although the oxygen levels go below 2.3 $mg \cdot L^{-1}$ in the central basin and the species does not survive during the anoxic period, it persists in Lake Erie because the species reaches sexual maturity and lays eggs during other times of the year.

Temperature/precipitation.
Freshwater ostracodes may show an indirect relationship to climatic parameters (Forester, 1983; Delorme, unpublished data) through salinity and probably more specifically through solute composition. There are several species which are restricted to a geographical area, apparently controlled by climate. Delorme and Zoltai (1984) and Delorme (1971c) indicate that such species as *Candona protzi* and *Cyclocypris globosa* are found only north of latitude 65°, and *Candona rectangulata* is restricted to north of latitude 62°. However, these species have been found as fossils in basal lacustrine sediments of southwestern Manitoba (49°23'N lat., 98°35'W long.) indicating a cold-climate, cold-water type of environment adjacent to a retreating ice-margin. Conversely, species found in sediments (7100 to 8610 years B.P.) at the Jiggle Lake site (67°41'N lat., 132°07'W long.) reflect a warmer climate like that found on the Canadian prairie today (Delorme and Zoltai, 1984).

CONCLUSIONS

The use of Quaternary freshwater ostracodes is a recent advance in paleoenvironmental research. The reasons for this relatively new interest has been a confused systematics of the group and a lack of autecological data. The systematics are not as unaccommodating as they may appear with many good references now available. For Canada and the northern United States, autecological data are now available.

A carefully designed model is used to interpret Holocene paleolimnologic and paleoclimatic conditions. These results are acceptable within certain statistical limits. The most important aspect of these studies and their results is that they provide trends in water chemistry, changes in the physical habitat, and paleoclimate. These trends and changes may be equivalent to temporal or regional variation. Some workers are still hesitant to accept objective paleoenvironmental results, but more and more there is the acceptance that the fossils did exist at the time the lacustrine deposits were being deposited and that their distribution may have been controlled, in their environments, by physical and chemical factors. We are attempting to understand these relationships in the modern fauna. As pointed out by Smol (1987) and many other paleolimnologists, the best approach is the multidisciplinary approach, each approach complimenting the other.

Current concerns regarding climatic

change in Canada are centered in two regions: the prairies (Arthur, 1988) and coastal areas including the Great Lakes. The agricultural economy of the prairie region fluctuates with climate. Long-term climatic records are not available for this region, therefore, those wishing objective data going back farther in time for research on drought scenarios must rely on climatic proxy data. We are now at the point in this research where we can begin to produce such data from the Holocene lacustrine record. Changes in climate may affect water levels in coastal regions (Sanderson, 1987; Martec Limited, 1987; P. Lane and Associates Limited, 1988). Despite control structures on the Great Lakes system, fluctuations in water levels related to climate do occur. Extensive water level fluctuation data for the Great Lakes are not available. Therefore, lacustrine sediments, with their inherent fossils, should be able to provide objective water level and paleoclimatic data.

REFERENCES

Arthur, L.M., 1988, The implication of climate change for agriculture in the prairie provinces: Climate Change Digest, Environment Canada, Downsview, Ontario, CCD 88-01, 12 p.

Benson, R.H. and MacDonald, H.C., 1963, Postglacial (Holocene) ostracodes from Lake Erie: University of Kansas Paleontological Contributions, Article 4, p. 1-16.

Cvancara, A.M., Clayton, L., Bickley, W.B., Jr., Jacob, A.F., Ashworth, A.C., Brophy, J.A., Shay, C.I., Delorme, L.D. and Lammers, G.E., 1971, Paleolimnology of late Quaternary deposits Seibold site, North Dakota: Science, v. 171, p. 172-174.

Delorme, L.D., 1964, A checklist of Pleistocene and Recent freshwater ostracods in Canada: Sterkiana, no. 14, p. 39-44.

Delorme, L.D., 1965, Pleistocene and post-Pleistocene Ostracoda of Saskatchewan, Ph.D. thesis, University of Saskatchewan, Saskatoon.

Delorme, L.D., 1967, Field key and methods of collecting freshwater ostracodes in Canada: Canadian Journal of Zoology, v. 45, p. 1275-1281.

Delorme, L.D., 1968, Pleistocene freshwater Ostracoda from Yukon, Canada: Canadian Journal of Zoology, v. 46, p. 859-876.

Delorme, L.D., 1969, On the identity of the ostracode genera *Cypriconcha* and *Megalocypris*: Canadian Journal of Zoology, v. 47, p. 997-999.

Delorme, L.D., 1970a, Freshwater ostracodes of Canada, part I; Subfamily Cypridiniae: Canadian Journal of Zoology, v. 48, p. 153-169.

Delorme, L.D., 1970b, Freshwater ostracodes of Canada, part II; Subfamilies Cypridopsinae, Herpetocypridinae, and family Cyclocyprididae: Canadian Journal of Zoology, v. 48, p. 253-266.

Delorme, L.D., 1970c, Freshwater ostracodes of Canada, part III; Family Candonidae: Canadian Journal of Zoology, v. 48, p. 1099-1127.

Delorme, L.D., 1970d, Freshwater ostracodes of Canada, part IV; Families Ilyocyprididae, Notodromadidae, Darwinulidae, Cytherideidae and Entocytheridae: Canadian Journal of Zoology, v. 48, p. 1251-1259.

Delorme, L.D., 1971a, Freshwater ostracodes of Canada, part V; Families Limnocytheridae, Loxoconchidae: Canadian Journal of Zoology, v. 49, p. 43-64.

Delorme, L.D., 1971b, Paleoecological determinations using Pleistocene freshwater ostracodes: Centre de Recherche, Pau-SNPA Bulletin, v. 5, supplement, p. 341-347.

Delorme, L.D., 1971c, Paleoecology of Holocene sediments from Manitoba using freshwater ostracodes, *in* Turnock, A.C., ed., Geoscience Studies in Manitoba: Geological Association of Canada, Special Paper 9, p. 301-304.

Delorme, L.D., 1972, Groundwater flow systems, past and present: 24th International Geological Congress, Montreal, Canada, Section 11, p. 222-226.

Delorme, L.D., 1978, Distribution of freshwater ostracodes in Lake Erie: Journal of Great Lakes Research, v. 4, p. 216-220.

Delorme, L.D., 1982, Lake Erie oxygen, the prehistoric record: Canadian Journal of Fisheries and Aquatic Sciences, v. 39, p. 1021-1029.

Delorme, L.D., 1990a, Ostracoda, *in* Thorp, J.H. and Covick, A.P., eds., Ecology and Classification of North American Freshwater Invertebrates: Academic Press, Inc., New York.

Delorme, L.D., 1990b, The preparation of lacustrine sediment samples from cores for use in dating and paleolimnology, *in* Mudroch, A. and MacKnight, S.D., CRC Handbook of Techniques for Aquatic Sediment Sampling: CRC Press, Florida.

Delorme, L.D., Duthie, H.C., Esterby, S.R., Smith, S.M. and Harper, N.S., 1986, Prehistoric inferred pH changes in Batchawana Lake, Ontario from sedimentary diatom assemblages: Archiv für Hydrobiologie, Bd. 108, p. 1-22.

Delorme, L.D. and Zoltai, S.C., 1984, Distribution of an arctic ostracod fauna in space and time: Quaternary Research, v. 21, p. 65-73.

Delorme, L.D., Zoltai, S.C., and Kalas, L.L., 1977, Freshwater shelled invertebrate indicators of paleoclimate in Northwestern Canada during the late glacial: Canadian Journal of Earth Sciences, v. 14, p. 2029-2046.

Duncan, G.A. and LaHaie, G.G., 1979, Size analysis procedures used in the sedimentology laboratory, NWRI; manual: National Water Research Institute, Hydraulics Division, Burlington, Ontario, 55 p.

Forester, R.M., 1983, Relationship of two lacustrine ostracode species to solute composition and salinity; implications for paleohydrochemistry: Geology, v. 11, p. 435-438.

Furtos, N.C., 1933, The Ostracoda of Ohio: Ohio Biological Survey Bulletin 29, v. 5, p. 413-524.

Hoff, C.C., 1942, The ostracodes of Illinois, their biology and taxonomy: Illinois Biological Monograph, v. 19, 196 p.

Holloway, D.L. and Delorme, L.D., 1987, Specific gravity determinations of sediments from Manitoba lakes (cores 127 to 132), and additional cores from Alberta (core 123), Saskatchewan (cores 120b, 124 to 126), and Ontario (core 122): National Water Research Institute, Burlington, Ontario, Technical Note AED-87-2, 33 p.

Kesling, R.V., 1951a, Terminology of ostracod carapaces: Contributions to the Museum of Paleontology, University of Michigan, v. 9, p. 93-171.

Kesling, R.V., 1951b, The morphology of ostracod molt stages: Illinois Biological Monograph, v. 21, 324 p.

Kesling, R.V., 1952, Doubling in size of ostracod carapaces in each molt stage: Journal of Paleontology, v. 26, p. 772-780.

Kesling, R.V., 1954, Ornamentation as a character in specific differentiation of ostracods: Contribution to the Museum of Paleontology, University of Michigan, v. 12, p. 13-21.

Klassen, R.W., Delorme, L.D. and Mott, R.J., 1967, Geology and paleontology of Pleistocene deposits in southwestern Manitoba: Canadian Journal of Earth Sciences, v. 4, p. 433-447.

Martec Limited, 1987, Effects of a one metre rise in mean sea-level at Saint John, New Brunswick and the lower reaches of the Saint John River: Climate Change Digest, Environment Canada, Downsview, Ontario, CCD 87-04, 8 p.

Moore, R.C., 1961, ed., Treatise on invertebrate paleontology, part Q; Arthropoda 3, Crustacea, Ostracoda: Geological Society of America, University of Kansas Press, Q442 p.

Müller, O.F., 1776, Zoologiae Danicae produmus, seu animalium Daniae et Norvegiae indigenarum. Characteres, nomina, et synonyma imprimis popularium: Havniae, Typis Hallageriis, p. 198-199.

Pennak, R.W., 1953, Ostracoda (seed shrimps), *in* Freshwater Invertebrates of the United States: Ronald Press Co., New York, p. 410-421.

P. Lane and Associates Limited, 1988, Preliminary study of the possible impacts of a one meter rise in sea level at Charlottetown, Prince Edward Island: Climate Change Digest, Environment Canada, Downsview, Ontario, CCD 88-02, 7 p.

Sanderson, M., 1987, Implications of climatic change for navigation and power generation in the Great Lakes: Climate Change Digest, Environment Canada, Downsview, Ontario, CCD 87-03, 20 p.

Sars, G.O., 1915, Entomostraca of Georgian Bay: Contribution of Canadian Biology (1911-1914), no. 39b, art. 13, p. 221-222.

Sars, G.O., 1926, Freshwater Ostracoda from Canada and Alaska: Report on the Canadian Arctic Expedition (1913-1918), v. 7, pt. 1, 22 p.

Smol, J.P., 1987, Freshwater Algae: Geoscience Canada, v. 14, p. 208-217.

Staplin, F.L., 1963a, Pleistocene Ostracoda of Illinois, Part 1. Subfamilies Candoninae, Cyprinae, general ecology, morphology: Journal of Paleontology, v. 37, p. 758-797.

Staplin, F.L., 1963b, Pleistocene Ostracoda of Illinois, part 2. Subfamilies Cyclocyprinae, Cypridopsinae, Ilyocyprinae; families Darwinulidae and Cytheridae. Stratigraphic ranges and assemblage patterns: Journal of Paleontology, v. 37, p. 1164-1203.

Tressler, W.L., 1959, Ostracoda, *in* Edmondson, W.T., ed., Fresh-water Biology: John Wiley and Sons, New York, p. 657-734.

Turner, L.J. and Delorme, L.D., 1988, ^{210}Pb dating of lacustrine sediments from Hamilton Harbour (Cores 137, 138, 139, 141, 142, 143): National Water Research Institute, Burlington, Ontario, Technical Note LRB-88-9, p. 1-42.

Westgate, J.A., Chen, F.-J. and Delorme, L.D., 1987, Lacustrine ostracodes in the Late Pleistocene Sunnybrook diamicton of southern Ontario, Canada: Canadian Journal of Earth Sciences, v. 24, p. 2330-2335.

Wong, K.T., Nriagu, J.O. and Coker, R.D., 1984, Atmospheric input of heavy metals chronicled in lake sediments of the Algonquin Provincial Park, Ontario, Canada: Chemical Geology, v. 44, p. 187-201.

Originally published in
Geoscience Canada Volume 16 Number 2
(June 1989)

Non-marine Molluscs

Barry B. Miller
Department of Geology
Kent State University
Kent, Ohio 44242

Andrew F. Bajc
Ontario Geological Survey
Ministry of Northern Development & Mines
77 Grenville Street
Toronto, Ontario M7A 1W4

SUMMARY

Molluscs are among the most ubiquitous fossils found in Quaternary non-marine sediments. They are found in a wide variety of deposits including: loess, cave, sinkhole and fissure deposits; fluvial, lacustrine, glaciolacustrine and paludal sediments; and spring-fed travertine accumulations.

In spite of problems associated with identification of some species from shell morphology alone and a general inadequacy in our understanding of the factors controlling modern species distribution, Quaternary studies of non-marine molluscs have resulted in some significant contributions. In North America fossil molluscs have been used to: (1) reconstruct former stream confluences; (2) provide a basis for biostratigraphic zonation of sedimentary sequences in the Lake Michigan, Superior and Huron basins; and (3) reconstruct local habitat and climatic conditions. Isotopic data obtained from the analysis of molluscan shell has provided information about the environment in which the molluscs lived. Amino acid studies of the shell protein have provided information about the age of fossil assemblages and a method for estimating post depositional effective temperature histories of fossil sites.

One of the basic problems in Quaternary biogeography concerns the timing, the nature, and the magnitude of biotic changes that accompanied glacial advances and retreats. The ubiquity of non-marine molluscs in Quaternary sediments provides an excellent opportunity to study effects of both short and long scale climatic fluctuations on at least one important part of the non-marine biota, the molluscs.

INTRODUCTION

Molluscs have been reported from a diversity of environments. Sediments known to preserve non-marine molluscs include supraglacial lake deposits (Tuthill *et al.*, 1964); windblown silt (Leonard and Frye, 1960); cave, sinkhole and fissure sediment accumulations (Guilday *et al.*, 1978); alluvial floodplain (Miller and Kay, 1981), back-water channel (Miller, 1983), littoral and lagoonal lacustrine sediments (Karrow *et al.*, 1975; Bajc, 1986), accumulations of spring-fed travertine (Hansel and Johnson, 1986), and peat (Miller and Thompson, 1987). Human activities have accumulated significant shell concentrations in Indian middens (Parmalee and Bogan, 1986).

The widespread occurrence of fossil non-marine molluscs in a variety of environments, has frequently led to their inclusion in reports dealing with Quaternary sites, even though they usually were not the main focus of the study. For example, the panhandle of Oklahoma – southwestern Kansas contains one of the best documented molluscan successions in North America. It includes a sequence of faunas that range in age from the Pliocene to the Holocene, that were collected over a period of three decades by C.W. Hibbard and his associates. The molluscs, which usually formed the most abundant faunal element, were retained during wet-sieving of sediments for the fossil microvertebrate remains: the main focus of the study.

In spite of the large amount of important research that remains to be done in North American Quaternary malacology, few paleontologists have been attracted to this field of study. The reasons for this are probably comparable to those given by Morgan and Morgan (1980) for the paucity of paleoentomologic researchers, *viz.*, it requires ... "a blend of paleontology, stratigraphy"... "taxonomy and zoogeography, so it is rare that any prospective researcher starts in this field fully qualified in all aspects". Add to this the sheer tedium of sorting through large quantities of material for shells, the few well-organized systematic reference collections and taxonomic keys, and the absence of ecological and distributional data for many species, has probably discouraged potential students and driven former workers from the field.

Although these difficulties have hindered, they have not prevented malacology from being an important contributor to North American Quaternary studies. The sections that follow present a short review of Quaternary malacology in North America.

HISTORICAL DEVELOPMENT

One of the first references to Canadian Quaternary non-marine molluscs was made in 1861 by Robert Bell who reported on collections made between 1838 and 1861 from several mollusc-bearing deposits in the eastern Great Lakes and St. Lawrence valley (Coleman, 1922). Among these are reports of freshwater molluscs enclosed in calcareous nodules from Green Creek near Ottawa; molluscs contained within fluvial gravels of an ancestral Niagara River near Niagara Falls; and mollusc remains recovered from clay pit excavations near Montreal. Subsequent reports of fossil molluscs from the north shore of Lake Superior were published after Bell's visit to the area around 1860 (Coleman, 1922).

Baker's (1920a) classic publication *Life of the Pleistocene or Glacial Period* presents a compilation of the early studies dealing with fossil molluscs in eastern North America, with special reference to the interglacial and post-glacial biota of the Great Lakes region. Reference to interglacial shells at Toronto (Lemon, 1898), and general discussions on fossil land snails in Quaternary paleoecology (Billups, 1902), are among the many significant historical citations found in this publication. As early as 1884, molluscan faunal studies from Canadian archaeological sites were being used to reconstruct cultural and natural environments (Matthew, 1884). Other early work, dealing primarily with taxonomy, distribution, and ecology of modern and fossil molluscs include publications by Sterki (1920) and Baker (1920b). Quaternary molluscan studies of post-glacial and interglacial sites in Saskatchewan were undertaken by Russell (1934) in an attempt to compare distributions of modern populations in the province with their fossil counterparts.

Wittaker (1921) described molluscan faunas from marl deposits in the Ottawa area. The molluscan succession of Holocene basin deposits in Ontario and Quebec were the subject of a series of studies by Shallom (1965), Clowers (1966), Gibson (1967), Warner (1968), and Bickel (1970). These investigations utilized substrata and water depth, chemistry, temperature, and pH data in the interpretation of molluscan paleoecology. Harris and Pip (1973) studied fossil molluscs in southwestern Alberta in an attempt to reconstruct post-glacial climatic change from distribution shifts of extant taxa. A recent study of gastropods from the interglacial Don Formation at Toronto revealed a fossil fauna similar to that found in southern Ontario today (Kerr-Lawson, 1985).

Molluscan studies of Quaternary assemblages from the Ontario (Karrow *et al.*, 1972), Huron (Miller *et al.*, 1985), Agassiz (Ashworth and Cvancara, 1983), and Superior basins (Bajc, 1986) have focussed on the potential for biostratigraphic zonation of these faunas. Data from the Michigan basin suggests faunal changes similar to those observed in the Huron basin during the past 11,000 years B.P. (Miller, unpublished).

A number of modern studies have focussed on the analysis of stable isotopes in molluscan shell for deducing past temperatures, evaporative effects, and productivity levels of fossil environments (Webber and LaRocque, 1964; Stuiver, 1968; Fritz and Poplawski, 1974; Fritz *et al.*, 1975; Yapp, 1979; Godwin, 1985).

Much recent work is currently being directed toward the use of amino acids in shell protein to infer ages of fossil assemblages and estimating post-depositional effective temperature histories of fossil sites (Miller and Hare, 1980).

FIELD TECHNIQUES AND SOURCE OF FOSSILS

The quality of preservation of fossil molluscs may vary greatly. Individual shells from some sites may retain the external organic periostracum layer and original shell colour patterns (Figure 1). Preservation of snail eggs in some deposits is not uncommon. Tompa (1976) reported calcified eggs of *Vallonia* from middle Pleistocene sediments associated with a 700,000 years B.P. ash in the Meade County area of southwestern Kansas. Fossil eggs (Figure 2) have also been recovered from magnetically reversed silts at Whites Branch in west-central Indiana (Miller *et al.*, 1987). At the other end of the preservation spectrum is complete destruction of the shell by leaching leaving only a mold of the exterior and/or interior.

Access to natural exposures of fossil assemblages may occur as a consequence of some geologic processes, *e.g.*, wave action along the shores of a lake (Gutschick and Gonsiewski, 1976) or the erosive action of a stream that might expose fossils in its cut bank (Wayne, 1981). Many sites containing molluscs become accessible as a result of such human activities as sand and gravel extraction operations (Leonard, 1953).

In the field, fossil molluscs are often bleached white on the surface of exposures and may stand out in sharp contrast against the colour of the enclosing sediments. All but the smallest (<2 mm) shells can be seen easily with the unaided eye. To avoid a bias that would favour larger individuals, bulk samples are usually collected at selected stratigraphic intervals of the exposure for systematic analysis.

Extraction of molluscs from the enclosing sediment usually starts by drying the sample to remove most of the interstitial water. The sample is then immersed in a container of cold water. After several hours of soaking the sediments will usually disaggregate, and many shells, insects, seeds, and plant fragments may float to the surface where they are skimmed off with a teastrainer. The fossils remaining in the water-sediment slurry are then poured through a stacked series of sieves (0.071 mm to 2.0 mm mesh diameter). Washing under a gentle spray of water flushes the fine silt and clay through the sieves and breaks down small clods of sediment that did not completely disaggregate during the soaking.

If this simple procedure was insufficient to free all shells, the sample may be redried, soaked or boiled in a water-soluble deflocculant such as Quaternary-O or sodium hexametaphosphate or placed in an ultrasonic cleaner. This concentrate of shell, organic matter, and the coarse-grained fraction from the sample is then dried, labelled and stored in vials for later examination and sorting under a low-power microscope. The molluscs are sorted on the basis of shell morphology with forceps or a fine sable brush and stored in gelatin capsules or vials.

IDENTIFICATION

Most taxa encountered as fossils in North American Quaternary deposits are referable to extant species. Consequently, reliable identification of fossil molluscs requires access to good collections of identified modern representatives and familiarity with malacological literature. Often taxonomic specialists are needed to confirm identification.

Figure 1 Anguispira alternata *from the Chinatown Silt (Illinoian) Bellville, Illinois showing preserved colour pattern in shell.*

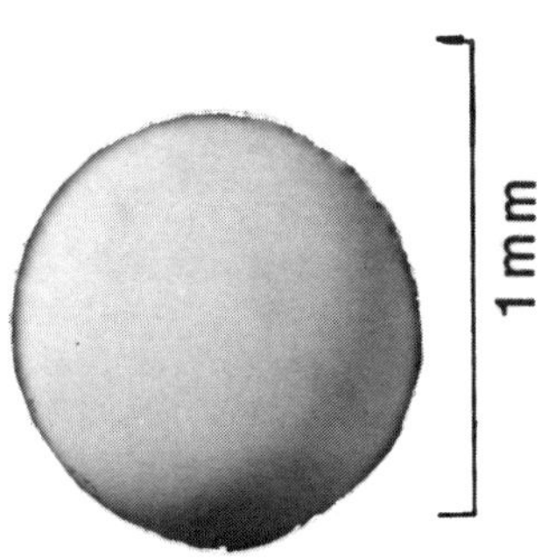

Figure 2 *Fossil calcified egg from magnetically reversed silts (>730,000 years B.P.) Whites Branch, Indiana (Miller* et al., *1987).*

Figure 3 *Angular parietal fold (upper part of aperture) and lower and basal lamellae in* Gastrocopta cf. Gastrocopta armifera *from middle Pleistocene sediments, Little Sioux, Iowa. The number, shape, and distribution of these apertural structures are important features used for the identification of species within certain families of gastropods.*

Identification may require the careful scrutiny of minute details in shell microstructure at high magnifications (100 ×). In some genera, (*e.g.*, *Strobilops, Vertigo, Gastrocopta, Carychium*, and *Planorbula*), examination of folds, lamellae, and denticles (Figure 3) that may extend varying distances within the shell may be necessary for species determination. X-radiography (Miller, 1957) or immersion of the opaque shell in oil (refractive index 1.60), which makes the shell translucent to transmitted light (Miller, 1968), provide non-destructive methods for studying internal structures.

Taxa that require the examination of soft parts, (usually the genitalia), for species identification, are of little value in fossil specimens. Although many aquatic species commonly encountered in Quaternary assemblages belong to families in which the shell alone is frequently insufficient to permit specific identification, (for example, many species of the families Succiniidae, Lymnaeidae, Physidae and Hydrobiidae), there are far more taxa that can be confidently identified to the species level on the basis of shell characters alone.

FACTORS CONTROLLING FOSSIL MOLLUSC ASSEMBLAGES

The suitability of a fossil molluscan assemblage for Quaternary environmental reconstruction is determined by factors related to: natural abundance of taxa in living populations; processes of deposition and preservation; effects of contamination by older or younger shells; adequacy of the sample; and specimen identification (Sparks, 1964). Further consideration of this problem can be found in Pip (1988), Lozek (1982), Birks and Birks (1980), Crisman (1978), Miller (1978), Kerney (1977), LaRocque (1966), and Sparks (1961).

Natural Abundance of Taxa.

The relative abundance of various taxa in their natural habitat is an important consideration for paleoecologic reconstructions. Some freshwater forms, such as *Valvata tricarinata, Lymnaea stagnalis, Sphaerium striatinum,* and *Pisidium casertanum* are ubiquitous, whereas others, such as *Acella haldemani, Pisidium cruciatanum,* and *Pisidium henslowanum* are invariably scarce, even in favourable, present-day environments. The single shell of a rare or uncommon taxon in any paleoecological analysis has quite a different significance than the solitary occurrence of a common species.

Allochthonous and Autochthonous Assemblages.

Most Quaternary molluscan assemblages are composed of taxa that have been brought together after death from a variety of habitats. Air trapped in the shells of gastropods frequently permit them to float long distances. Species diversity in an open river deposit, therefore, may be quite high because it could represent an accumulation of shell from a wide spectrum of terrestrial and aquatic habitats indiscriminately brought together during high-water periods. In contrast, a small, spring-fed lake, may yield only a limited number of taxa that were living in the water column near the sampling site.

Interpretation of mixed assemblages, frequently requires invoking the "rule of relative abundance", which assumes that, if all other factors are equal, species native to a given environment (autochthones) should be more abundant than exotic (allochthonous) taxa brought in from outside (LaRocque, 1966). Sparks (1964) has suggested that autochthonous elements in a fossil assemblage can often be recognized by the quality of preservation of the periostracum and the absence of surface abrasion. An allochthonous origin for a mollusc assemblage can be inferred when only a small percentage of the sample shows absence of wear and fragmentation.

Allochthonous assemblages redeposited by fluvial processes are likely to have been subjected to some degree of sorting. The composition of the assemblage may be determined by the size, weight, and surface morphology of the shells in the fossil deposit (Pip, 1988). As stream competence increases, the larger, heavier, and more streamlined shells will tend to settle out, while the smaller, lighter molluscs will remain in suspension and be swept away to a position downstream. Generally, loosely coiled gastropods have the greatest potential for redeposition because of their low mass to volume ratio and high sphericity. In contrast, the streamlined shape of clam valves are more resistant to transport after disarticulation.

Typically, collections of fossils are assumed to be the same age unless evidence warrants rejection of this hypothesis. Taylor (1965), for example, utilized the worn and etched condition of most of the shells from the Rezabeck local fauna, Lincoln Co., Kansas, in contrast with the normal fresh condition of shells from other nearby Quaternary age sites in the area, as evidence for probable reworking of this assemblage. A wide range of values obtained from amino acid racemization analyses of shells from the same taxa and stratigraphic level, would suggest that the fossils include reworked shells from different age deposits. Failure to recognize the possibility of contamination from older deposits may result in misleading ecological, biogeographic, and age interpretation.

The reworking of post-glacial mollusc-bearing sediments within the Great Lakes basins is a serious potential problem given the history of water level changes, particularly transgressive phases that have reworked sediments from older and lower lake levels. Within the Huron basin, alluvial deposits transitional in age between Algonquin and Nipissing phases, may well contain reworked fossils from the oldest Algonquin sediments (Miller *et al.*, 1985). Along the north shore of Lake Superior, the potential for reworking of older shells also exists as continuously falling post-glacial water levels resulted in the insetting of progressively younger terraces into older lacustrine and alluvial deposits (Bajc, 1986).

Preservation of Fossil Molluscs.

The preservation of fossil molluscs can be related to both the biological and chemical attributes of the depositional environment as well as the strength of the individual shell structures. In non-calcareous environments, molluscs are quickly leached after deposition. Generally, fossil molluscs are absent from large areas of the Canadian Shield containing acidic overburden and bedrock. The scarcity of fossil mollusc sites along the northern parts of the Huron basin has been attributed to this phenomenon (Miller *et al.*, 1985). Calcareous drift on Precambrian bedrock is responsible for the preservation of fossil molluscs along the north shore of Lake Superior (Bajc, 1986), east of Lake Nipigon (Kristjansson, pers. comm.), and in the Fort Frances-Rainy River area (Bajc and Gray, 1987).

Solution features are commonly found in molluscs recovered from sand

and gravel deposits because high groundwater flow rates through permeable sediments greatly enhance the leaching process. Calcareous, low permeability (fine-grained) sediments, tend to favour preservation of gastropods and pisidiid clams whereas, better drained sediments favour preservation of the larger unionid clams.

Differential destruction of species occurs during transportation and deposition, sediment compaction from burial, and the laboratory extraction of fossils from the enclosing matrix. Factors that contribute to differential destruction of molluscan shell include relative weakness of: larger *versus* smaller shells, thin *versus* thick shells, and loosely-coiled and elongate *versus* tightly coiled, compact shells (Sparks, 1964). Over-representation of small species, juveniles and the broken early whorls of larger forms, often results from the operation of these factors. Molluscan paleoecologists must be aware of such problems to avoid overestimating the abundance of a particular taxon or group of taxa.

USES OF FOSSIL NON-MARINE MOLLUSCS

Stratigraphy.

As Quaternary North American non-marine molluscs have apparently undergone little change in shell morphology, they have proven useful indicators of environmental change that locally have stratigraphic significance. Miller *et al.* (1985) have shown that deposits of the Algonquin, Stanley, and Nipissing lake phases of the Huron basin are characterized by distinctive assemblages of molluscs that relate to climatic changes,

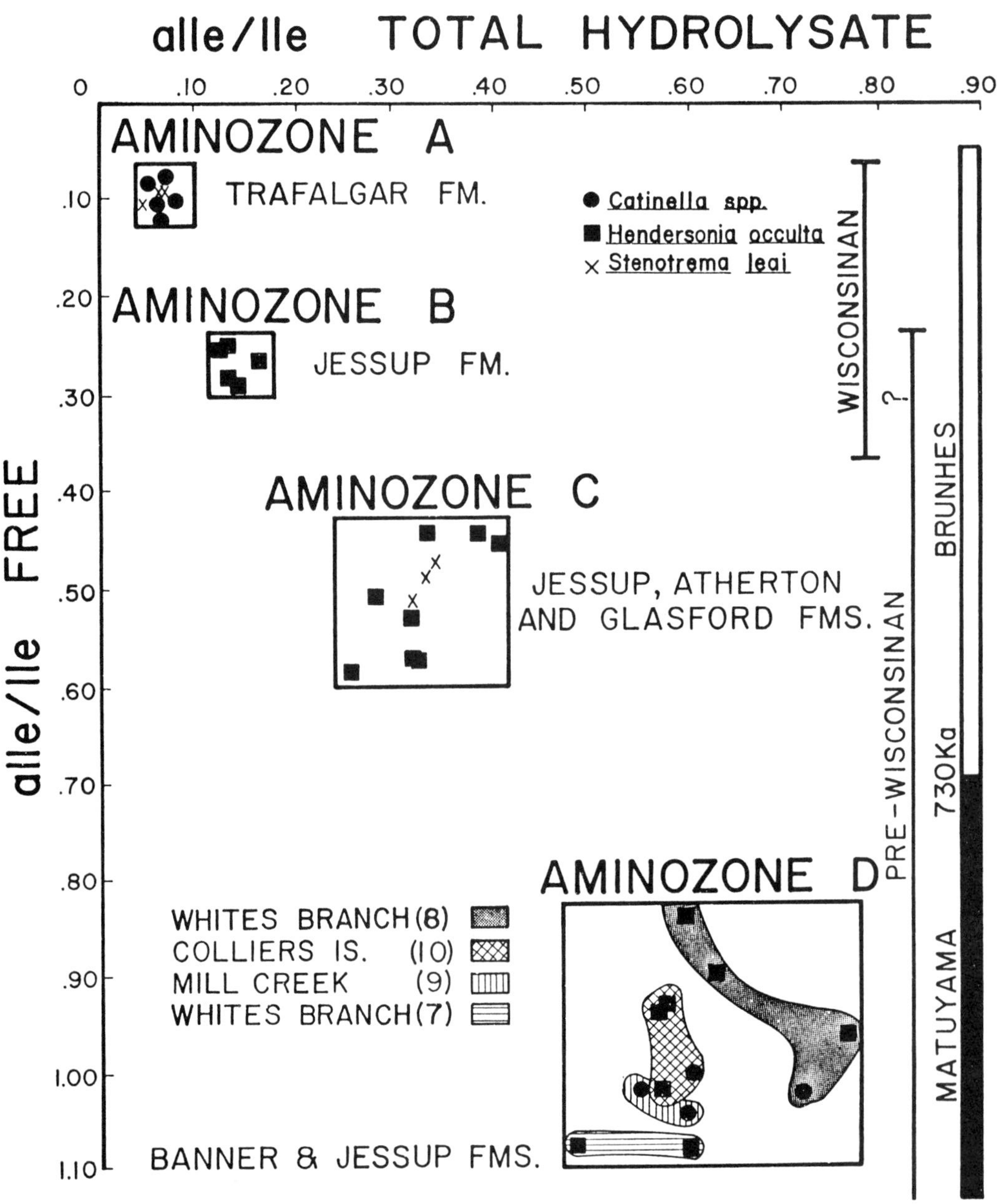

Figure 4 *Plot of free* versus *total alloisoleucine/isoleucine from fossil mollusc shell protein. The ratio of free to total hydrolysate increases asymptotically with age. The sites with the highest ratios are magnetically reversed and should be older than the Brunhes-Matuyama boundary or >730,000 years B.P. (Miller* et al., *1987).*

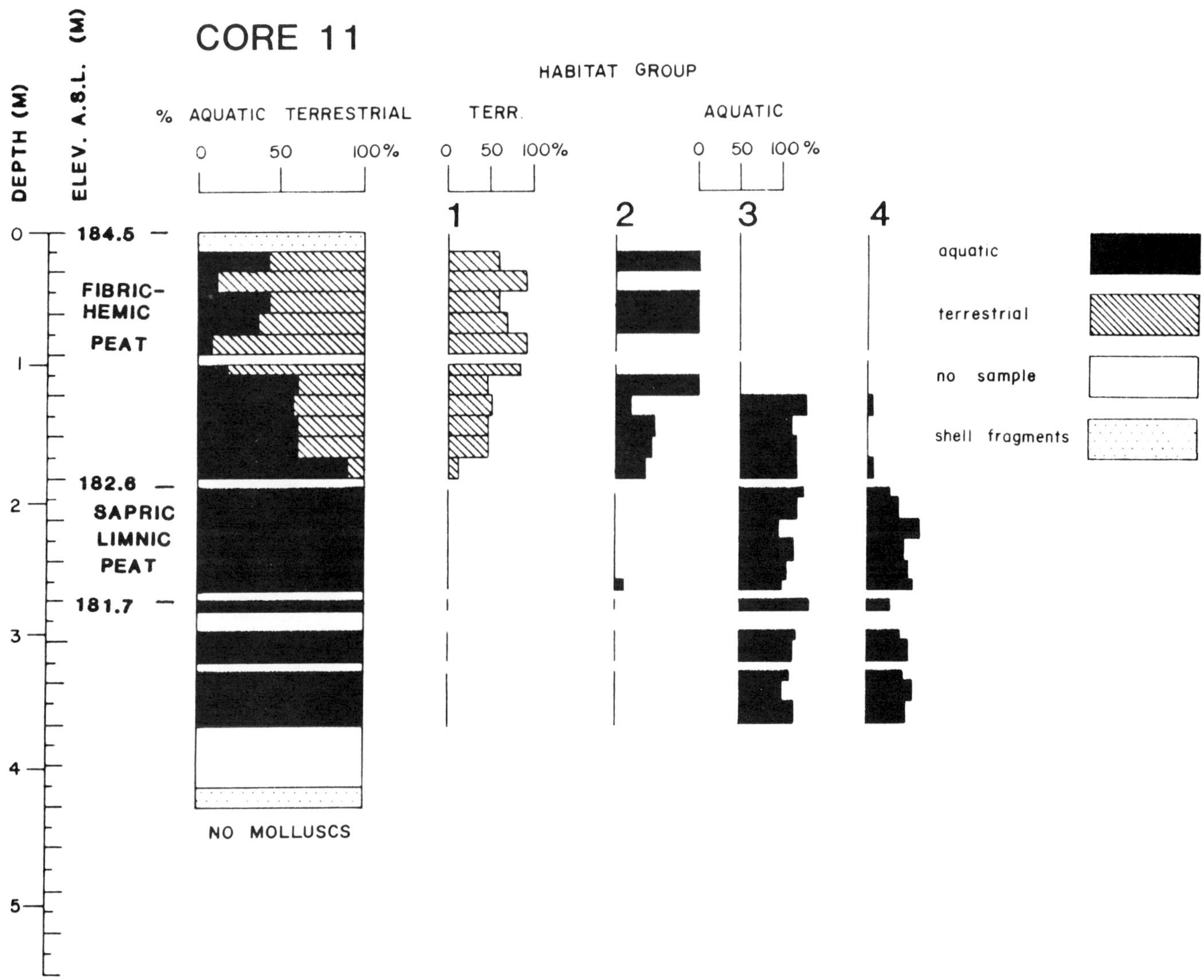

Figure 5 *Plot showing the change in abundance of molluscan habitat groupings in core 11 from Cowles Bog, Indiana (Miller and Thompson, 1987).*
Habitat group 1 *includes terrestrial species living in wooded, damp lowland areas, close to water, with moist areas available throughout the year.*
Habitat group 2 *contains species associated with bodies of water subject to significant seasonal drying and with areas of submerged vegetation.*
Habitat group 3 *includes species associated with eutrophic, perennial bodies of water that are not subject to significant seasonal drying.*
Habitat group 4 *includes species that live in perennial unpolluted water bodies containing aquatic vegetation.*

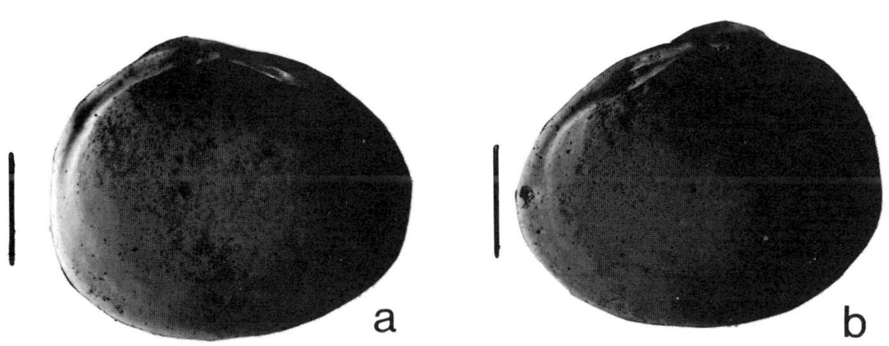

Figure 6 *Right and left valve of* Pisidium casertanum, *a species which can tolerate a broad range of environmental conditions. Scale bars equal 1 mm.*

Figure 7 *Right valve of* Sphaerium occidentale, *one of the most amphibious species of North American bivalves (Clarke, 1981). Scale bar equals 1 mm.*

retreat of ice margins, and isostatic rebound which altered the drainage. The widespread occurrence of well-preserved non-marine molluscs in a variety of environments has led to their use in stratigraphic studies utilizing amino acid racemization ratios obtained from shell protein (Miller and Hare, 1980). Miller *et al.* (1987) have shown that changes in the ratio of alloisoleucine to isoleucine (alle/Ile) in fossil gastropods from west-central Indiana could be used to differentiate four distinct groupings (Figure 4). The lowest alle/Ile values were from shells collected from sites within the Trafalgar Formation (*ca.* 20,000 years B.P.), while the highest values were associated with the magnetically reversed (>730,000 years B.P.) Jessup and Atherton Formations.

Habitat Reconstruction.

Most interpretations of Quaternary non-marine molluscs start with the tacit assumption that fossils with shell morphology similar to those of living forms represent the same species, and therefore, probably responded in the same manner to environmental factors as their living counterparts (Taylor, 1960, 1965, 1988). Some support for the belief that many non-marine molluscs have been evolutionary conservatives during the Quaternary comes from examination of a large number of fossil faunas in which the same ecologic association of species reappear. These assumptions have frequently provided the basis for reconstruction of local paleohabitat and paleoclimatic conditions.

Analysis of molluscs for environmental reconstructions usually involves sorting of extant species into habitat associations in which they most frequently occur. These data may be plotted as relative abundance histograms. Miller and Thompson (1987) used this approach to graphically illustrate habitat changes in a core from the Cowles Bog area of northern Indiana that they interpreted as responses to water-level oscillations (Figure 5).

Although most species may be found in a range of habitats today, many of them seem to occur more frequently in only a few. This point is illustrated by Pip (1986) in an ecological study of 38 aquatic gastropods species in central Canada. She found that most of the species were not restricted to a single habitat, but many were far more common in lakes and ponds that in river and

HABITAT	Taxa No.	Taxa %	SIGNIFICANT SPECIES
FLOODPLAIN WOODED WITH AREAS OF DENSE SHADE, LOCALLY WET TO DAMP.	13	22%	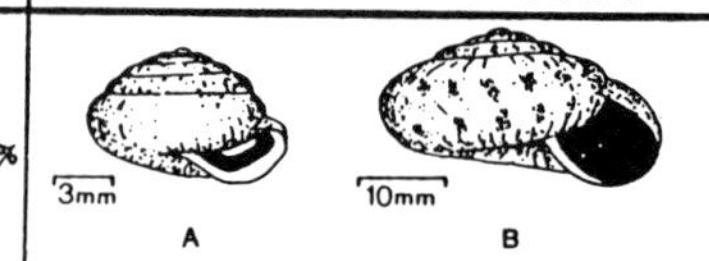
VALLEY SLOPES OPEN WOODLAND AND SHRUBS, SHADED AREAS..	10	18%	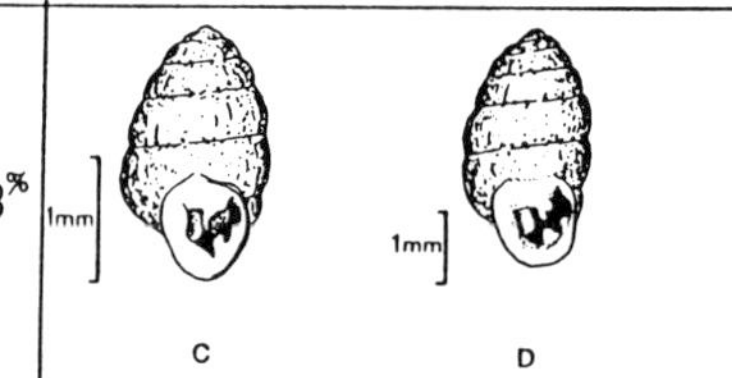
UPLAND GRASSLAND PRAIRIE: SCATTERED TREES – SHRUBS.	3	10%	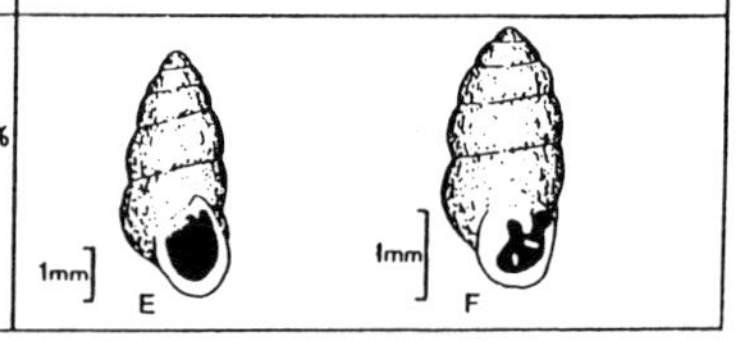

Figure 8 *Kanopolis local fauna terrestrial habitats inferred from molluscs (Miller and Kay, 1981). The total number of species and their abundance expressed as a percentage of the total number of individuals counted is also included. The significant species illustrated are:* ***(A)*** Stenotrema leai, ***(B)*** Anguispira alternata, ***(C)*** Gastrocopta contracta, ***(D)*** Gastrocopta armifera, ***(E)*** Pupoides albilabris, *and* ***(F)*** Gastrocopta cristata.

HABITAT	Taxa No.	Taxa %	SIGNIFICANT SPECIES
EPHEMERAL WATER BODIES ON FLOODPLAIN.	5	2%	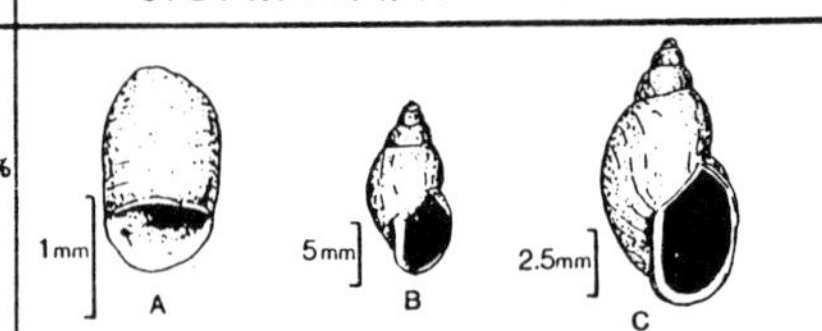
SMALL BODIES OF WATER ON FLOODPLAIN: NOT SUBJECT TO SIGNIFICANT SEASONAL DRYING.	7	24%	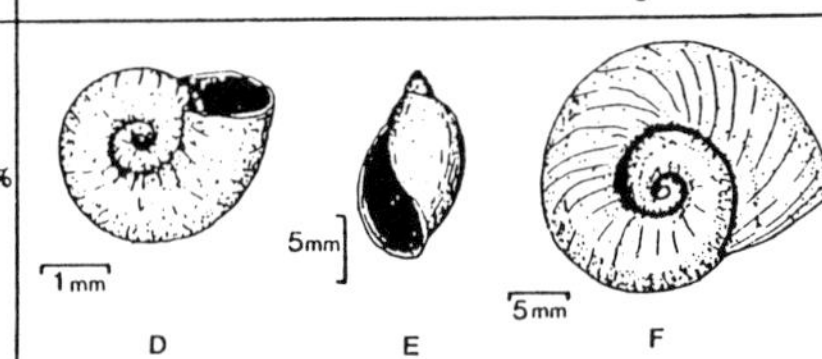
PERENNIAL STREAM OR RIVER.	3	24%	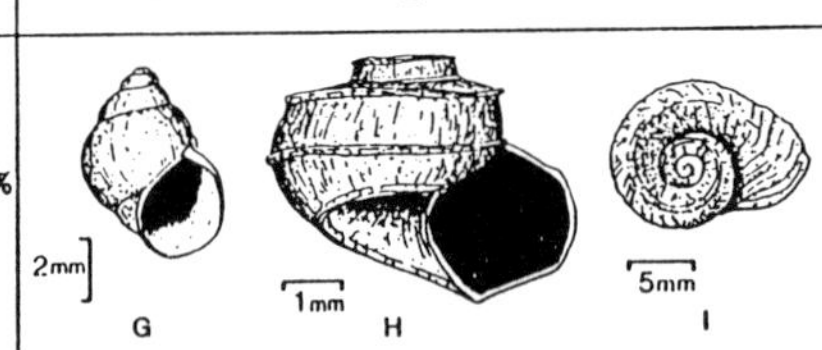

Figure 9 *Kanopolis local fauna aquatic habitats inferred from molluscs (Miller and Kay, 1981). The total number of species and their abundance expressed as a percentage of the total number of individuals counted is also included. The significant species listed are:* ***(A)*** Ferrissia fragilis, ***(B)*** Stagnicola caperata, ***(C)*** Fossaria obrussa, ***(D)*** Gyraulus parvus, ***(E)*** Physa gyrina, ***(F)*** Promenetus exacuous kansasensis, ***(G)*** Amnicola integra, ***(H)*** Valvata tricarinata, *and* ***(I)*** Helisoma anceps.

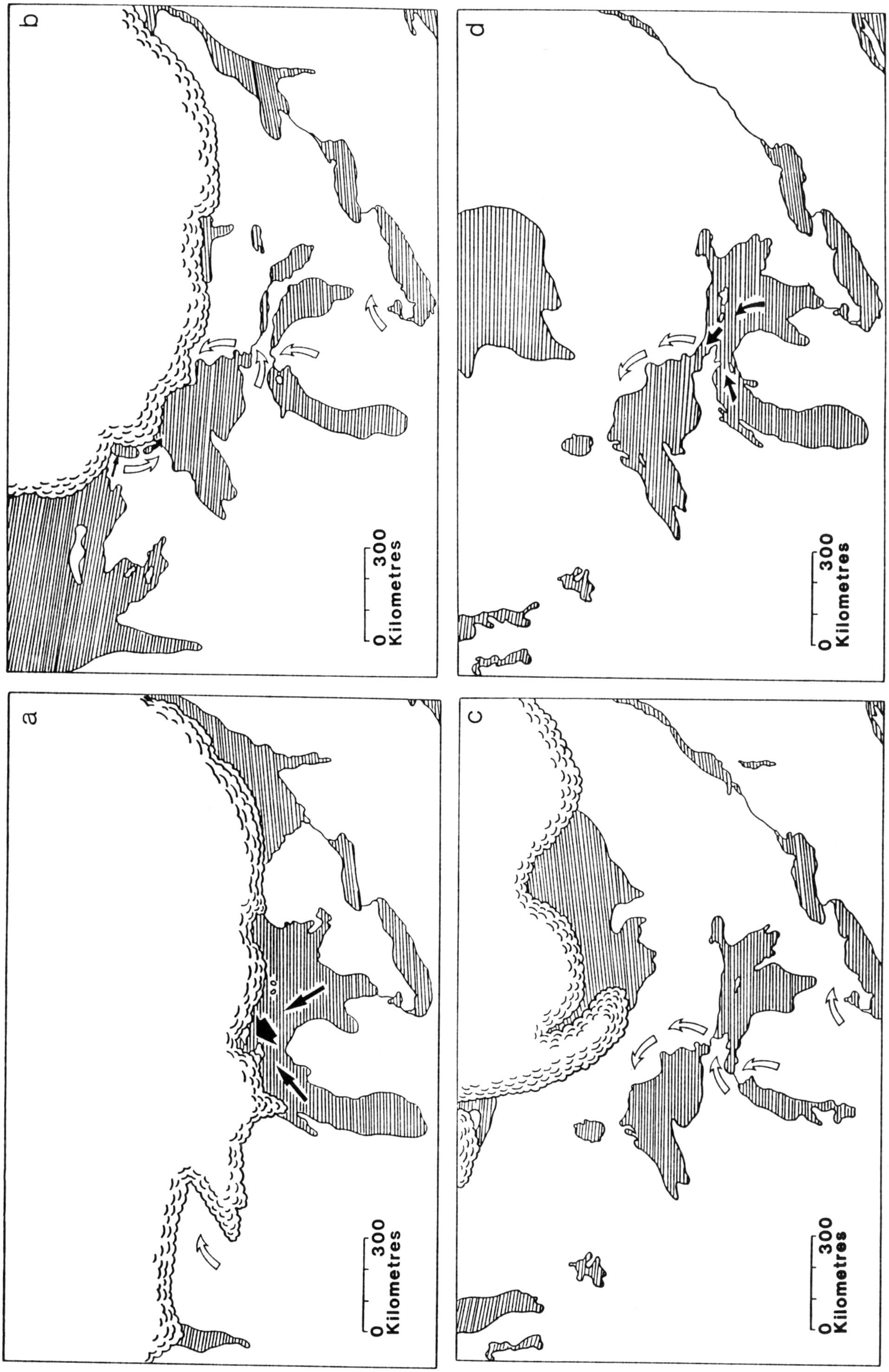

Figure 10 *Major mollusc migration pathways in the Great Lakes drainage basins. Solid arrows indicate migration pathways of aquatic molluscs and open arrows indicate migration pathways of terrestrial molluscs.*

(a) *Migration of Algonquin aquatic fauna into the southeastern corner of the Superior basin between 12,000 and 10,000 years B.P.*

(b,c) *migration of terrestrial molluscs from southern Michigan into southwestern Ontario and from the upper peninsula of Michigan to the north shore of Lake Superior between 10,000 and 8000 years B.P.*

(d) *Migration of Michigan and Huron basin aquatic faunas into the Superior basin between 7000 and 4000 years B.P. during the Nipissing phase of the upper Great Lakes.*

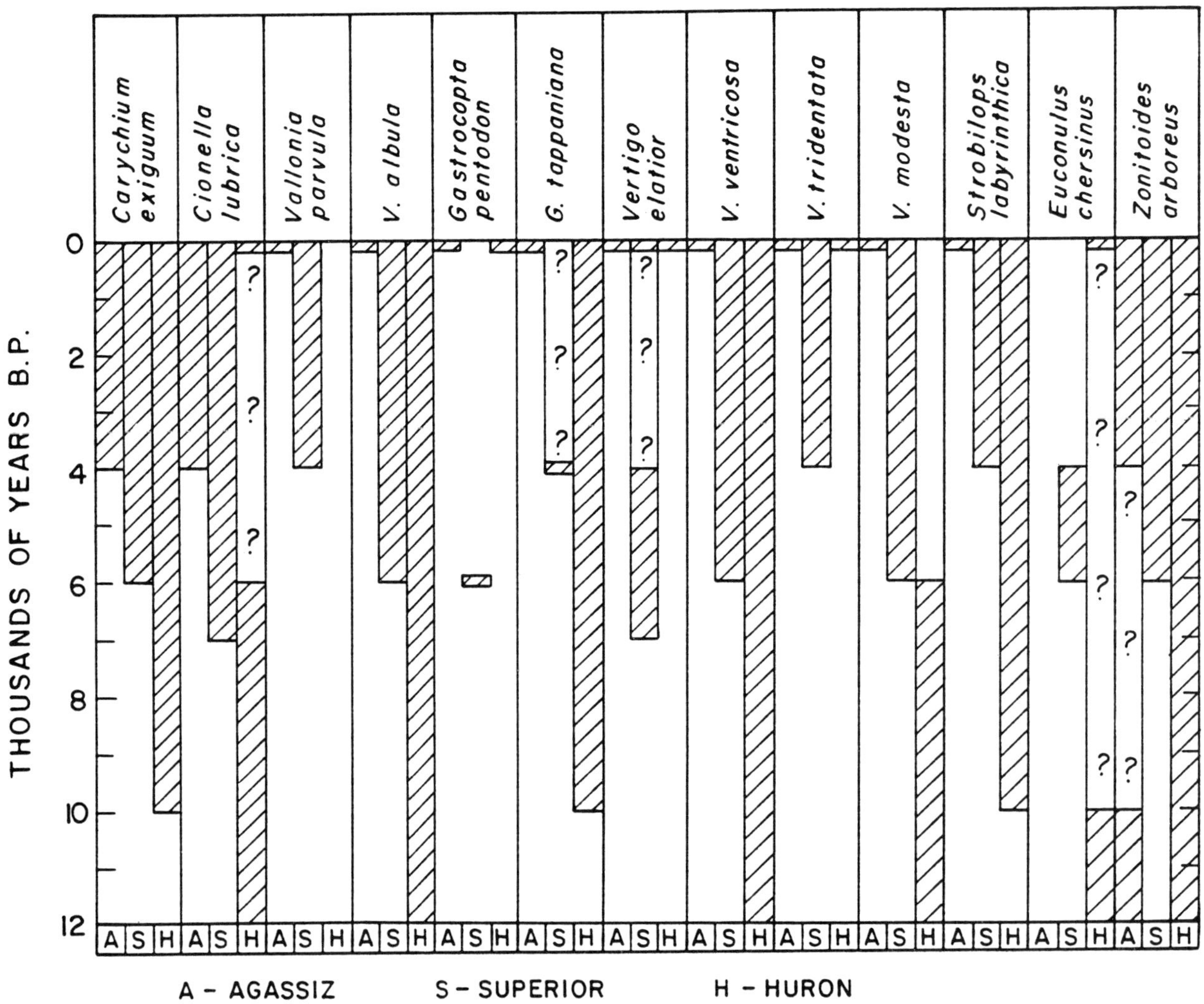

Figure 11 *Temporal occurrences of aquatic molluscs in the Agassiz, Superior, and Huron basins for those taxa that occur first on the north shore of Lake Superior between 7000 and 4000 years B.P. Note the overwhelming westerly migration of aquatic molluscs (Bajc, 1986).*

creek habitats. While many species are generalists and can tolerate a broad range of environmental conditions (*e.g.*, *Pisidium casertanum*, Figure 6) other species, for example, *Discus macclintocki*, are limited to a few sites in northeastern Iowa which are kept cool during the summer months by cold-air drainage (Frest and Fay, 1980a). *Sphaerium occidentale*, (Figure 7) is a species that is found only in aquatic habitats that dry up for parts of each year. It is more amphibious than any other North American bivalve and often occurs among damp leaves (Clarke, 1981).

Some species develop special shell modifications that reflect habitat conditions. The development of a septum that partially closes the aperture in some species of *Ferrissia* living in habitats subjected to periodic drought has been reported by Basch (1963) and Richardot (1978). Fossil septate individuals of *F. fragilis* in the Kanopolis local fauna near Ellsworth, Kansas, therefore, were considered *prima facie* evidence for the presence of temporary water bodies near the depositional site (Miller and Kay, 1981). Six habitats were inferred from the gastropods of the Kanapolis local fauna (Figures 8 and 9). The general compatibility between interpretations based on molluscs, and those derived from the study of other fossil organisms in the assemblage (*e.g.*, Hall, 1980), suggests that this approach is valid.

Migration Patterns and Dispersal Routes.

In his review of the distribution of fossil freshwater molluscs, Yen (1947) suggested several mechanisms by which dispersal can occur. River and lake currents can transport molluscs effectively. Often, molluscs attach themselves to

Figure 12 ***(opposite page - top)***
Temporal occurrences of terrestrial molluscs in the Agassiz, Superior, and Huron basins for those taxa that occur first on the north shore of Lake Superior between 7000 and 4000 years B.P. Note the overwhelming westerly migration of terrestrial molluscs (Bajc, 1986).

Figure 13 ***(opposite page - lower left)***
Oxygen-18 isotope data from fossil shell collected at 13 sites along Lake Huron. The molluscs range from about 4000 to 11,000 years B.P. The change in the oxygen isotope values from Lake Algonquin to Nipissing age sites is equivalent to about a 1.5°C increase in temperature (Godwin, 1985).

Figure 14 ***(opposite page - lower right)***
Distribution map showing modern geographic range (dots) of Physa skinneri. *The triangle indicates the fossil occurrence of this species from Illinoian sites in southwestern Kansas (Miller, 1966).*

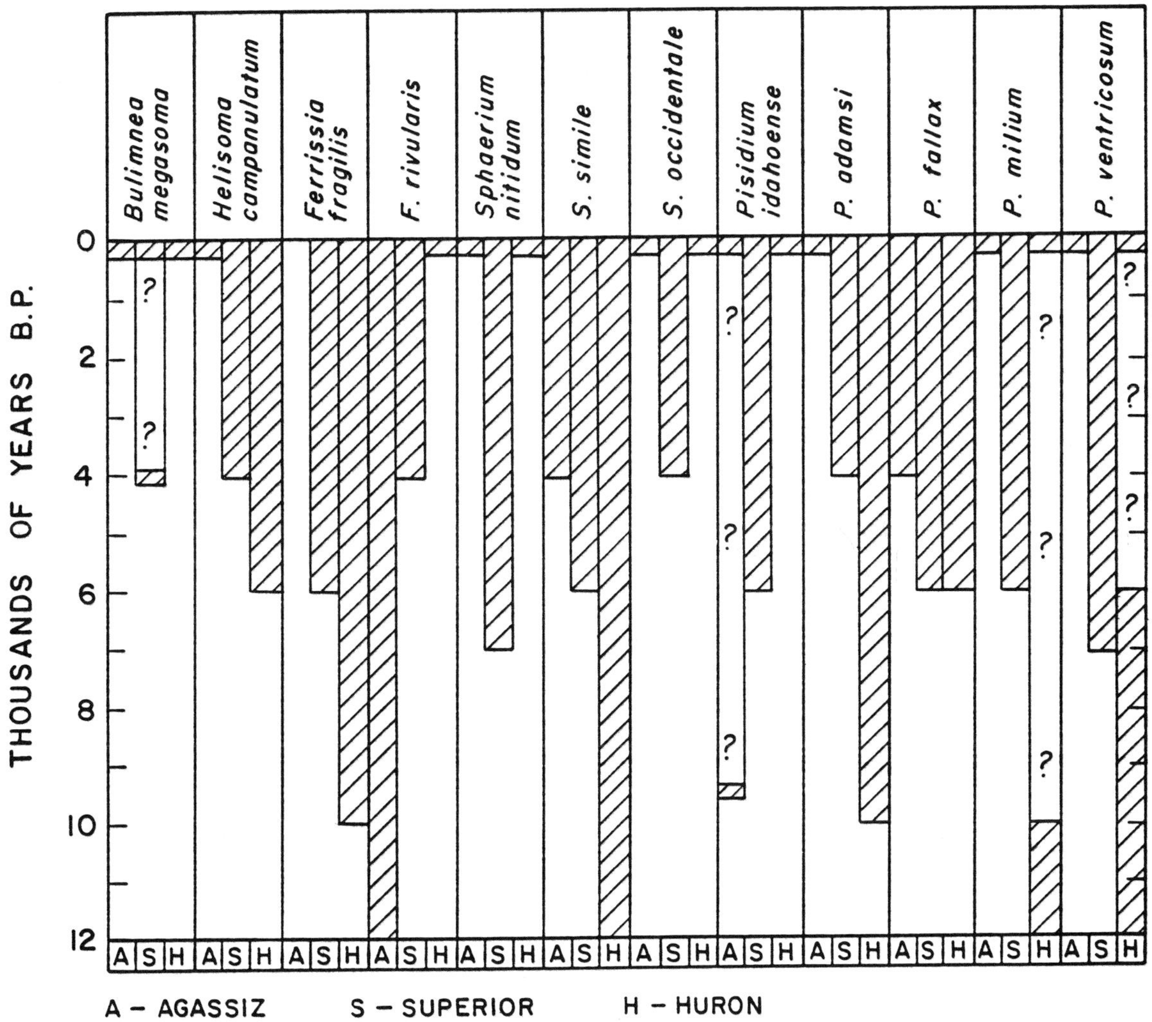

Bulimnea megasoma
Helisoma campanulatum
Ferrissia fragilis
F. rivularis
Sphaerium nitidum
S. simile
S. occidentale
Pisidium idahoense
P. adamsi
P. fallax
P. milium
P. ventricosum
THOUSANDS OF YEARS B.P.
0
2
4
6
8
10
12
?
A S H
A – AGASSIZ
S – SUPERIOR
H – HURON

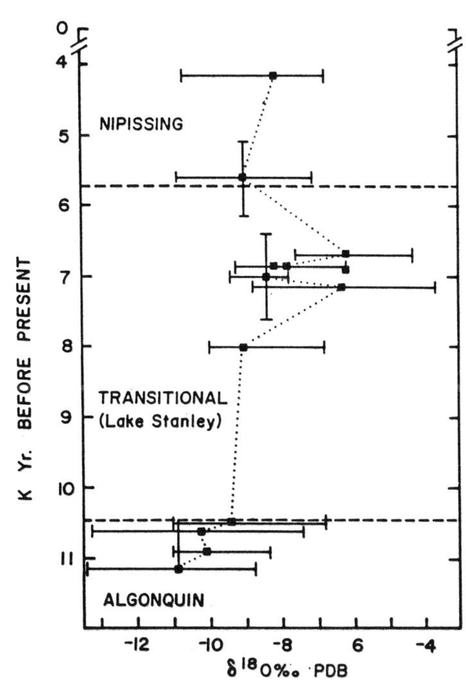

K Yr. BEFORE PRESENT
0
4
5
6
7
8
9
10
11
NIPISSING
TRANSITIONAL
(Lake Stanley)
ALGONQUIN
-12
-10
-8
6
-4
δ18O‰ PDB

floating objects and have been observed on aquatic insects, amphibians, birds, and fish (Kew, 1893; Cooke, 1895). Molluscs have also been found floating within ice on rivers and lakes; some species have been known to survive after being frozen for several months. Malone (1965) suggested that snails could also be dispersed by being regurgitated from the crop of water birds. The dispersal of many species of unionid clam is highly dependent upon fish migrations (Taylor, 1988). During their glochidial stage, the clams are parasitic upon a host fish and therefore, are easily spread.

The dispersal of terrestrial gastropods tends to be a much slower process. Common dispersal mechanisms include: natural locomotive processes; organic agents, such as mammals and birds; and physical agents, such as wind and water. Factors such as natural abundance; resistance to desiccation, drowning and cold; and the fertility of taxa play important roles in dispersal (Oughton, 1948).

Ice-marginal positions and intra-basin confluence appear to have imposed constraints on mollusc migration patterns within the Great Lakes basins. Between 12,000 and 10,000 years B.P., the Huron, Michigan, and southeast corner of the Superior basins were confluent and formed Lake Algonquin (Figure 10a). Freshwater molluscs followed the retreating ice margin in the Superior basin and reached the north shore by 9500 years B.P. (Figure 10b). Aquatic mollusc assemblages (9000-8000 years B.P.), recovered from the north shore of the Superior basin are remarkably similar to those found in Algonquin age sediments of the Huron basin (12,000-10,000 years B.P.)(Bajc, 1986). Basin confluence appears to be the primary controlling factor responsible for this similarity. Between 10,000 and 6000 years B.P., the Michigan and Huron basins stood at the Chippewa and Stanley low-water levels, respectively, and the transitional levels while the Superior basin remained perched above the lower lakes at the Minong through Houghton water levels. It was during this time that important migrations of terrestrial molluscs from the upper peninsula of Michigan to the north shore of Lake Superior occurred (Figures 10b and 10c). By 6000 years B.P., confluence of the upper Great Lakes was again restored due to isostatic processes, the water body standing at the Nipissing shoreline (Figure 10d). Basin confluence resulted in rapid dispersal of aquatic taxa and hence, a great increase in diversity in Nipissing faunas of the Huron basin and the introduction of 16 new taxa of aquatic molluscs to the north shore of the Superior basin (Bajc, 1986).

It is also possible to suggest directions of intra-basin migrations by plotting the temporal occurrences of various taxa in adjoining basins. For example, if one plots the temporal occurrences within the Agassiz, Superior, and Huron basins of aquatic taxa that first occur in the Superior basin between 7000 to 4000 years B.P., a better sense of migration is achieved (Figure 11). The majority of migrations appear to be in the westerly direction, however, some exceptions exist (*e.g.*, *Ferrissia rivularis* and *Pisidium idahoense*). Terrestrial taxa appear to demonstrate similar trends to those documented by the aquatic molluscs (Figure 12).

Climatic Interpretations.

Although evidence directly implicating climate as a prime factor in controlling the distribution of molluscs is difficult to obtain, there are observations concerning distribution of some terrestrial gastropods which strongly suggest that temperature may be a major control. Miller (1975) reviewed field observation and laboratory experiment data that suggested temperature extremes and available moisture as two of the major factors controlling non-marine gastropod distributions. In a more recent study, Frest and Fay (1980a) have shown that three species of terrestrial gastropods first described from Pleistocene loess deposits of the American midwest, *Vertigo hubrichti, V. briarensis,* and *Discus macclintocki*, can now be found living as isolated relict colonies in the Driftless Area of northeastern Iowa. The molluscs at these sites are associated with "cold air slopes" which buffer the maximum summer temperature to about 10°C and provide a climatically acceptable refugia for species that were widely distributed in the region during the Pleistocene.

Unfortunately, no standard methods of analyses are now in use to infer climatic data from molluscan assemblages. Some workers have tried to make qualitative statements about climate by dividing molluscs, on the basis of their modern distribution patterns, into what are considered climatically significant groupings (Harris and Pip, 1973). Changes in relative abundance of these species-groupings are interpreted as responses to climatic change (Zakrewski and Kolb, 1982). In other studies, an area of sympatry for the majority of the extant faunal elements is established, and the climate of the overlap area is considered to be an approximation of climatic conditions locally at the time the fossil assemblage lived (Taylor, 1960, 1965; Miller, 1966).

Some recent workers have tried to relate variation in shell morphology to climatic parameters. Baerreis (1980) produced a climatic reconstruction for the interval between 6350 and 10,500 years B.P., at the Cherokee Site in northwestern Iowa, in which climatic interpretations were derived from equations that regressed variation in shell size of extant land snails against a series of climate variables.

Godwin (1985) used changes in the ^{18}O isotopic composition in freshwater shell carbonate to calculate water temperature changes in the Huron basin between 11,000 and 4000 years B.P. Her calculations suggest a 2.5°C increase in water temperature between the pro-glacial, main Algonquin lake phase and the transitional, low-water, Lake Stanley and a drop of 1°C between Lake Stanley and the Nipissing levels (Figure 13).

Amino acid epimerization rates of isoleucine from shell protein of non-marine gastropods have been used to estimate the average post-depositional integrated thermal history in a region. Unpublished data for molluscs from Late Wisconsinan sites in Ohio and Indiana, suggest an average temperature during the past 20,000 years that was about 1.5°C lower than the present mean annual temperature at these sites.

Pleistocene assemblages frequently include a greater diversity of species than now exist in the area of the fossil sites (Miller, 1976). These faunas usually include associations of species with modern geographic ranges that may approach, but do not reach the area of the fossil localities (Figure 14). The Peoria loess, in parts of northern Missouri, south central Illinois and southern Indiana, present well-documented examples of molluscan species associations that have no modern analogues (Frest and Fay, 1980b). The most parsimonious explanation that can account for the former co-existence of species that now have such disjunct geographic distributions is to assume that these assemblages lived at a time that combined cooler, moister summers (allowing boreal and montane species to extend their range southward and to lower elevations), with shorter, less severe winters (permitting southern forms to range further north during a longer growing season).

Disjunctive molluscan assemblages persisted in the southern Great Plains until about 10,000 years B.P. (Miller, 1975). In the Huron basin disjunct molluscan faunas continued on until 6000 to 8000 years B.P., when climatic warming probably led to a local extirpation of the northern species (Miller *et al.*, 1985).

The relative ease with which large numbers of fossil molluscs can be collected from sites, together with their increasing use as a source of stratigraphic information from amino acid studies of shell protein, virtually guarantees that they will continue to be an important component of Quaternary stratigraphic and paleoecologic studies.

ACKNOWLEDGEMENTS

This paper is published with the permission of Dr. V.G. Milne, Director of the Ontario Geological Survey, Ministry of Northern Development and Mines. Some of the data presented in this paper are based on the results of field mapping programs in the Hemlo and Fort Frances areas of Ontario. Other unpublished research reported in this paper was supported by National Geographic Society Grant Number 3224-85. We would like to thank Kymberly Foster and Jennifer John for typing the manuscript and the following for their help in drafting: Robert Maschue for figure 4; Martin Layman for figures 5 and 13; Harvey Geizer for figures 8 and 9; and Nadia Bahar for figures 11 and 12. The authors would like to thank Dr. G. Mackie, Dr. T. Frest and Dr. B. Warner for their constructive reviews of this manuscript.

REFERENCES

Ashworth, A.C. and Cvancara, A.M., 1983, Paleoecology of the southern part of the Lake Agassiz basin, *in* Teller, J.T. and Clayton, L., eds., Glacial Lake Agassiz: Geological Association of Canada, Special Paper 26, p. 133-156.

Baerreis, D.A., 1980, Habitat and climate interpretation from terrestrial gastropods at the Cherokee Site, *in* Anderson, D.C. and Semken, H.A., Jr., eds., The Cherokee Excavations: Academic Press, New York, p. 101-122.

Bajc, A.F., 1986, Molluscan paleoecology and Superior basin water levels, Marathon, Ontario, M.Sc. thesis, University of Waterloo, 271 p.

Bajc, A.F. and Gray, P.A., 1987, Quaternary geology of the Rainy River area, District of Rainy River: Ontario Geological Survey, Map P. 3065, Geological-Series Preliminary Map, scale 1:50,000. Geology 1986.

Baker, F.C., 1920a, The life of the Pleistocene or glacial period, as recorded in the deposits laid down by the great ice sheets: University of Illinois Bulletin 41, v. 17, 476 p.

Baker, F.C., 1920b, Pleistocene Molluscs from Indiana and Ohio: Journal of Geology, v. 28, p. 439-457.

Basch, P., 1963, A review of the recent freshwater limpet snails of North America (Mollusca: Pulmonata): Harvard University, Museum of Comparative Zoology, Bulletin, v. 129, p. 401-461.

Bickel, E.D., 1970, Pleistocene nonmarine Molluscs of the Gatineau Valley and Ottawa areas of Quebec and Ontario, Canada: Sterkiana, v. 38, p. 1-50.

Billups, A.C., 1902, Fossil land snails: The Nautilus, v. 16, p. 50-52.

Birks, H.J.B. and Birks, H.H., 1980, Quaternary Paleoecology: Edward Arnold Publishers, London, 289 p.

Clarke, A.H., 1981, The freshwater molluscs of Canada: National Museum of Natural Sciences, Ottawa, 446 p.

Clowers, S.R., 1966, Pleistocene Mollusca of the Box Marsh deposit, Adamston Township, Renfrew County, Ontario, Canada: Sterkiana, v. 22, p. 31-59.

Coleman, A.P., 1922, Glacial and post-glacial lakes in Ontario: Toronto University Studies, Ontario Fisheries Research Laboratory, Publication no. 10, p. 1-31.

Cooke, A.H., 1895, Molluscs: MacMillan and Co., New York, p. 1-460.

Crisman, T.L., 1978, Reconstruction of past lacustrine environments based on the remains of aquatic invertebrates, *in* Walker, D. and Guppy, J.C., eds., Biology and Quaternary Environments: Australian Academy of Science, Canberra, p. 69-101.

Frest, T.J. and Fay, L.P., 1980a, Relict land snails from the Driftless Area, Iowa, with implications for Pleistocene climates: Geological Society of America, Abstracts with Programs, v. 12, p. 429.

Frest, T.J. and Fay, L.P., 1980b, Peoria loess mollusc faunas and Woodfordian biomes of the upper midwest: American Quaternary Association, 6th Biennial Meeting, Abstracts and Programs, p. 82.

Fritz, P., Anderson, T.W. and Lewis, C.F.M., 1975, Late Quaternary climatic trends and history of Lake Erie from stable isotope studies: Science, v. 190, p. 267-269.

Fritz, P. and Poplawski, S., 1974, O^{18} and C^{13} in the shells of freshwater molluscs and their environments: Earth and Planetary Science Letters, v. 24, p. 91-98.

Gibson, G.G., 1967, Pleistocene nonmarine Molluscs of the Richardson Lake deposit, Clarendon Township, Pontiac County, Quebec, Canada: Sterkiana, v. 25, p. 1-36.

Godwin, A.G., 1985, Stable isotope analyses on postglacial fluvial and terrestrial molluscs from the Kincardine area of southern Ontario, Unpublished M.Sc. thesis, University of Waterloo, 233 p.

Guilday, J.C., Hamilton, H.W., Anderson, E. and Parmalee, P.W., 1978, The Baker Bluff cave deposit, Tennessee, and the late Pleistocene faunal gradient: Bulletin of the Carnegie Museum of Natural History, No. 11, 67 p.

Gutschick, R.C. and Gonsiewski, J., 1976, Coastal geology of the Mt. Baldy area, Indiana Dunes National Lakeshore, south end of Lake Michigan: Geological Society of America, North-Central Section, Fieldtrip Guidebook, p. 40-90.

Hall, S., 1980, Paleoenvironmental synthesis of Hominy Creek Valley: pollen and land snail evidence, *in* Henry, D.O., ed., The Prehistory and Paleoenvironment of Hominy Creek Valley: 1978 Field Season: University of Tulsa, Laboratory of Archaeology, Contributions in Archaeology No. 6, p. 44-55.

Hansel, A.K. and Johnson, W.H., 1986, Late Quaternary record of the Chicago outlet area, *in* Hansel, A.K. and Johnson, W.H., eds., Quaternary Records of Northeastern Illinois and Northwestern Indiana: American Quaternary Association, Ninth Biennial Meeting, Field Guide Trip 5, Illinois State Geological Survey, Champaign, Illinois.

Harris, S.A. and Pip, E., 1973, Molluscs as indicators of late and post-glacial climatic history in Alberta: Canadian Journal of Zoology, v. 21, p. 209-213.

Karrow, P.F., Anderson, T.W., Clarke, A.H., Delorme, L.D. and Sreenivasa, M.R., 1975, Stratigraphy, paleontology, and age of Lake Algonquin sediments in southwestern Ontario, Canada: Quaternary Research, v. 5, p. 49-87.

Karrow, P.F., Clarke, A.H. and Herrington, H.B., 1972, Pleistocene molluscs from Lake Iroquois deposits in Ontario: Canadian Journal of Earth Sciences, v. 9, p. 589-595.

Kerney, M.P., 1977, British Quaternary nonmarine Mollusca: a brief review, *in* Shotten, F.W., ed., British Quaternary Studies: recent advances: Oxford University Press, p. 31-42.

Kerr-Lawson, L.J., 1985, Gastropods and plant macrofossils from the Quaternary Don Formation (Sangamonian Interglacial), Toronto, Ontario, M.Sc. Thesis, University of Waterloo, 202 p.
Kew, H.W., 1893, The dispersal of shells: Kegan Paul, French Trubner, and Company, Ltd., London, 291 p.
LaRocque, A., 1966, Pleistocene Mollusca of Ohio: Division of the Geological Survey Bulletin, v. 62, Part 1, p. 1-111.
Lemon, J.H., 1898, Interglacial shells at Toronto: The Nautilus, v. 12, p. 6-7.
Leonard, A.B., 1953, Molluscan faunules in Wisconsinan loess at Cleveland, Ohio: American Journal of Science, v. 251, p. 369-376.
Leonard, A.B. and Frye, J.C., 1960, Wisconsinan molluscan faunas of the Illinois Valley region: Illinois State Geological Survey, Circular 304, 32 p.
Lozek, V., 1982, Mollusca analysis, *in* Berglund, B.E., ed., Paleohydrological Changes in the Temperate Zone in the Last 15,000 Years. Subproject B. Lake and Mire Environments, v. 3: International Geological Correlation Programme Project 158, p. 149-163.
Malone, C.H., 1965, Dispersal of aquatic gastropods via the intestinal tract of water birds: The Nautilus, v. 78, p. 135-139.
Matthew, G.F., 1884, Discoveries at a village of the Stone Age at Bocabec, N.B.: Natural History Society of St. John, New Brunswick, Bulletin, v. 1, p. 6-29.
Miller, B.B., 1966, Five Illinoian molluscan faunas from the southern Great Plains: Malacologia, v. 4, p. 173-260.
Miller, B.B., 1968, *Planorbula campestris* (Dawson) Gastropoda: Planorbidae) from the Cudahy fauna (Kansas) of Meade County, Kansas and the status of the subgeneric categories of *Planorbula*: Malacologia, v. 6, p. 253-265.
Miller, B.B., 1975, A sequence of radiocarbon-dated Wisconsinan nonmarine molluscan faunas from south-western Kansas-northwestern Oklahoma, *in* Smith, G.R. and Friedland, N.E., eds., Studies on Cenozoic Paleontology and Stratigraphy: University of Michigan, Papers on Paleontology, v. 12, p. 9-18.
Miller, B.B., 1976, Late Cenozoic molluscan succession in the Meade County, Kansas area: Kansas Geological Survey, Guidebook Series 1, p. 73-85.
Miller, B.B., 1978, Nonmarine molluscs in Quaternary paleoecology: Malacological Review, v. 11, p. 27-38.
Miller, B.B., 1983, Late Quaternary fluvial history of the lower Cuyahoga River in northeastern Ohio: Journal of Great Lakes Research, v. 9, p. 96-104.
Miller, B.B., Karrow, P.F. and Mackie, G.L., 1985, Late Quaternary molluscan faunal changes in the Huron basin, *in* Karrow, P.F. and Calkin, P.E., eds., Quaternary Evolution of the Great Lakes: Geological Association of Canada, Special Paper 30, p. 95-107.
Miller, B.B. and Kay, W.T., 1981, Gastropods from the Kanopolis local fauna (Yarmouthian?), Ellsworth County, Kansas: Journal of Paleontology, v. 55, p. 65-71.
Miller, B.B., McCoy, W.D. and Bleuer, N.K., 1987, Stratigraphic potential of amino acid ratios in Pleistocene terrestrial gastropods: an example from west-central Indiana: Boreas, v. 16, p. 133-138.
Miller, B.B. and Thompson, T.A., 1987, Environmental changes in the Cowles Bog area, Indiana Dunes National Seashore, during the past 4500 years: Symposium '87, Wetlands/Peatlands, (August 23-27, 1987), Edmonton, Alberta, Proceedings, p. 341-348.
Miller, G.H. and Hare, P.E., 1980, Amino acid geochemistry: integrity of the carbonate matrix and potential of molluscan fossils, *in* Hare, P.E., Hoering, T.C. and King, K.M., Jr., eds., Biochemistry of Amino Acids: J. Wiley and Sons, New York, p. 415-443.
Miller, R.R., 1957, Utilization of x-rays as a tool to systematic zoology: Systematic Zoology, v. 6, p. 29-40.
Morgan, A.V. and Morgan, A., 1980, Beetle bits - the science of paleoentomology: Geoscience Canada, v. 7, p. 22-29.
Oughton, J., 1948, A zoogeographical study of the land snails of Ontario: University of Toronto, Biological Series, v. 47, 128 p.
Parmalee, P.W. and Bogan, A.E., 1986, Molluscan remains at the Clinch River breeder reactor plant site, Roane County, Tennessee: American Malacological Bulletin, v. 41, p. 25-37.
Pip, E., 1986, The ecology of freshwater gastropods in the Canadian region: The Nautilus, v. 100, p. 56-66.
Pip, E., 1988, Differential attrition of molluscan shells in freshwater sediments: Canadian Journal of Earth Sciences, v. 25, p. 68-73.
Richardot, M., 1978, Ecological factors inducing aestivation in the freshwater limpet *Ferrissia wautieri* (Basommatophora: Ancylidae). III. Density levels and food supply. General conclusions: Malacological Review, v. 11, p. 47-58.
Russell, L.S., 1934, Pleistocene and post-Pleistocene molluscan faunas of southern Saskatchewan (with description of a new species of *Gyraulus* by F.C. Baker): Canadian Field-Naturalist, v. 48, p. 34-37.
Shallom, L.J., 1965, Pleistocene molluscan fauna of the Lac Blanc deposit, Matapedia County, Quebec, Canada: Sterkiana, v. 19, p. 41-54.
Sparks, B.W., 1961, The ecological interpretation of Quaternary nonmarine Mollusca: Linnean Society of London, Proceedings, v. 172, p. 71-80.
Sparks, B.W., 1964, Non-marine mollusca and Quaternary ecology: Journal of Animal Ecology, v. 33, p. 87-98.
Sterki, V., 1920, Marl deposits in Ohio and their fossil Mollusca: Ohio Journal of Science, v. 20, p. 173-184.
Stuiver, M., 1968, Oxygen-18 content of the atmospheric precipitation during the last 11,000 years in the Great Lakes Region: Science, v. 162, p. 994-997.
Taylor, D.W., 1960, Late Cenozoic molluscan faunas from the High Plains: United States Geological Survey, Professional Paper 337, 103 p.
Taylor, D.W., 1965, The study of the Pleistocene nonmarine molluscs in North America, *in* Wright, H.E. and Frey, D.G., eds., The Quaternary of the United States: Princeton University Press, p. 597-611.
Taylor, D.W., 1988, Aspects of freshwater mollusc ecological biogeography: Palaeogeography, Palaeoclimatology, Palaeoecology, v. 62, p. 511-576.
Tompa, A., 1976, Fossil eggs of the land snail genus *Vallonia* (Pulmonata:Vallonidae): The Nautilus, v. 90, p. 5-7.
Tuthill, S.J., Clayton, L. and Laird, W.M., 1964, A comparison of fossil Pleistocene molluscan fauna from North Dakota with a Recent molluscan fauna from Minnesota: American Midland Naturalist, v. 71, p. 344-362.
Warner, D.J., 1968, Pleistocene gastropods of a lake deposit, Rimouski County, Quebec, Canada: Sterkiana, v. 31, p. 1-14.
Wayne, W.J., 1981, Kansan periglacial environment, east-central Nebraska: American Journal of Science, v. 281, p. 375-389.
Weber, J.N. and LaRocque, A., 1964, Carbon isotopic composition of lacustrine gastropods from pondweed environments: Journal of Paleontology, v. 38, p. 965-967.
Wittaker, E.J., 1921, The fossil molluscan faunas of the marl deposits of the Ottawa district: Geological Survey of Canada, Bulletin 33, p. 59-77.
Yapp, C.J., 1979, Oxygen and Carbon isotope measurements of land snail shell carbonate: Geochimica et Cosmochimica Acta, v. 43, p. 629-635.
Yen, T.C., 1947, Distribution of fossil freshwater mollusks: Geological Society of America, Bulletin, v. 58, p. 293-298.
Zakrzewski, R.J. and Kolb, K.K., 1982, Late Kansan faunas from central Kansas: Transactions of Kansas Academy of Sciences, v. 85, p. 200-215.

Originally published in
Geoscience Canada Volume 16 Number 3
(September 1989)

Beetles[1]

Alan V. Morgan
Department of Earth Sciences
and Quaternary Sciences Institute
University of Waterloo
Waterloo, Ontario N2L 3G1

Anne Morgan
304 Faraday Court
Waterloo, Ontario N2L 6A6

SUMMARY

A number of Quaternary and Late Tertiary sequences in North America have been analyzed for fossil insect remains. Most of the remains are from Coleoptera (beetles), and these can be used as indicators of environmental and climatic changes. Fragments of Coleoptera are present in many Quaternary deposits, but unlike most pollen grains, many of the exoskeletal remains can be identified specifically. Beetles appear to respond rapidly to climatic change illustrating thermal variations which are often not easily recognized, or detected, by other conventional indicators.

This paper is a revised version of a manuscript which originally appeared in *Geoscience Canada* entitled "Beetle Bits — The Science of Paleoentomology" (Morgan and Morgan, 1980a). It provides updated references and summarizes some of the techniques and results of two decades of work in Canada, parts of the United States and England.

INTRODUCTION

Whenever one talks about the subject of paleoentomology, even in academia, one is met with blank stares and a faintly bemused expression. Further elaboration of the topic signifying that it is the study of fossil insects (or, in the case of our own studies, fossil beetles) still elicits comments which usually revolve around the former four from Liverpool. To most paleontologists, beetles rank almost at the bottom of a scale headed by the better known zonal fossils; the ammonites, trilobites and graptolites. Everyone has seen a beetle; — the active field geologist perhaps under rocks or logs somewhere in the Rockies or Arctic, more sedentary persons may have observed them squashed on sidewalks, or under debris in the garden. If you are really unfortunate they may infest the old wood in your home, shelter in your carpets and furnishings, or even infest your food supplies. Beetles are ubiquitous; they surround us in practically every environment except the extreme high Arctic and in the oceans of the world. Beetles, or more scientifically, Coleoptera, come in a bewildering variety of shapes, sizes, colours and numbers. One in five living things is an insect, and well over a third of all insect species are beetles, or, put in another way, of the over 1,300,000 described species on the planet, about one quarter belong to the family Coleoptera.

The differences in beetles are enormous. In size alone, the longest beetle, *Xylorycites satyrus* is over 1000 times the length of the smallest member of the family Trichopterygidae (Ptiliidae). Individuals of this family would probably have a hard time in a confrontation with some protozoans, since many of the unicellular organisms are larger than the beetles themselves. The largest beetles in terms of volume are the "Goliath" beetles (*Goliathus* spp.), ranging up to 100 g. Still others have remarkable abilities; a capability to produce "cold light" as in the case of the Lampyridae (the so-called "fireflies"); chemical defense methods, particularly those of *Brachinus* which produce quinones in a defensive spray at almost 100°C; or the pupation of some Buprestids, recorded at a minimum of 47 years.

Beetles have a reasonable pedigree in the geological column, being first recorded in the Permian (Moore *et al.*, 1952). Thus, they have not yet reached the longevity of the class Trilobita, but they have outlived the last dinosaurs by about 65 million years and have a 50 million year head start on the mammals.

Despite their immense diversity, fossil beetles have received little study to date. The reason for the lack of interest is fairly clear. Insect remains are not particularly common in the fossil record with the exception of well-known localities such as Florissant, the Green River Shales, some mid-Tertiary deposits in Germany, Montana and British Columbia, and various amber deposits in Europe and North America. The vast majority of fossil remains (except for the amber deposits) consist of compressed and sometimes carbonized impressions of fragments or, more rarely, whole insects. Morphological details are poorly preserved and colour patterns usually obscured or non-existent in these fossil arthropods. In addition, the phylogeny of many mid-Tertiary and older insect groups is poorly understood.

The formidable problems associated with the study of older insect remains are somewhat reduced in the examination of Late Tertiary and Quaternary insects. The latter consist of the original chitinous parts, thus colour patterns, hairs and scales can be well preserved, even on specimens which date back two or more million years.

Almost any sediment containing organic debris is likely to preserve insect remains. The nature of the preservation depends upon the the initial environment of deposition, the rapidity with which the organism was buried, the depth of burial, post-depositional changes in the sediment and the position of past and present water tables in respect to the stratigraphic position of the site. Probably the two most important factors are oxidation of the organic horizon containing the fossils and the degree of compression undergone by the sequence. Insect fragments are remarkably resilient to chemical attack, but they are relatively brittle, and stresses caused by overburden pressures or cyclic wetting and drying tend to fragment specimens to the point where they become almost unidentifiable.

Materials which are known to preserve arthropod remains include silts and organic horizons within littoral facies of lacustrine sequences, organic lenses in fluvial deposits, and, to a lesser degree, peat bogs or swamps. Abundant and well-preserved insects have also been found in the well-known tar seeps of Peru, eastern Europe and southern California (Churcher, 1966; Angus, 1973; Miller, 1982). Drier environments do not preclude the preservation of arthropods since the chitinous exoskeletons remain unaltered, and insect

1 Quaternary Entomology Laboratory Contrib. 86

fossils have been described from wood rat middens in cave deposits in the arid southwestern United States (Ashworth, 1973; Spilman 1976; Morgan *et al.*, 1983), and equally well-preserved remains have been recovered from organic horizons in permafrost of the Arctic (Matthews, 1968, 1974; Matthews *et al.*, 1986). Human influences have also provided numerous environments favourable for preservation in such features as hand dug wells and ditches, refuse pits, grave sites and even granaries (Kenward, 1976). Anthropogenic modification of the natural environment can also be detected by fossil insect faunas well beyond the modern (post-industrial revolution) time frame normally thought of in respect to such activity (Osborne, 1988).

Although sites with potential for paleoentomological research are quite common, it is surprising that relatively few workers have ventured into this field. The reasons for the apparent lack of enthusiasm are varied. Initially, there appeared to be a widespread (but fallacious) belief that beetles, or other insect remains, could not be identified from disarticulated fragments. The nature of paleoentomological research is almost a deterrent in itself, being a blend of paleontology, stratigraphy, entomological taxonomy and zoogeography, so it is rare that any prospective researcher starts in this field fully qualified in all aspects. Other factors detrimental to beetle studies involve the tedium of sorting through large quantities of sediment to extract the fragments, lack of well-organized collections and taxonomic keys, uncertainty about the ecology and zoogeography of many species and the time taken to process a single site (about one site/year).

In spite of the problems, individuals or small research groups have become established in Britain, Canada, Germany, The Netherlands, Sweden, the United States and the Soviet Union; and paleoentomology, in terms of Quaternary studies, is becoming a more generally recognized subject.

HISTORICAL DEVELOPMENT OF PALEOENTOMOLOGY

The first references to fossil insects were made by the natural historians of the last century (Fisher, 1863). These early records only served to alert readers to the occurrence of coleopterous fragments in peats. Those organic deposits, which we now know to be of Pleistocene or Holocene age, were variously attributed to deposits of the Noachian flood, or later, the Ice Age (Pengelly, 1870). It should be noted that Charles Darwin's mentor, Charles Lyell, in his classic text on the "Antiquity of Man" referred to "forest beds" of organic debris interspersed between the English East Anglian drift sequences (Godwin-Austen, 1851). These deposits contained plants, vertebrates and insects. Lyell (1863) states (p. 216): ... "The insects, so far as they are known, including several species of *Donacea*, are, like the plants and freshwater shells, of living species."

A little over one century ago, the high cliff sections along the north shore of Lake Ontario, east of Toronto, attracted geologists interested in recording fossil remains from the detrital peats of the Scarborough Bluffs. G.J. Hinde (1878) discovered fossil beetle fragments in these deposits, initiating a series of studies which progressed for almost 50 years under Chamberlain, Coleman and Scudder.

Samuel H. Scudder described insect fragments from the Scarborough Formation sent to him by Coleman (Scudder, 1895, 1900). He obviously had problems with the identification of many of the beetle fragments. In a series of papers, he described the coleopterous fauna of the Scarborough deposits as consisting of seventy extinct, and two extant, species. Having worked in this research field for almost two decades, it is very easy to empathize with Scudder's reactions to the poorly preserved materials that he was presented with. His frustration with identifications is reflected in the specific names given in his publications describing the Scarborough fauna.

Table 1 Beetles recognized by Scudder from the "Interglacial clays near Scarboro', Ontario".

	"Extant" forms
GROUP A:	*Gyrinis confinis*
	Acidota crenata
	"Extinct" forms
GROUP B:	*Platynus halli*
	Platynus hindei
GROUP C:	*Bembidium glaciatum*
	Platynus interglacialis
	Anthonomus fossilis
GROUP D:	*Elaphrus irregularis*
	Pterostichus abrogatus
	Bembidium vestigium
GROUP E:	*Platynus delapidatus*
	Coelambus derelictus
	Phloeosinus squalidens
GROUP F:	*Olophrum dejectum*
	Lathrobium frustum
	Coelambus infernalis
	Bembidium expletum
	Bembidium damnosum

Table 1 illustrates what we have inferred to be Scudder's gradually deteriorating trust of the designations given to the beetles. This was probably directly proportional to the state of preservation of the "identified" fragments. We have taken the liberty of regrouping this partial list of Scudder's identifications to reflect his presumed frustration at matching the fossils with modern counterparts.

One can certainly empathize with Scudder in his problems of specific identification; many of the over 30,000 species of beetles in North America were undescribed at the time of his work and, from our own experience, the preservation of the Scarborough fossils is generally not good.

Group A (Table 1) are species identified by Scudder which are extant. Only one of the two identifications is correct to the species level. *Acidota crenata* has been identified from the Scarborough Formation (Morgan and Morgan, 1980b). Group B to F are all "extinct" species. Group B represents a normal taxonomic approach; *Platynus hindei* and *Platynus halli* are respectively named after G.J. Hinde (from whom Scudder obtained most of the materials from Scarborough) and the New York state paleontologist, James Hall. Group C represents species which Scudder seems to have named in recognition of their "interglacial" nature; an apparent function of his assumption that these are extinct interglacial forms. Group D reflects a degree of concern that the specimens do not quite match modern species, even though the generic identification is probably correct. Group E names start to show the poorly preserved nature of the specimens, and Group F finally exposes Scudder's real feelings about the difficulties of identification!

We recognize Scudder's reluctance to name the specimens provided by Hinde, however, our own work has shown that not only are the majority of Scudder's specific determinations incorrect, but unfortunately a number of the generic determinations are completely inaccurate. Examples of these major mis-identifications are given in Figure 1. Possibly part of Scudder's problems in specific determinations may have been because of the scientific logic of the day which postulated intense speciation during the "Ice Ages". We now know that this does not appear to have been the case at least in the Arctic and cool temperate zones of both Europe, North America and portions of the Soviet Union.

The first studies which identified Pleistocene fossil insects as extant forms on a faunal basis were written in Europe during the first half of this century. Foremost among these were papers by Bell (1922), Henriksen (1933), Zeuner (1934), Thorkelsson (1935) and Lindroth (1942, 1948). Although the Scandinavian workers (principally Lindroth) retained an interest in the fossil insect studies, the next major research group to emerge was at the University of Birmingham, England, in the late 1950s and early 1960s. G.R. Coope, P.J. Osborne and F.W. Shotton commenced a series of publications establishing insect faunas as useful indices for a variety of Pleistocene and Holocene environmental conditions (Coope, 1959; Coope *et al.*, 1961; Coope, 1970). Not only have they shown the potential for climatic and ecological reconstruction using Pleistocene beetles, but Coope, his co-workers and students, have refined the climatic curves for the last interglacial-glacial cycle in Britain, with a remarkable degree of accuracy (Coope *et al.*, 1971; Morgan, 1973; Coope, 1977). In the late 1960s and early 1970s the first North American beetle papers appeared (Matthews, 1968; Coope, 1968a; Ashworth and Brophy, 1972; Morgan, 1972). A series of papers describing insect faunas recovered from the tar pits and other Tertiary and Quaternary deposits in California had been produced by W.D. Pierce during the 1940s and 1950s, but a large number of the identifications have been shown to be erroneous (Miller and Peck, 1979; Miller *et al.*, 1981), and they have not been considered in this discussion.

During the period from 1970-1990, several hundred publications have appeared describing many sites of different ages from many areas of the world. Some of the more relevant North American references are cited below. Comprehensive listings are available from commercial or government reference computer databases. The authors can be contacted for further details.

EXTRACTION TECHNIQUES

In the field, fossil insects may be observed on bedding planes in peat or within fine silts which have thin detrital organic horizons. If insects are observed they are sometimes worth isolating because associated fragments may be together on the stratum. Generally, splitting of organic debris is not conducted because it is time consuming, and because the insects observed often are the largest and most brightly coloured, thus introducing a sampling bias. Rather, bulk samples (approximately 5-10 kg per 5 cm interval) are collected for laboratory treatment.

The standard technique for extracting fossil insect fragments has been described many times first by Coope (1968b), by many others in the intervening years, and most recently by Morgan (1988a). The technique is illustrated in the flow chart reproduced as Figure 2.

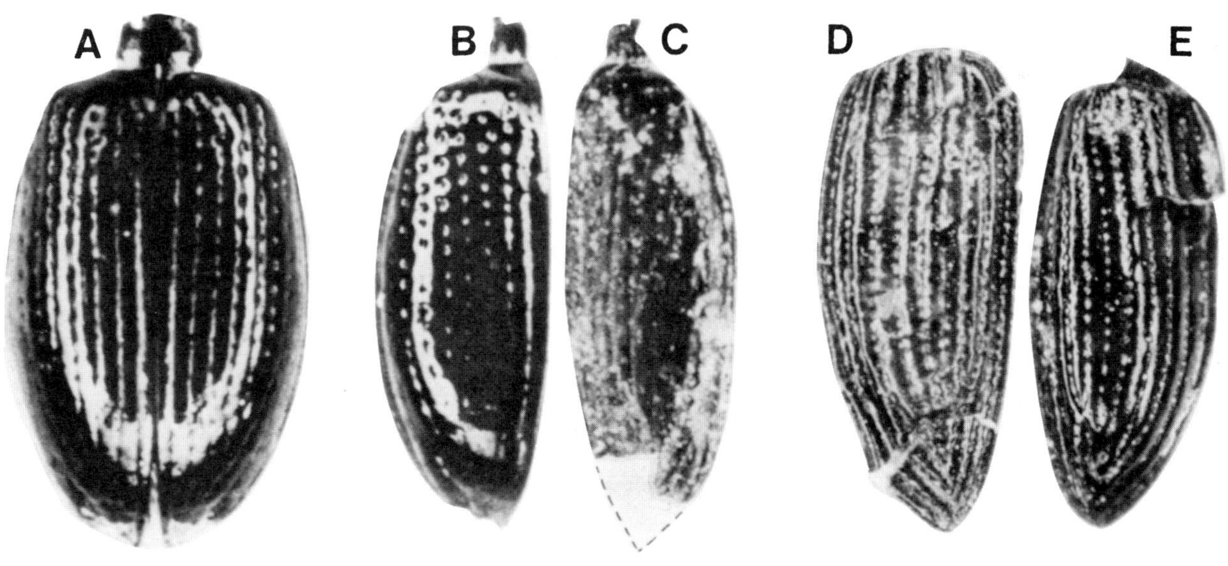

Figure 1 *Comparison of fossil specimens identified by S.H. Scudder, with other fossil material. The scale bar equals 1 mm except where specified.*

(**a**) *Partially fused elytra of* Dyschirius *cf.* globulosus *Say from Two Creeks, Wisconsin, about 12,000 years B.P.*

(**b**) *Single (left) elytron of* Dyschirius, *also from Two Creeks, Wisconsin (Morgan and Morgan, 1979).*

(**c**) *Single elytron identified by Scudder as "*Helophorus rigescens*!!". The specimen is hand-labelled (by Scudder?). "Clay beds near Cleveland, Ohio" and numbered 14,505. Unfortunately, the specimen is not an* Helophorus *species; indeed, it does not even belong to the family Hydrophilidae. The specimen is a member of the Carabidae (genus* Dyschirius*); specific determination is impossible.*

(**d**) *A left elytron, once again identified and labelled (by Scudder?) as "*Loricera? lutosa*! (det. Sc.) Scarboro"; and on a second label "Clay beds Scarboro Ca." and numbered 14,559. Although this was identified as a member of the family Carabidae, it is definitely a member of the Scarabaeidae, almost certainly of the genus* Aegialia, *and probably* Aegialia lacustris *LeC.*

(**e**) *Single (right) elytron of* Aegialia lacustris *from Two Creeks, Wisconsin.*

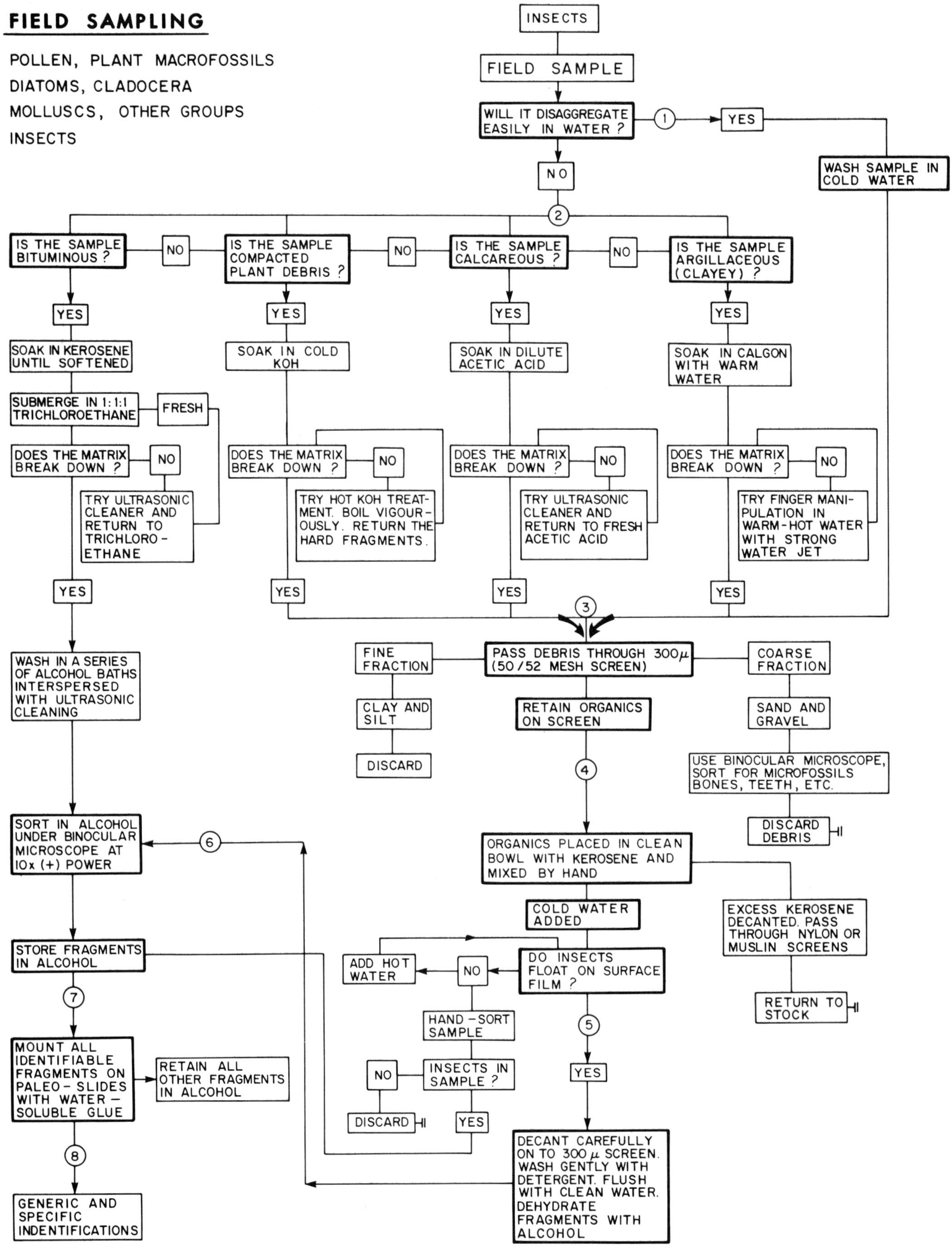

Figure 2 *Flow chart for insect separation and sorting techniques. (After Morgan, 1988a).*

The basic procedure involves placing the field sample in a polyethylene bowl and washing the organic debris through a 300 micron (No. 50 or 52) mesh sieve. The fine silt and clay fraction passes through while organic material is retained. Organic debris trapped on the sieve is placed in a clean polyethylene bowl. Kerosene is added and kneaded into the sample, the excess decanted, and cold water added. The insect fragments tend to rise to the surface on the water/kerosene interface. Variants on this theme can be used; for example, using other flotants to supplement the kerosene process such as a calcium chloride solution, or heptane. After the kerosene/other/separation the flotant is carefully decanted into a clean sieve, washed thoroughly with detergent, and rinsed in clean water until free of detergent. The sample is dehydrated in alcohol and sorted in alcohol under the binocular microscope at 8-10 × magnification. Insect fragments (and other associated fossils) can be picked out with fine forceps and stored in alcohol until they are mounted (Figure 3). Heads, thoraces and elytra are mounted on standard (white background) micropaleontological slides with water-soluble gum tragacanth. Semi-transparent specimens (thin elytra or thoraces, caddis and dipteran remains, together with other arthropod fragments) can be mounted in polyvinyl lactophenol mounting medium on glass microscope slides with a cover slip.

IDENTIFICATION

The identification of fossil fragments requires general familiarity with a wide spectrum of beetle families, a good systematic collection and sympathetic taxonomists. Modern taxonomic keys rely heavily upon features seen on articulated specimens; the disarticulated fossil material has to be looked at in an entirely different way, often utilizing features which are obscured, or not readily apparent under the regular light microscope. In this respect the paleoentomologist can utilize shape and macro-ornamentation of various sclerites (Figure 4), and by using the petrographic microscope, examine micro-ornamentation at magnifications of about 150 ×. This type of analysis is particularly useful when looking at carabid (ground) beetles. The genus *Bembidion*, for example, includes over 150 species in Canada and Alaska (probably over 400 species in North America) which can often be recognized to sub-generic level by gross morphological features, and then narrowed considerably by the examination of micro-ornamentation (Figure 5). One of the most powerful tools in this respect is the scanning electron microscope. The first SEM photographs of fossil beetles were taken at the University of Birmingham (Morgan, 1969, 1970). Detailed comparison of fossil and modern weevils of the same species revealed identical features at high magnifications (Figure 6). Besides revealing details of beetle microsculpture, the SEM is also capable of illustrating the remarkable preservation of parts which remain articulated in fossil specimens (Figure 7). The SEM is still used today for illustrating specific fossil fragments, but pre-treatment may damage rare specimens (*e.g.*, fragments which tend to charge readily and often must be coated). Colour patterns cannot be seen under the SEM, and photomicrographs are difficult to compare to specimens photographed under the light microscope because of the detail revealed by the SEM. Undoubtedly innovations, such as environmental chambers and low Kv operation, will allow more frequent use of this technique. We frequently use antistatic solutions and carbon coating to reduce charging, and to keep the specimen in a more natural state.

The key characters for a species, recognized by all modern taxonomists, are the beetle genitalia. All fossil abdomens are carefully dissected to see if they have retained the genitalia. Unfortunately, this search is often unsuccessful; the specimen may be female, parthenogenetic, or the genitalia have been lost. Figure 8 illustrates one particularly well-preserved example recovered from material in a borehole at Parkhill, Ontario, at a depth of 9 m. Associated organic detritus has been dated at 10,870 ± 90 years B.P. (WAT-376). After comparison with modern specimens the aedeagus was identified as *Gyrinus bifarius* Fall, a species found in the boreal zone of North America.

Figure 3 *Beetle and other arthropod fragments photographed in alcohol under reflected light. Generic identifications often can be made at this stage. The strongly punctate specimen* (**a**) *is part of the right elytron of the sedge-eating chrysomelid beetles* Donacia *or* Plateumaris*;* (**b**) *is a striped elytral fragment of the diving beetle* Colymbetes*; and* (**c**) *is the right elytron of the ground beetle* Bembidion*. These specimens are from Lockport Gulf, NY, and are about 11,000 years old (Miller and Morgan, 1982).*

USES OF FOSSIL INSECTS

The applied aspects of fossil insect research can be related to three major fields; archeology, geology (incorporating paleoenvironmental and paleoclimatological reconstruction) and zoogeography. Of course, there are areas of overlap between each of these disciplines, but we would like to give some examples of how each of these areas has incorporated the results of beetle research.

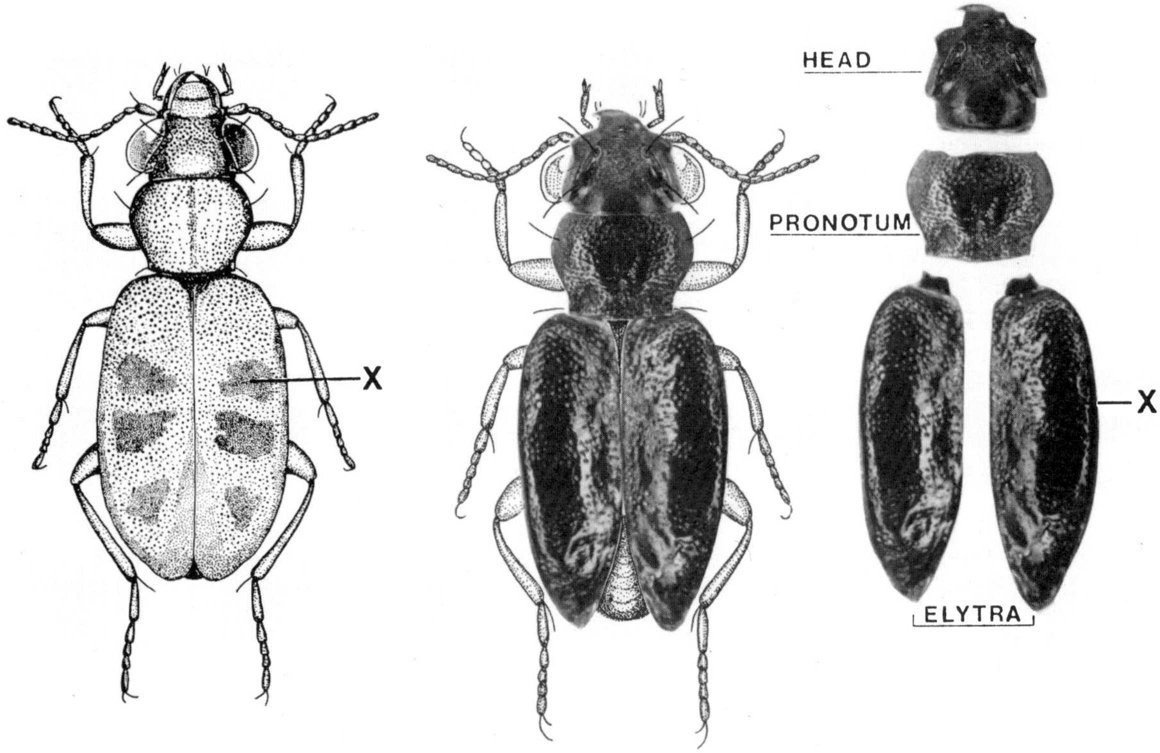

Figure 4 *Macro-ornamentation* (e.g., *the "mirror" areas illustrated as* **X** *on the elytra) provide diagnostic clues to help narrow the number of species during identification. These patterns, together with overall size and shape, the distinctive punctae on the head and thorax, enable the specimen to be identified as* Asaphidion yukonense *Wickham. The individual fragments can be put together to provide a picture of the whole beetle. The fossil fragments are from the Two Creeks site, Wisconsin (Morgan and Morgan, 1979); the closest modern records are from the Nordegg area, Alberta, but the species may be regarded as a typical western North American boreal form.* A. yukonense *was resident in eastern Canada (southwest Ontario) until about 10,000 years B.P., and has been recorded north of Lake Superior in the Holocene.*

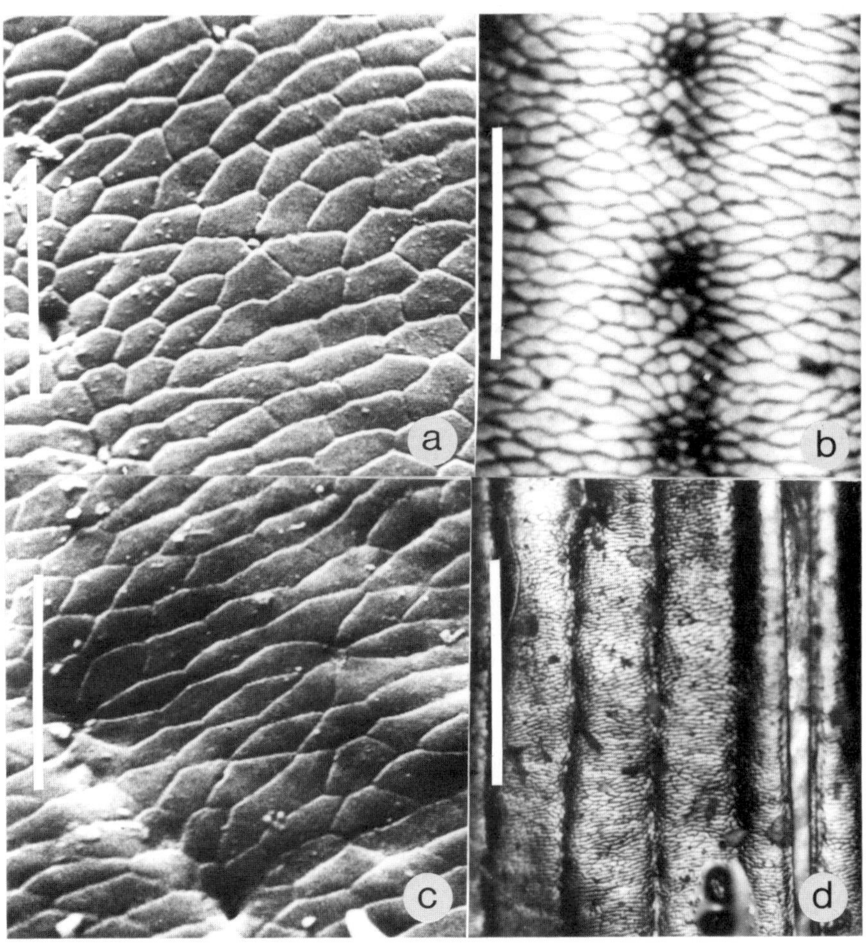

Figure 5 *Micro-ornamentation in the carabid genus* Bembidion *photographed under polarized light.* Bembidion fellmani *(Mnh.) A.* (**a**) *scanning electron micrograph and* (**b**) *polarized light.* Bembidion crenulatum *(Sahlb.) C.* (**c**) *scanning electron micrograph and* (**d**) *polarized light. The specimens are from Four Ashes, Staffordshire, England, and are greater than 30,500 years B.P. Note more traverse "brick-like" mesh in c and d (Morgan, 1970). Scale bars equal approximately 8μm for a, b, 16μm for c, and 90μm for d.*

Archeology.
Archeological sites, especially in Britain, have been the subject of a number of papers describing associated insect faunas (Osborne, 1973). These include a Bronze Age ceremonial shaft near Wilsford, Wiltshire (Osborne, 1969), a Roman well at Barnsley Park, Gloucestershire (Coope and Osborne, 1967), a Viking-period site at York (Kenward, 1975) and a Medieval site, also at York (Buckland *et al.*, 1974). These fossil faunas associated with early occupation sites provide much interesting information, such as the type and extent of insect infestations in ancient granaries or on cultivated crops. To a lesser degree, we can determine the directions of early trading systems, or even the distribution of disease-carrying organisms such as bed bugs and fleas. Some recent papers on aspects of arthropod remains in archeological sites in Europe, Greenland and Africa are given in Buckland and Sadler (1989), Osborne (1988) and Mumcuoglu and Zias (1988), and more can be found from the references therein.

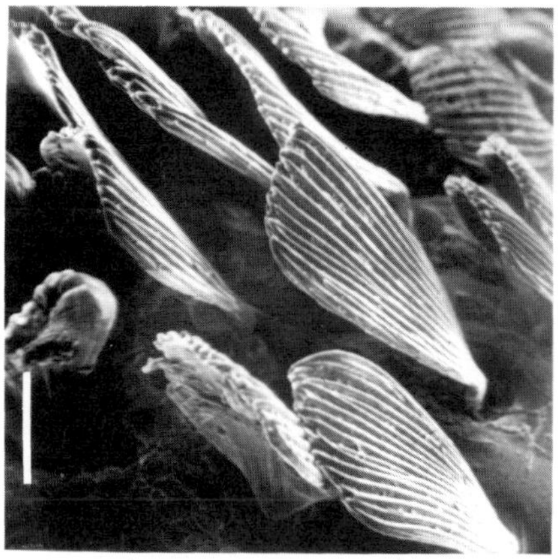

Figure 6 *Scales of the weevil* Cryptorrhyncidius lapathi *Linnaeus. This scanning electron micrograph illustrates one of three different sets of diagnostic scales on the elytron of this curculionid. These can be matched exactly with scales on fossil counterparts of the same species (Morgan, 1970). Scale bar equals 10μm approximately.*

Deposits (not necessarily associated with archeological sites) belonging to Mesolithic, Neolithic and younger periods (from about 10,000 years B.P.) have been examined at a number of locations and show good evidence on the nature of the forest cover and agricultural practices (Kelly and Osborne, 1964; Osborne, 1965; Morgan, 1976). The examination of insect faunas associated with archeological sites in North America has not been conducted on the scale seen in western Europe. Insect fragments, recovered in the late 1970s from the Lubbock (Texas) site, are all contaminants, probably brought into the archeological horizons by parasitic Hymenoptera (Morgan, unpublished data). Elias (1986) has described a fossil insect assemblage from a 13,000 year old archeological site south of Denver, Colorado. At Parkhill, Ontario, deposits close to a paleo-Indian locality have been processed for environmental data. Unfortunately, the borehole locality is approximately 400 m from the human occupation site, but is believed to be of the same age, based upon stratigraphic interpretation.

Geology.
Perhaps the most dramatic use of fossil insects has been in the reconstruction of paleoenvironments and paleoclimates during the Pleistocene Epoch. A series of increasingly severe climatic fluctuations during the Tertiary eventually culminated in major ice advances in both hemispheres during Quaternary time. The magnitude of most of these advances and retreats is unknown because of the complexity of the terrestrial record coupled with the poor preservation of many of the sequences. Insects have been utilized in examining deposits of earliest Pleistocene, Pliocene or possibly even Miocene sites in northern Canada and Alaska (Matthews, 1974, 1976; Morgan *et al.*, 1979) and Greenland (Böcher, 1989). Other sites of Middle and Late Pleistocene age have been examined in Western Europe (Coope, 1970, 1977; Lemdahl, 1985; Hufnagel, 1987), and various parts of North America (many references summarized most recently in Morgan, 1987, 1988a; and Schwert and Ashworth, 1988). Most fossil insect work has been concentrated on the time span involved in the last interglacial/glacial cycle. Much of our research has dealt with sites centred on southern Ontario, but encompassing adjacent American

A

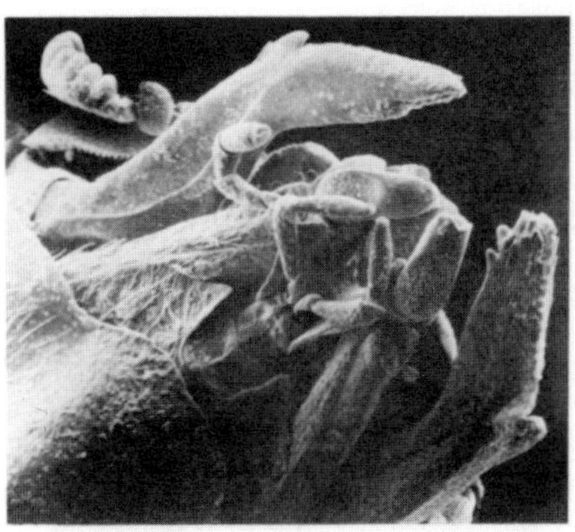

B

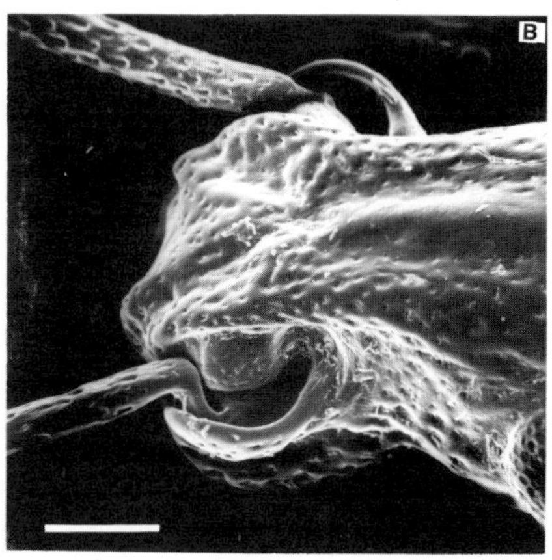

Figure 7 *Detail on heads of fossil specimens illustrated by the SEM. Scale bar = 300μm.*
(**a**) *Ventral view of a specimen of the genus* Heterocerus. *This is a fossil specimen recovered from the McKittrick tar seeps in California. The tar was dissolved, and the specimen cleansed in successive baths of alcohol. A* Cybister *water beetle from the same horizon was dated by tandem accelerator at close to 8000 years B.P., nevertheless, specimens still have hairs, articulated palps, and attached antennae.*
(**b**) *Dorsal view of* Otiorrhyncus nodosus *Mueller. The specimen is from Four Ashes, England, (Morgan, 1973) from a sequence dated at 36,350 years B.P. The antennae of the weevil are still in the sockets. The left eye is at the lower right.*

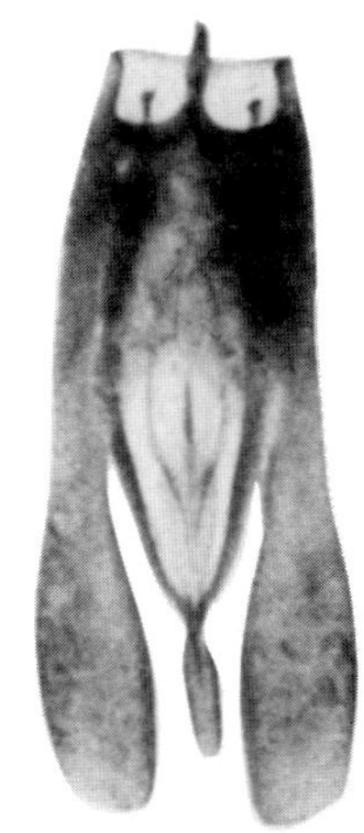

Figure 8 *Aedeagus of* Gyrinus bifarius *Fall. The specimen is from Parkhill Creek, southwest Ontario, and is approximately 10,800 years old. (×85 approx.)*

states. In this way, we have made a start in reconstructing climatic curves for restricted geographic areas, similar to those developed by Coope and his co-workers in southern Britain (Coope *et al.*, 1971; Coope, 1977; Atkinson *et al.*, 1986). Before embarking on a précis of the uses of Coleoptera in the determination of climatic change, we should discuss why these insects are believed to be so useful.

As Coope (1970, 1979) has frequently pointed out, a large number of beetle species are quite specific about their living requirements. For example, scolytids (bark beetles) are often restricted to certain types of trees, frequently occupying specific locations on the tree under attack, *e.g.*, the lower bole and root region or the mid-portion of the bole, or the growing tips of branches. Certain scarabaeids (dung beetles) are confined to the dung of specific animals. Other species from different families have only been collected in specific habitats such as beaver lodges. Many carabid (ground) beetles have decided preferences for different types of substrates, often having modified sclerites to take advantage of differing lithologies. Finally, many phytophagous (plant-eating) beetles, representing a number of families, utilize specific host plants, although they frequently are not present over the whole range of the plant species, but often stop as though confined by certain climatic factors (Morgan, 1970).

There are, of course, many species of beetles whose ecological requirements are very poorly understood, and also large numbers which are eurythermic, extending in Europe from the Mediterranean coast to North Cape, and in North America from the Gulf of Mexico to the Tuktoyaktuk Peninsula. Nevertheless, two important principles have been established by an examination of a large number of fossil faunas: (1) Ecological assemblages have remained relatively constant through time with groups of genera and species apparently occupying the same niches for hundreds of thousands, or even millions of years; and (2) beetles appear to be rapid colonizers and are able to take advantage of their mobility and position in the food chain to move quickly into a region once conditions become suitable for life. For example, in recently deglaciated terrain, carnivorous beetles can prey upon spiders and flies, which in turn depend upon lichens or algae for sustenance. In this way, they are not dependent upon higher plants which in turn are often restricted from moving into a region because of unsuitable substrate development.

Perhaps the most remarkable use of these principles has been in the English Midlands where a short-lived, but intense, warmer episode punctuated the Siberian-like cold of the Middle Devensian (Middle Wisconsinan/Weichselian). One locality, a small gravel pit at Four Ashes near Wolverhampton, produced a number of organic lenses which revealed deposition from the last (Ipswichian/Sangamonian) Interglacial through to the late Devensian (A.V. Morgan, 1973). More importantly, the Four Ashes sequence indicated the presence of a thermophilous insect fauna between faunas which contained Scandinavian and Siberian species (Morgan, 1970, 1973). Microhabitat and microclimatic differences could not be readily invoked, since the substrate was the same for all the sites, and since they were all deposited within an area of some 400 m by 200 m in relatively flat terrain (vertical amplitude less than 5 m). This observation

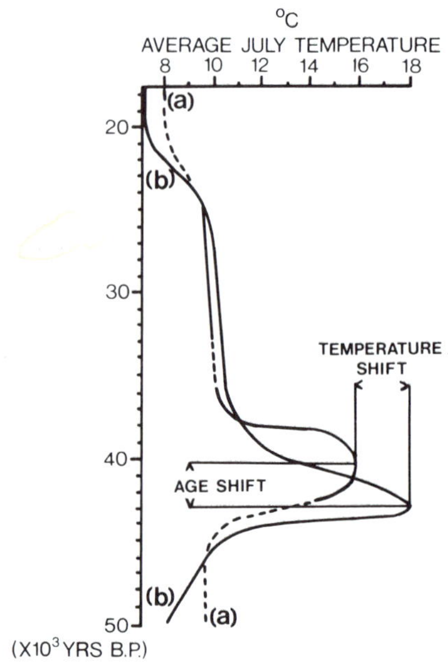

Figure 9 *Paleotemperature curve for the Middle Devensian in Central England. The short-lived climatic amelioration illustrated in curve* (**a**) *(Coope* et al., *1971) has been modified to curve* (**b**) *(Coope, 1977). The magnitude, longevity and timing in the peak of warmth has gradually been refined with research into additional sites.*

verified the results of a number of other sites of similar ages, but found in scattered geographic locations. These data, along with data from other areas, were incorporated in a paper on climatic reconstruction of Central England in Middle and Late Devensian times (Coope *et al.*, 1971). The preliminary climatic curve (Figure 9) shown in this publication was modified by subsequent research (Coope, 1975, 1977). The beetles changed from arctic boreomontane species, to thermophilous species largely representative of the North German plain (*i.e.*, warmer than southern England), and then reverted once again to treeline/tundra species. Other remarkably rapid shifts in insect faunas have been recorded from late-glacial (Late Wisconsinan) sites in many different parts of the British Isles (Coope, 1977), but our work in southern Ontario has not yet produced a similar continuity of the climatic record. Rather, we have established isolated points for a climatic curve, and such examples are discussed below.

The first locality is the long-famous Sangamonian site of the Don Valley Brickpit. For about a century, the Don Formation has been examined for its warm climate flora and the fossils of beetles, which have remained both generically and specifically unidentified. We have examined and identified a small fauna which shows that the insects reflect a climate which is almost identical to that of Toronto today. The presence of a diverse group of beetles and associated caddisflies (Williams and Morgan, 1977) indicates a mixed deciduous and coniferous woodland bordering a well-vegetated, slowly moving river flowing into a lake with a water level well above that of the present day Lake Ontario. The July average temperature would have been about 20-21°C. A site with similar climatic implications, and possibly of the same age, has recently been described from the region southwest of Kitchener-Waterloo. The Innerkip site (Pilny and Morgan, 1987) has a moderately well-preserved beetle fauna with species (not recorded elsewhere until the early Holocene) in common with the Don. Many of the species from the Innerkip site today only reside in southern Ontario.

The second locality to the east of Toronto, referred to earlier in the historical discussion, is at the Scarborough Bluffs.

An insect fauna recovered from the lower part of the Bluffs indicates a vastly different climatic regime from that of the Don. The Scarborough Formation, of Early Wisconsinan age, rests on top of the Don Formation (Karrow, 1967; Karrow and Morgan, 1975). Insects collected from the deltaic Scarborough Sands and the lacustrine Scarborough Silts and Clays are typical of the northern boreal forest or open tundra, close to treeline. The habitat reflected by the insects appears to be a patchy spruce environment with open ground areas. The July average temperature would have been about 11-12°C.

Although the thermal environments of the Don and Innerkip, and Scarborough, sites were drastically different, we cannot interpret the rapidity of change which took place during the deposition of the sequences since it lies beyond conventional radiocarbon dating. It would be fascinating to have good data from the terrestrial record on just how rapidly the climate of the last interglacial deteriorated — a prospect which may await us in the not so distant geological future. Such knowledge would present vital information for our species and would have an incredible fiscal impact on industries and governments who are currently trying to come to grips with the economic and social realities of global change.

It is equally interesting to examine the poorly preserved insect faunas of the Early and Middle Wisconsinan and to contemplate the long timeframes with far more severe climates which are indicated by tundra insect assemblages in southern Ontario. (Morgan, 1972; Warner *et al.*, 1988). The prospect of increasingly harsh climates, the disappearance of deciduous trees, followed by conifers and the advent of a bleak tundra with active permafrost and a slowly advancing ice sheet is one which can only be welcomed by Alberta and OPEC! Such information would also be invaluable in helping address possible counterbalances to the predicted atmospheric warming suggested in various Global Circulation Models (GCMs) in respect to the "greenhouse effect".

Recent concerns about the rapidity of CO_2 buildup, and the potential implications of climate warming have led to renewed interest in understanding the proxy data record preserved in Quaternary sediments. Most research has been conducted by palynologists, and unfortunately the beetle record seems to have been overlooked by many. Nevertheless, it was over twenty years ago, in 1970, when the British Quaternary Association discussed the perceived climatic ameliorations of latest Pleistocene time at the University of Birmingham. Out of those discussions came the realization that there had been very rapid climatic ameliorations at *ca.* 13,500-13,000 years B.P. and at 10,200 years B.P. (Coope *et al.*, 1971; Coope and Brophy, 1972). These movements were later shown to be caused by fluctuations of the polar front in the North Atlantic, but they had been detected in the coleopteran record despite relatively complacent records in the pollen stratigraphy from sites of similar ages. The Allerod/Dryas fluctuations are, of course, recorded in pollen sequences in western Europe, and, more recently, believed recorded in eastern Canada (Anderson, 1983; Mott *et al.*, 1986).

Oscillations in pollen spectra in Ohio and Indiana (Shane, 1987) attributed to a "Younger Dryas" effect in the mid-continent are not matched by climatic fluctuations in the beetle records nor are they matched by the oxygen isotope record (Fritz *et al.*, 1987). The oxygen isotope records in western Europe clearly reflect the movement of the polar front (Eicher and Siegenthaler, 1976), and one must look for another explanation for pollen spectral shifts in the North American continental interior. Morgan *et al.* (1982), Morgan, (1987) and Fritz *et al.* (1987) have suggested that such shifts are not connected to regional climatic change, but rather due to lake effects. We now understand that these are likely attributable to discharges of cold glacial meltwater, and these, in turn may have triggered movements in the polar front in the North Atlantic.

With few exceptions (Elias, 1985; Morgan *et al.*, 1985; Schwert and Ashworth, 1985; Nelson and Carter, 1987) Holocene insect sites have not yet been examined in any great detail, but there is much to be done. Insect preservation in these younger sites is generally good to excellent, the sites can be well dated and undoubtedly there is much to be learned in terms of zoogeography, archeology, and climate reconstruction.

Zoogeography.

One of the most exciting facets of this research is that of zoogeography, although this subject would seem to be furthest removed from the geological applications of paleoentomology. Understanding the full modern range of a species, like the life history and habitat ecology, is of paramount importance in the interpretation of past environmental conditions. These elements provide a classical case of the importance of basic "natural history observations" which went out of vogue around the turn of the last century. The lack of importance attached to such observations, which are often not regarded as "science", has done a grave disservice to our understanding of the knowledge bases required for the correct interpretation of past conditions. We recognized at the start of paleoentomological research in Canada that modern range distributions and habitat data ("non-applied" observations) were essential to our understanding of past ecological interactions. Over the past two decades, we have gradually and systematically collected from northern North America, from Alaska to Newfoundland, and have built up modern reference materials which are invaluable to our work. The Waterloo collection now consists of more than 60,000 Coleoptera, principally from northern Canada and Alaska, and is an excellent working reference for paleoentomological research.

Most people tend to regard beetles as being secretive and possibly somewhat lethargic, but anyone observing beetles on a warm summer day will quickly realize that large numbers of species can move extremely rapidly both on the ground and in the air. In the previous section, we alluded to the fact that beetles were rapid colonizers and were well adapted to certain ecological niches. The climatic fluctuations of the Pleistocene forced beetles to move vast distances so that their present distributions are often quite dissimilar to their distributions during the Pleistocene. For example, the present Canadian insect fauna would only have been able to colonize most of Canada during the period following ice retreat. With the exception of the Yukon, parts of the Northwest Territories, possibly portions of the east coast of Labrador and some of the islands in the Gulf of St. Lawrence, ice retreat has taken place less than 15,000 years ago. Substantial portions of Quebec and smaller areas of the Northwest Territories were ice-covered as recently as 7,000 years ago. Figure 10 illustrates the distribution of four beetle species which have moved considerable distances during the Pleistocene.

Diacheila polita Falderman (a ground beetle) is today confined to the extreme northwest of Canada, with a range expanding westward through Alaska, Kamchatka and eastern Siberia, eventually terminating on the east side of the Kola Peninsula. This beetle is flightless and usually inhabits open peaty soil on the tundra. It has been found in southern Ontario in the Scarborough Formation, in the St. Lawrence Valley of southern Quebec, and in the period approximately contemporaneous with the maximum development of ice it was a common resident in the area west of the Michigan lobe in Illinois and Iowa (Morgan and Morgan, 1982, 1986; Baker *et al.*, 1986). The youngest reported date for the species in eastern North America is about 12,100 years B.P. (Barnosky *et al.*, 1988). Comments on the North American zoogeography of this species are provided in Morgan and Morgan (1980b, 1982) and Morgan (1987, 1988a). Figure 10 illustrates the known and postulated movement of the species through time in North America; Figure 11 shows the known modern range.

Helophorus arcticus Brown (a water scavenger beetle) provides a good example of the additional knowledge which has been gained since the original "Beetle Bits" paper was written a decade ago. *H. arcticus* was first described by Brown (1937) and, in 1980, was only known from three modern collecting localities (Churchill, Manitoba; at Lake Harbour (Baffin Island) and at Hebron, Labrador). As a result of deliberate searches, the species was again collected at Churchill in 1983 and was found in northeastern Ungava Bay in 1986 (Figure 11). The ecology is not fully understood, but generally the beetle appears to be associated with granular littoral substrates at, or relatively close to, treeline. In all modern finds, the species is situated close to cold ocean water. The fossil record also indicates that it may have lived near cold water bodies in the form of large glacial lakes, meltwater spillway routes, or the sea. Fossil specimens of *H. arcticus* were first found in the Scarborough Formation of southern Ontario (Morgan, 1972), and later in possibly contemporaneous Early Wisconsinan deposits at Woodbridge, northwest of Toronto (Williams *et al.*, 1981). It has subsequently been recovered from localities outside the maximum Late Wisconsinan ice limit in Iowa (Baker *et al.*, 1986), as well as in postglacial sequences in southern Ontario and Quebec (Mott *et al.*, 1981). Recent publications (Smetana, 1985; Morgan, 1989) summarize what is known about this enigmatic species. *H. arcticus* and *Diacheila polita* have been recognized as beetles with contrasting colonizing potentials, and this has been expanded upon elsewhere (Morgan and Morgan, 1982; Morgan, 1987).

Anotylus gibbulus (Eppelsheim), a rove beetle, was found in supposed Sangamonian (last) Interglacial deposits at the Don Valley Brickpit (Toronto) in 1974. This species is of interest to entomologists because it is not living in North America (or even Western Europe) at present. The core area of *A. gibbulus* appears to be the Caucasus Mountains in Southern Russia. One other modern locality near Vladivostock has also been described (Hammond *et al.*, 1979). The reason for the contraction in range is not clear, because *A. gibbulus* was also present in Central England until approximately 35,000 years ago (Figure 11). It is possible that the range of this beetle (often associated with dung from large herbivores) may have been curtailed by the demise of the large mammals in latest Pleistocene time.

Aphodius holdereri Reitter (a lamellicorn beetle) is another species which is now living far from fossil sites of Middle Wisconsinan age. This dung beetle is exceptionally common in certain deposits of Central England. In spite of an extremely distinctive morphology, the fossil specimens were not matched to living counterparts until the early 1970s (Coope, 1973). The somewhat restricted distribution in the region of Tibet, north of the Himalayas, is shown in Figure 11.

Although we have only used four examples above, many more could be

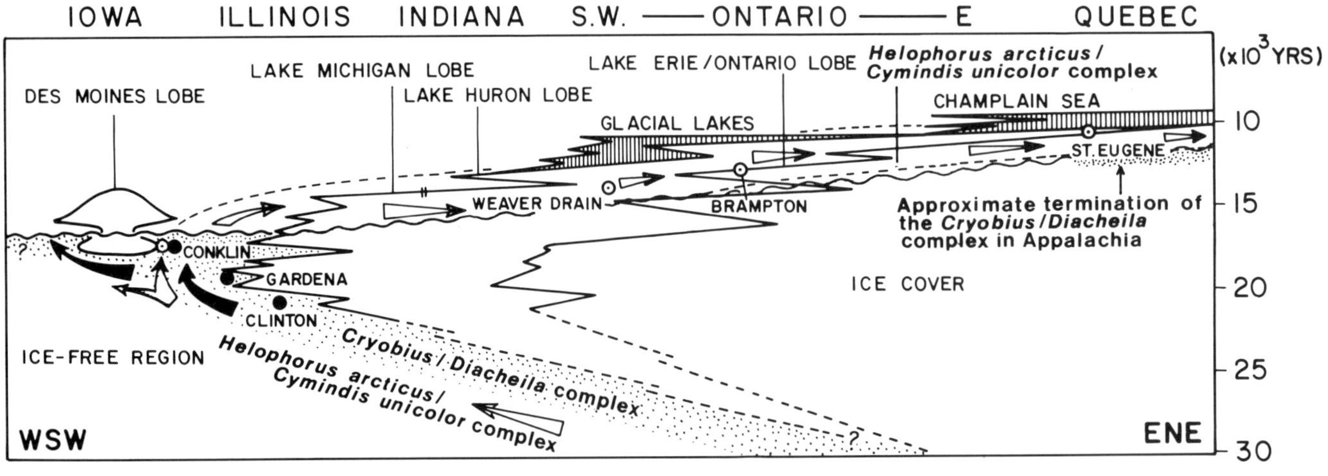

Figure 10 *Changes in geographic range of* Diacheila polita *Fald. through time.* D. polita *occurs at sites along the ice margin at maximum glaciation in Illinois and Iowa (black spots). The species is recorded in eastern Ontario and in Quebec in presumed Early Wisconsinan time, eventually becoming extirpated at the ice margin between 16,000 and 15,000 years B.P. west of the Michigan lobe, and somewhat later in Appalachia (wavy line). Unlike members of the* Helophorus arcticus *complex,* D. polita *failed to recolonize eastern Canada following ice retreat. It survived maximum Wisconsin glaciation in the Beringian "refugium", and is today a resident of the northwest of North America.*

cited illustrating movements in distribution which amount to thousands of kilometres between fossil and modern locations, even for relatively recent Late Wisconsinan sites. The reader is referred to publications on *Cicindela* (Morgan and Freitag, 1982; Nagano *et al.*, 1982), *Holoboreaphilus* (Morgan *et al.*, 1984), *Blethisa* (Morgan *et al.*, 1986) and *Polygraphus* (Morgan, 1988b) as examples of North American fossil and modern range comparisons. As this research area expands, it should be possible to further illustrate the movement of selected species, or even whole faunal assemblages through time and to reconstruct the shifts of beetle populations in front of the advancing and retreating Wisconsinan Ice Sheet.

ACKNOWLEDGEMENTS

We are grateful to the Natural Sciences and Engineering Research Council of Canada for TEAM and Operating grants to both authors. These funds support on-going research in paleoentomology at Waterloo and have provided the sole support for basic research in both fossil insects studies and modern collecting. We would like to acknowledge Dr. David Adam (United States Geological Survey at Menlo Park) for assistance with the SEM photographs of the McKittrick specimens. Our sincere thanks also go to friends and colleagues at the Biosystematics Research Unit, Agriculture Canada, Ottawa, for their help over the years. Much of our work could not have been done without their assistance.

REFERENCES

Anderson, T.W., 1983, Preliminary evidence for late Wisconsinan climatic fluctuations from pollen stratigraphy in Burin Peninsula, Newfoundland: Geological Survey of Canada, Paper 83-1B, p. 185-188.

Angus, R.B., 1973, Pleistocene *Helophorus* (Coleoptera, Hydrophilidae), from Borislav and Starunia in the western Ukraine, with a re-interpretation of M. Lomnicki's species, description of a new Siberian species, and comparison with British Weichselian faunas: Royal Society of London, Philosophical Transactions, v. 265B, p. 299-326.

Ashworth, A.C., 1973, Fossil beetles from a fossil woodrat midden in Western Texas: Coleopterists Bulletin, v. 27, p. 139-140.

Ashworth, A.C. and Brophy, J.A., 1972, A Late Quaternary fossil beetle assemblage from the Missouri Coteau, North Dakota: Geological Society of America, Bulletin, v. 83, p. 2981-2988.

Atkinson, T.C., Briffa, K.R., Coope, G.R., Joachim, M.J. and Perry, D.W., 1986, Climatic calibration of coleopteran data, *in* Berglund, B.E., Handbook of Holocene Palaeoecology and Palaeohydrology: John Wiley and Sons Ltd., Toronto, p. 851-859.

Baker, R.G., Rhodes, R.S., Schwert, D.P., Ashworth, A.C., Frest, T.J., Hallberg, G.R. and Janssens, J.A., 1986, A full-glacial biota from southeastern Iowa: Journal of Quaternary Science, v. 1, p. 91-107.

Barnosky A.D., Barnosky, C.W., Nickmann, R.J., Ashworth, A.C., Schwert, D.P. and Lantz, S.W., 1988, Late Quaternary paleoecology at the Newton site, Bradford Co., Northeastern Pennsylvania: *Mammuthus columbi*, palynology and fossil insects, *in* Laub, R.S., Miller, N.G. and Steadman, D.W., eds., Late Pleistocene and Early Holocene Paleoecology and Archaeology of the Eastern Great Lakes Region: Buffalo Society of Natural Sciences, Bulletin, v. 33, p. 173-184.

Bell, A., 1922, On the Pleistocene and later Tertiary British insects: Annual Report of the Yorkshire Philosophical Society, (1921), p. 41-51.

Böcher, J., 1989, Boreal insects in northernmost Greenland: Palaeoentomological evidence from the Kap Kobenhavn Formation (Plio-Pleistocene), Peary Land: Fauna Norvegica (Norwegian Journal of Entomology), Series B, v. 36, p. 37-43.

Brown, W.J., 1937, The Coleoptera of Canada's eastern Arctic: The Canadian Entomologist, v. 69, p. 106-111.

Buckland, P.C., Greig, J.R.A. and Kenward, H.K., 1974, York: An Early Medieval site: Antiquity, v. 47, p. 25-33.

Buckland, P.C. and Sadler, J.P., 1989, A biogeography of the human flea *Pulex irritans* L.: (Siphonaptera:Pulicidae): Journal of Biogeography, v. 16, p. 115-120.

Churcher, C.S., 1966, The insect fauna from the Talara tar-seeps, Northwestern Peru: Canadian Journal of Zoology, v. 44, p. 985-993.

Coope, G.R., 1959, A late Pleistocene insect fauna from Chelford, Cheshire: Royal Society of London, Philosophical Transactions, v. 151B, p. 70-86.

Coope, G.R., 1968a, Insect remains from silts below till at Garfield Heights, Ohio: Geological Society of America, Bulletin, v. 79, p. 753-756.

Coope, G.R., 1968b, An insect fauna from Mid-Weichselian deposits Brandon, Warwickshire: Royal Society of London, Philosophical Transactions, v. 254B, p. 425-456.

Coope, G.R., 1970, Interpretations of Quaternary insect fossils: Annual Review of Entomology, v. 15, p. 97-120.

Coope, G.R., 1973, Tibetan species of dung beetle from Late Pleistocene deposits in England: Nature, v. 245, p. 335-336.

Coope, G.R., 1975, Mid-Weichselian climatic changes in western Europe, re-interpreted from coleopteran assemblages, *in* Suggate, R.P. and Creswell, M.M., eds., Quaternary Studies: Royal Society of New Zealand, p. 101-108.

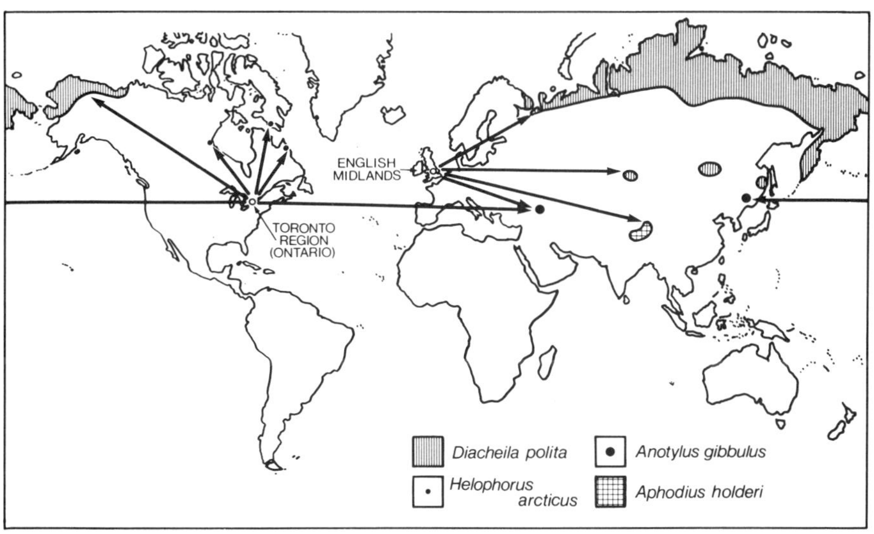

Figure 11 *Changes in geographic ranges of selected species. Modern distributions are shown and the general location of fossil localities in North America.*

Coope, G.R., 1977, Fossil coleopteran assemblages as sensitive indicators of climatic changes during the Devensian (last) cold stage: Royal Society of London, Philosophical Transactions, v. 280B, p. 313-340.

Coope, G.R., 1979, Late Cenozoic fossil Coleoptera: Evolution, biogeography and ecology: Annual Review of Ecology and Systematics, v. 10, p. 247-267.

Coope, G.R. and Brophy, J.A., 1972, Late glacial environmental changes indicated by coleopteran succession from North Wales: Boreas, v. 1, p. 97-142.

Coope, G.R. and Osborne, P.J., 1967, Report on the coleopterous fauna of the Roman well at Barnsley Park, Gloucestershire: Transactions of the Bristol and Gloucestershire Archaeological Society, v. 86, p. 84-87.

Coope, G.R., Morgan, A. and Osborne, P.J., 1971, Fossil Coleoptera as indicators of climatic fluctuations during the Last Glaciation in Britain: Palaeogeography, Palaeoclimatology, Palaeoecology, v. 10, p. 87-101.

Coope, G.R., Shotton, F.W. and Strachan, I., 1961, A Late Pleistocene fauna and flora from Upton Warren, Worcestershire: Royal Society of London, Philosophical Transactions, v. 244B, p. 379-421.

Eicher, U. and Siegenthaler, U., 1976, Palynological and oxygen isotope investigations on late-glacial sediment cores from Swiss lakes: Boreas, v. 5, p. 109-117.

Elias, S.A., 1985, Paleoenvironmental interpretations of Holocene insect fossil assemblages from four high-altitude sites in the Front Range, Colorado, U.S.A.: Arctic and Alpine Research, v. 17, p. 31-48.

Elias, S.A., 1986, Fossil insect evidence for Late Pleistocene paleoenvironments of the Lamb Spring site, Colorado: Geoarchaeology: An International Journal, v. 1, p. 381-387.

Fisher, O., 1863, On the Brick-pit at Lexden, near Colchester: Quarterly Journal of the Geological Society of London, v. 19, p. 393.

Fritz, P., Morgan, A.V., Eicher, U. and McAndrews, J.H., 1987, Stable isotope, fossil Coleoptera and pollen stratigraphy in Late Quaternary sediments from Ontario and New York State: Palaeogeography, Palaeoclimatology, Palaeoecology, v. 58, p. 183-202.

Godwin-Austen, R.A.C., 1851, On the superficial accumulation of the coasts of the English Channel: Quarterly Journal of the Geological Society of London, v. 7, p. 118-136.

Hammond, P., Morgan, A. and Morgan, A.V., 1979, On the *gibbulus* group of *Anotylus*, and fossil occurences of *Anotylus gibbulus* (Staphylinidae): Systematic Entomology, v. 4, p. 215-221.

Henriksen, K.L., 1933, Undersogelser over Danmark-Skanes Kvartaere Insektfauna: Videnskabelige Meddelelser fra Danske Naturhistorisk Forening i Kjobenhavn, v. 96, p. 77-355.

Hinde, G.J., 1878, Glacial and interglacial strata of Scarboro' Heights and other localities near Toronto, Ontario: Canadian Journal of Science, Literature and History (New Series: defunct), v. 15, p. 388-413.

Hufnagel, M., 1987, Fossile Käferfaunen in Typischen Quartärprofilen aus Schleswig-Holstein und ihre bedeutung als Klimaindikatoren, unpublished Doctoral Dissertation, Christian Albrechts Universität, Kiel, West Germany, 148 p.

Karrow, P.F., 1967, Pleistocene geology of the Scarborough area: Ontario Department of Mines, Geological Report 46, p. 1-108.

Karrow, P.F. and Morgan, A.V., 1975, Quaternary stratigraphy of the Toronto area, *in* Telford, P.J., ed., Field excursions guidebook, Geological Association of Canada Annual Meeting, Waterloo, Ontario: University of Waterloo, p. 161-180.

Kelly, M. and Osborne, P.J., 1964, Two faunas and floras from the alluvium at Shustoke, Warwickshire: Proceedings of the Linnaean Society, London, v. 176, p. 37.

Kenward, H.K., 1975, The biological and archaeological implications of the beetle *Aglenus brunneus* (Gyllenhal) in ancient faunas: Journal of Archeological Science, v. 2, p. 63-69.

Kenward, H.K., 1976, Further archaeological records of *Aglenus brunneus* (Gyll.) in Britain and Ireland, including confirmation of its presence in the Roman Period: Journal of Archeological Science, v. 3, p. 275-277.

Lemdahl, G., 1985, Fossil insect faunas from Late-Glacial deposits in Scania (South Sweden): Ecologia Mediterranea, v. 11, p. 185-191.

Lindroth, C.H., 1942, Ett subfossilfynd av *Calosoma sycophanta* L.: Popular Biology Review, v. 2, p. 1-7.

Lindroth, C.H., 1948, Interglacial insect remains from Sweden: Arsbok Sveriges Geologiska Undersökning, C, v. 42, p. 1-29.

Lyell, C., 1863, The Geological Evidences of the Antiquity of Man with remarks on theories of the origin of species by variation: John Murray, (Spottiswoode and Co.,), London, 520 p.

Matthews, J.V., Jr., 1968, A paleoenvironmental analysis of three Late Pleistocene coleopterous assemblages from Fairbanks, Alaska: Quaestiones Entomologica, v. 4, p. 202-224.

Matthews, J.V., Jr., 1974, Quaternary environments at Cape Deceit (Seward Peninsula, Alaska): Evolution of a tundra ecosystem: Geological Society of America, Bulletin, v. 85, p. 1353-1384.

Matthews, J.V., Jr., 1976, Insect fossils from the Beaufort Formation: geological and biological significance: Geological Survey of Canada, Paper 76-1B, p. 217-227.

Matthews, J.V., Jr., Mott, R.J. and Vincent, J.-S., 1986, Preglacial and Interglacial environments of Banks Island: Pollen and plant macrofossils from Duck Hawk Bluffs and related sites: Géographie physique et Quaternaire, v. 40, p. 279-298.

Miller, R.F. and Morgan, A.V., 1982, A postglacial coleopterous assemblage from Lockport Gulf, New York: Quaternary Research, v. 17, p. 258-274.

Miller, S.E., 1982, Quaternary insects of the California asphalt deposits: Third North American Paleontological Convention, Proceedings, v. 2, p. 377-380.

Miller, S.E., Gordon, R.D. and Howden, H.F., 1981, Re-evaluation of Pleistocene scarab beetles from Rancho la Brea, California (Coleoptera:Scarabaeidae): Proceedings of the Entomological Society of Washington, v. 83, p. 625-630.

Miller, S.E. and Peck, S.B., 1979, Fossil carrion beetles of Pleistocene California asphalt deposits, with a synopsis of Holocene California Silphidae (Insecta: Coleoptera:Silphidae): San Diego Society of Natural History, Transactions, v. 19, p. 85-106.

Moore, R.C., Lalicker, C.G. and Fischer, A.G., 1952, Invertebrate Fossils: McGraw Hill, Toronto, 766 p.

Morgan, A., 1969, A Pleistocene fauna and flora from Great Billing, Northamptonshire, England: Opuscula Entomologica, v. 34, p. 109-129.

Morgan, A., 1970, Weichselian insect faunas of the English Midlands, unpublished Ph.D. Thesis, University of Birmingham, England, 188 p.

Morgan, A., 1972, The fossil occurrence of *Helophorus arcticus* Brown (Coleoptera: Hydrophilidae) in Pleistocene deposits of the Scarborough Bluffs, Ontario: Canadian Journal of Zoology, v. 50, p. 555-558.

Morgan, A., 1973, Late Pleistocene environmental changes indicated by fossil insect faunas of the English Midlands: Boreas, v. 2, p. 173-212.

Morgan, A. and Morgan, A.V., 1982, *Diacheila polita* Fald. (Coleoptera: Carabidae) and *Helophorus arcticus* Brown (Coleoptera: Hydrophilidae) in Late Wisconsinan deposits: Joint Meeting of the Entomological Societies of America, Canada and Ontario, Programme volume, p. 101.

Morgan, A., Morgan, A.V. and Elias, S.A., 1985, Holocene insects and paleoecology of the Au Sable River, Michigan: Ecology, v. 66, p. 1817-1828.

Morgan, A.V., 1973, The Pleistocene geology of the area north and west of Wolverhampton, Staffordshire, England: Royal Society of London, Philosophical Transactions, v. 265B, p. 233-297.

Morgan, A.V., 1976, An antler artifact from Hinksford, Straffordshire: Transactions of the South Staffordshire Archeological and Historical Society, v. 17, p. 85-87.

Morgan, A.V., 1987, Late Wisconsin and early Holocene paleoenvironments of east-central North America based on assemblages of fossil Coleoptera, *in* Ruddiman W.F. and Wright, H.E., Jr., eds., North America and its Adjacent Oceans During the Last Deglaciation: Geological Society of America, The Geology of North America, v. K-3, p. 353-370.

Morgan, A.V., 1988a, Late Pleistocene and Early Holocene Coleoptera in the lower Great Lakes Region, *in* Laub, R.S., Miller, N.G. and Steadman, D.W., eds., Late Pleistocene and Early Holocene Paleoecology and Archaeology of the Eastern Great Lakes Region: Buffalo Society of Natural Sciences, Bulletin, v. 33, p. 195-206.

Morgan, A.V., 1988b, *Polygraphus convexifrons* Wood (Coleoptera:Scolytidae); a range extension to Ungava Bay, Quebec, Canada: Coleopterists Bulletin, v. 42, p. 69-72.

Morgan, A.V., 1989, Late Pleistocene zoogeographic shifts and new collecting records for *Helophorus arcticus* Brown (Coleoptera:Hydrophilidae) in North America: Canadian Journal of Zoology, v. 67, p. 1171-1179.

Morgan, A.V. and Freitag, R., 1982, The occurrence of *Cicindela limbalis* Klug (Coleoptera: Cicindelidae) in a late-glacial site at Brampton, Ontario: Coleopterists Bulletin, v. 36, p. 105-108.

Morgan, A.V., Miller, R.F. and Morgan, A., 1982, Palaeoenvironmental reconstruction of southwestern Ontario between 10,000 and 11,000 yr B.P. using fossil insects as indicators: Third North American Paleontological Convention, Proceedings, v. 2, p. 381-386.

Morgan, A.V. and Morgan, A., 1979, The fossil Coleoptera of the Two Creeks Forest Bed, Wisconsin: Quaternary Research, v. 12, p. 226-240.

Morgan, A.V. and Morgan, A., 1980a, Beetle Bits — the science of paleoentomology: Geoscience Canada, v. 7, p. 22-29.

Morgan, A.V. and Morgan, A., 1980b, Faunal assemblages and distributional shifts of Coleoptera during the Late Pleistocene in Canada and the Northern United States: The Canadian Entomologist, v. 112, p. 1105-1128.

Morgan, A.V. and Morgan, A., 1986, A preliminary note on fossil insect faunas from central Illinois, *in* Follmer, L.R., McKenna, P. and King, J.E., eds., Quaternary Records of Central and Northern Illinois: American Quaternary Association, Ninth biennial meeting, University of Illinois, Urbana, Field Guide, p. 81-83.

Morgan, A.V., Morgan, A., Ashworth, A.C. and Matthews, J.V., Jr., 1983, Late Wisconsin fossil beetles in North America, *in* Wright, H.E., Jr., ed., Late Quaternary Environments of the United States: University of Minnesota Press, Minneapolis, p. 354-363.

Morgan, A.V., Morgan, A. and Carter, L.D., 1979, Paleoenvironmental interpretation of a fossil insect fauna from bluffs along the Lower Colville River, Alaska: United States Geological Survey, Circular 804-B, p. 41-44.

Morgan, A.V., Morgan, A. and Miller, R.F., 1984, Range extension and fossil occurrences of *Holoboreaphilus nordenskioeldi* (Mäklin) (Coleoptera: Staphylinidae) in North America: Canadian Journal of Zoology, v. 62, p. 463-467.

Morgan, A.V., Morgan, A., Nelson, R.E. and Pilny, J.J., 1986, Current status of knowledge on the past and present distribution of the genus *Blethisa* (Coleoptera: Carabidae) in North America: Coleopterists Bulletin, v. 40, p. 105-115.

Mott, R.J., Matthews J.V., Jr. and Anderson, T.W., 1981, Late-glacial paleoenvironments bordering the Champlain Sea based upon pollen and macrofossil evidence, *in* Mahaney, W.C., ed., Quaternary Paleoclimate: Geo Abstracts Ltd., Norwich, England, p. 129-171.

Mott, R.J., Grant, D.R., Stea, R. and Occhietti, S., 1986, Late-glacial climatic oscillation in Atlantic Canada equivalent to the Alleröd/Younger Dryas event: Nature, v. 123, p. 247-250.

Mumcuoglu, Y.K. and Zias, J., 1988, Head lice, *Pediculus capitis*, (Anoplura: Pediculidae) from hair combs excavated in Israel, and dated from the first century B.C. to the eighth century A.D.: Journal of Medical Entomology, v. 25, p. 545-547.

Nagano, C.D., Miller, S.E. and Morgan, A.V., 1982, Fossil tiger beetles (Coleoptera: Cicindelidae): Review and new Quaternary records: Psyche, v. 89, p. 339-346.

Nelson, R.E. and Carter, L.D., 1987, Paleoenvironmental analysis of insects and extralimital *Populus* from an early Holocene site on the Arctic Slope of Alaska, U.S.A.: Arctic and Alpine Research, v. 19, p. 230-241.

Osborne, P.J., 1965, The effect of forest clearance on the distribution of the British insect fauna: Proceedings of the XII International Congress of Entomology, London, p. 456-457.

Osborne, P.J., 1969, An insect fauna of Late Bronze Age date from Wilsford, Wiltshire: Journal of Animal Ecology, v. 38, p. 555-566.

Osborne, P.J., 1973, A late-glacial insect fauna from Lea Marston, Warwickshire: Proceedings of the Coventry and District Natural History and Science Society, v. 4, p. 209-213.

Osborne, P.J., 1988, A late Bronze age insect fauna from the river Avon, Warwickshire, England: its implications for the terrestrial and fluvial environment and climate: Journal of Archaeological Science, v. 15, p. 715-727.

Pengelly, W.C., 1870, On the submerged forest at Blackpool, near Dartmouth, South Devon: Geological Magazine, v. 7, no. 4, p. 164-167.

Pilny, J.J. and Morgan, A.V., 1987, Paleoentomology and paleoecology of a possible Sangamon age site near Innerkip, Ontario: Quaternary Research, v. 28, p. 157-174.

Schwert, D.P. and Ashworth, A.C., 1985, Fossil evidence for Late Holocene faunal stability in southern Minnesota (Coleoptera): The Coleopterists Bulletin, v. 39, p. 67-79.

Schwert, D.P. and Ashworth, A.C., 1988, Late Quaternary history of the northern beetle fauna of North America: a synthesis of fossil and distributional evidence: Memoir of the Entomological Society of Canada, v. 144, p. 93-107.

Schwert, D.P. and Morgan, A.V., 1980, Paleoenvironmental implications of a Late-glacial insect assemblage from Northwestern New York: Quaternary Research, v. 13, p. 93-110.

Scudder, S.H., 1895, The Coleoptera hitherto found fossil in Canada: Geological Survey of Canada, Contributions to Canadian Palaeontology, Ottawa, v. 2, pt. 1, p. 27-56.

Scudder, S.H., 1900, Additions to the coleopterous fauna of the interglacial clays of the Toronto district: Geological Survey of Canada, Contributions to Canadian Palaeontology, Ottawa, v. 2, pt. 11, p. 67-92.

Shane, L.C.K., 1987, Late-glacial vegetational and climatic history of the Allegheny Plateau and the till plains of Ohio and Indiana: Boreas, v. 16, p. 1-20.

Smetana, A., 1985, Revision of the subfamily Helophorinae of the Nearctic region (Coleoptera:Hydrophilidae): Memoir of the Entomological Society of Canada, 131.

Spilman, T.J., 1976, A new species of the fossil *Ptinus* from fossil wood rat nests in California and Arizona (Coleoptera: Ptinidae), with a postscript on the definition of a fossil: Coleopterists Bulletin, v. 30, p. 239-244.

Thorkelsson, T., 1935, A fossiliferous interglacial layer at Eillidaarvogur, Reykjavik: Greinar, v. 1, (Societas Scientarum Islandica), p. 78-91.

Warner, B.G., Morgan, A.V. and Karrow, P.F., 1988, A Wisconsinan interstadial Arctic flora and insect fauna from Clarksburg, southwestern Ontario, Canada: Palaeogeography, Palaeoclimatology, Palaeoecology, v. 68, p. 27-47.

Williams, N.E. and Morgan, A.V., 1977, Fossil caddisflies (Insecta:Trichoptera) from the Don Formation, Toronto, Ontario and their use in paleoecology: Canadian Journal of Zoology, v. 55, p. 519-527.

Williams, N.E., Westgate, J.A., Williams, D.D., Morgan, A. and Morgan, A.V., 1981, Invertebrate fossils (Insecta: Trichoptera, Diptera, Coleoptera) from the Pleistocene Scarborough Formation at Toronto, Ontario, and their paleoenvironmental significance: Quaternary Research, v. 16, p. 146-166.

Zeuner, F.E., 1934, Die Orthopteren aus der diluvialen nashornschicht von Starunia (Polnische Karpathen): Starunia, Krakow, Poland, v. 3, 17 p.

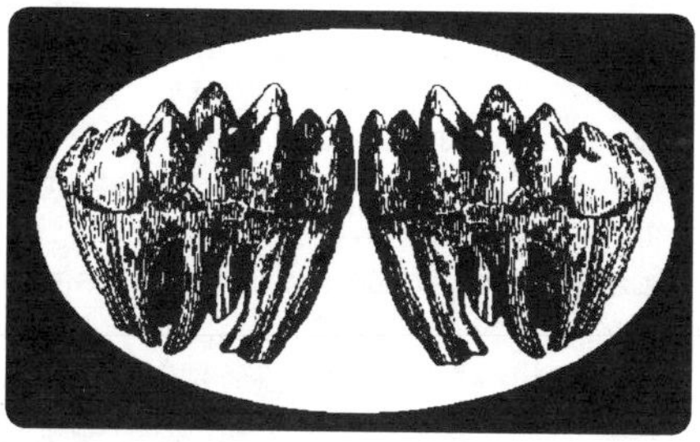

Vertebrates

C.S. Churcher
Department of Zoology
University of Toronto
Toronto, Ontario M5S 1A1

M.C. Wilson
Department of Geography
University of Lethbridge
Lethbridge, Alberta T1K 3M4

SUMMARY

The original fascination of scholars with Quaternary vertebrates was more related to their sometimes astonishing size and characteristics than to their value in paleoenvironmental reconstruction. The dominant place occupied by Quaternary vertebrates in the paleontological literature owes much to the high visibility of larger species as well as to our own affinity with the Mammalia. Nevertheless, it is clear that vertebrate fossils have enormous potential in the reconstruction of ancient climates and environments. This potential extends to the detection of unusual environmental parameters, such as extent and severity of seasonal stresses and even such a specific factor as depth of winter snow. Although many of the larger Quaternary mammals became extinct, modern analogues exist for a large number of fossil forms, allowing inferences of temperature, moisture, substrate, vegetative cover or presence of particular food plant species. Further, the study of small vertebrates, whether mammals or other species, allows paleontologists to make similar inferences based on animals that are relatively tied to one place and cannot migrate as do large ungulates or birds. Information obtained from these disparate sources can serve as a primary suite of proxy environmental data, as a cross-check on other proxy sources, or as an element to be used in complex transfer functions employing input from multiple proxy sources. As methods are refined, the potential for dating bones and teeth directly through radiocarbon, amino-acid racemization, electron spin resonance, uranium-series or other methods should ensure that vertebrate remains will also be of great value in establishing absolute time-sequences of local and regional environmental changes.

INTRODUCTION

Vertebrate paleontology has a relatively long pedigree as compared with other Quaternary paleoecological disciplines. Early interest in Quaternary vertebrates can, in part, be attributed to the large size and high visibility of many mammalian fossils, coupled with a fascination with "antediluvial giants" and other quasi-scientific pursuits. Association of artifacts with remains of extinct animals heightened interest before, and especially during, the early Darwinian era. Direct use of fossil vertebrates in reconstructing Quaternary paleoenvironments was well established by the late 19th century.

The subphylum Vertebrata comprises five major chordate groups, all amply represented by fossilized hard parts: Pisces (fishes), Amphibia (amphibians), Reptilia (reptiles), Aves (birds) and Mammalia (mammals). In terms of the modern world fauna, fishes are by far the most diverse and numerous group, but are mainly marine and thus of lesser importance in the Quaternary non-marine record. Early taxonomists treated Pisces as a single class, but currently the fishes are placed in three or more classes (*e.g.*, Agnatha, Chondrichthyes and Osteichthyes), while retaining as classes the other vertebrate groups listed above. Mammalian remains, because of their abundance, ubiquity and ease of preservation, dominate the literature on vertebrate paleontology. Mammoths and other "Quaternary giants" are awe-inspiring and have attracted the disproportionate attention of public and scientists alike. On the other hand, microvertebrate fossils are, in many cases, more informative about past environments than are the larger forms; these include the remains of such diverse groups as Rodentia (rodents), Chiroptera (bats), Insectivora (shrews and moles) and a host of small birds, amphibians and fishes.

Vertebrates have enormous potential in reconstructions of paleoclimates, substrate characteristics, vegetation cover, food chains, evolutionary changes, biostratigraphy and general paleoenvironments of a given time period. Vertebrate species display extraordinary size variation: mammals range in length from about 55 mm for the smallest shrew *(Microsorex hoyi)* to about 30 m for the blue whale *(Sibbaldus musculus)*. The animals' size and mobility are important considerations in paleoenvironmental reconstruction, so that interpretations can range from the local environmental level on the basis of microvertebrate fossils to broad regional interpretations based on the presence of large, mobile animals such as bovids (cattle, bison and relatives) or proboscideans (elephants and mammoths).

Vertebrate fossils are usually directly preserved hard parts, such as bones, teeth and occasionally muscle, hair and hornsheaths (Figure 1). Indirect evidence includes trace fossils or ichnites (footprints, skin casts or burrows), nests such as middens of woodrats *(Neotoma)* or dens of hyenas *(Crocuta* or *Hyaena)*, tooth marks or distinctive bone breakage patterns left by predators and scavengers, coprolites or droppings and owl pellets that may contain bones and hair of other vertebrates. Mummified remains of animals may be encountered in dry caves and other arid environments; examples are the North American camel *(Camelops hesternus)* or the Australian marsupial wolf *(Thylacinus cynocephalus)*. Frozen carcasses have been found in permafrost; they include the woolly mammoth *(Mammuthus primigenius)* and woolly rhinoceros *(Coelodonta antiquitatis)* from Siberia, and long-horned bison *(Bison priscus)* from Fairbanks, Alaska from frozen "muck" or organic silt overlying gold-bearing gravel (Guthrie, 1990; Harington, 1990). These provide more complete evidence of morphology, colour, preserved external and internal soft parts, and stomach contents. Similar soft-part preservation occurs in woolly rhinoceros remains from salt-, tannin- and petroleum-impregnated deposits in Poland (Kurtén, 1968) and in human remains from northern European bogs (Glob, 1969). The detection and specific identification of blood residue on stone tools (Loy, 1983) has opened up a new and important area of study.

This paper presents a general introduction to Quaternary vertebrate paleontology, particularly in the context of research undertaken in North America. The nature and potential sources of vertebrate remains, and field sampling techniques are discussed. Evolutionary and zoogeographic patterns can be deduced through analysis of fossil bones, and vertebrate remains play an important role in biostratigraphy. For further details on Quaternary vertebrates in North America see books or papers by Auffenberg and Milstead (1965), Graham (1986), Graham and Semken (1987), Harington (1984a,b), Hibbard (1958, 1970), Hibbard *et al.* (1965), Kurtén and Anderson (1980), Lundelius *et al.* (1983, 1987) and Semken (1983).

HISTORICAL DEVELOPMENT

Bones of large fossil mammals were viewed by Teutonic peoples as the remains of trolls from the underworld, by the Chinese as remains of devils or dragons and by the ancient Greeks as those of the Cyclops. Later, Christian mythology suggested that they were the remains of victims of Noah's Flood who failed to get into the ark. For example, a fossil skeleton of the large salamander *Andrias scheutzeri* was named *Homo diluvii testis*, literally "Man, witness of the flood." This interpretation, while fanciful, was an attempt to integrate fossil evidence with an hypothesized earlier climatic-environmental event, a "worldwide" flood that destroyed and altered the landscape and its biotas.

Detailed studies of vertebrate paleontology began with Baron Georges Cuvier and Alexandre Brongniart in France. They recognized a succession of vertebrate faunas, but saw them as separated by pervasive catastrophes and extinctions. "Catastrophism" of a sort is seen in current concepts of the importance of asteroidal impacts in causing mass extinctions. Specialized studies of vertebrate groups followed close on the heels of Cuvier's pioneering work. Smith (1817) was one of the first to recognize that there was a stratigraphic framework in which fossils acted as indicators of similar strata and thus relative ages and vertical relationships. The study of fishes, begun by Cuvier (1812), was continued by Agassiz (1833-1844), still a basic reference to the field. Reptiles and Quaternary mammalian fossils, especially "vanished giants" *sensu lato*, were studied in Great Britain by Buckland (1824, 1836), Mantell (1825), Owen (1846, 1849-1884) and Dawkins (1878), whose work was continued by Reynolds (1939).

North American Quaternary vertebrate explorations began in the late 18th century near Big Bone Lick, Kentucky, the first fossiliferous deposit with a recognized complex fauna in North America. The description of animals such as the extinct ground sloth by Thomas Jefferson (1799), mastodon *(Mammut americanum)* and giant bison *(Bison latifrons)* by Peale (1803a,b) attracted great public interest. Jefferson mistakenly believed that the ground sloth *Megalonyx* was a giant carnivore, on the evidence of its isolated huge claws. Some of the material from Big Bone Lick was transported to the fledgling Natural History Museum in London, England, ultimately to be studied by Owen or Dawkins. Rembrandt Peale, who was both an artist and gifted amateur naturalist, excavated the first complete skeleton of a mastodon at Newburgh, New York, with his brothers Charles and Titian, in 1801. Peale did a large oil painting of the excavation showing the bones, the lifting mechanism for removing the soil, and himself holding forth about the discovery.

During these early years, the reconstruction of paleoclimates and paleoecology was uncertain and rudimentary, but, in Canada, for example, J.W. Dawson (1878, 1883) recognized that the Gulf of St. Lawrence and the Ottawa Valley had been inundated by the Champlain Sea. Discoveries of vertebrate fossils (Harington and Occhietti, 1988), including fishes (Dawson, 1891), seals (Dawson, 1878; Leidy, 1856; Lyell, 1845) and whales (Billings, 1870; Dawson, 1883), played a role in reconstructing the Champlain Sea's environment. Even then, misidentification of one seal until the 1970s prevented geologists and biologists from realizing how cold this sea was (Harington, 1984b). Agassiz (1850) brought to North America an expertise in glacial sequences and land forms, learned in the Swiss Alps, as well as his experience with fossil fishes, and recognized a sequence of proglacial lakes in central North America partly on the basis of the fishes in the lake silts. By the late 19th Century, it was well understood that climatic change had been dramatic during the Pleistocene (Dawson, 1893; Wright, 1890).

During the late 19th and early 20th centuries, A.P. Coleman worked assiduously in southern Ontario and collected many vertebrate fossils, especially from the Toronto area (Coleman, 1901), with the specific goal of reconstructing ancient climates. Adding excursions to the

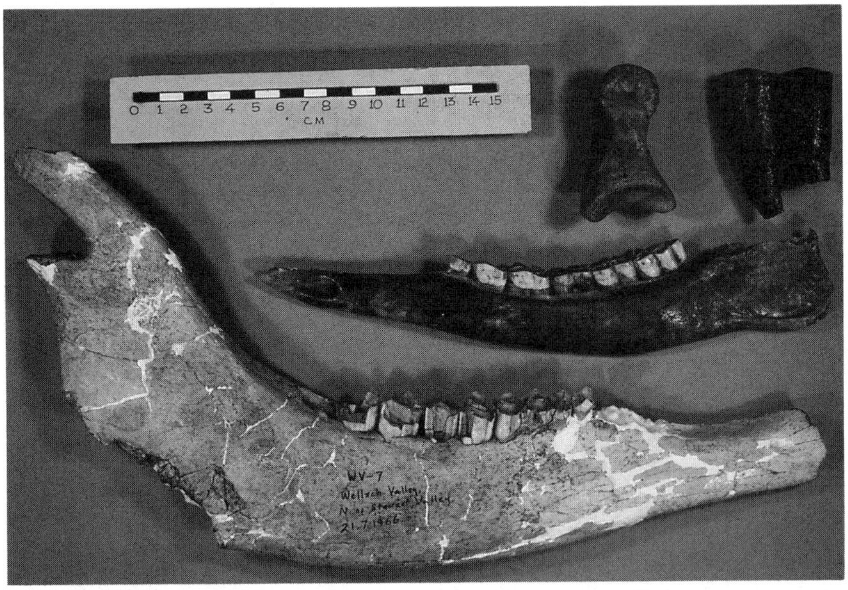

Figure 1 *Quaternary fossils from the South Saskatchewan River Valley, Alberta and Saskatchewan.* ***Top left:*** *extinct camel* (Camelops hesternus) *lower molar, Kansan age, Island Bluff, Medicine Hat;* ***top right:*** *giant sloth* (Nothrotheriops shastense) *toe bone (proximal phalanx), Kansan age, Low Bluff, Medicine Hat;* ***centre:*** *caribou* (Rangifer tarandus) *dentary, Sangamonian or Middle Wisconsinan age, Mitchell Bluff, Medicine Hat;* ***bottom:*** *camel* (Camelops minidokae) *dentary, Jaw Face Section, Wellsch Valley site, Stewart Valley.*

Prairies and Quebec, he formulated a sequence of glacial events that paralleled the sequence being described for the European Pleistocene. He showed that climatic change had involved cyclic glacial and interglacial events, with anywhere from two to five cycles represented. Many attempts have since been made to correlate vertebrate faunas with the glacial/interglacial sequences in both the Old and New Worlds (see Flint, 1971). The greatest problem with this biostratigraphic approach is that definition of faunal stages depends on the presumption of a well-established and dated geological chronostratigraphic equivalent. The utility of the concept of a "Kansan" or a "Nebraskan" fauna, for example, can readily be questioned as the geological stratigraphic names in their type areas have been abandoned (Richmond and Fullerton, 1986).

Early Quaternarists were primarily interested in the stratigraphic relationships of depositional units, and in fossils as indicators of age (and occasionally of environment). Later workers became more interested in the morphological succession of specific lineages of fossils, *e.g.*, horses (Scott, 1913; Simpson, 1951) or mammoths (Osborn, 1936-1942), and thus in the evolutionary sequences, phylogenies and dispersal histories of individual groups. Only later were biotas regularly seen as assemblages from which meaningful deductions about paleoenvironments could be made; this view depended upon uniformitarian thinking (Hutton, 1795) based in analogies from modern examples. In recent decades, bone *assemblages* have become the preferred time indicators, displacing individual index fossils from their previously pre-eminent perch as the key to the ages of strata.

Taphonomic Considerations.
The compound origin of a single death assemblage (thanatocoenosis) may not be obvious. Attempts to reconstruct communities are perturbed by differential transportation of skeletal elements, admixture of taxa from different communities, or by mixing through redeposition of fossils from deposits of different ages (Behrensmeyer, 1984; Efremov, 1940). Terrestrial vertebrate remains found in marine deposits illustrate processes of long-distance transport and reworking. Inclusion of an ungulate in an assemblage dominated by fish and other aquatic fossil forms, as well as rodents, might at first elicit less suspicion. The variety of possibilities available for such occurrences has given rise to the concept of "proximal" and "distal" taxa in mixed assemblages, the former representing animals resident in the area of deposition and the latter those that resided elsewhere and were included because of postmortem transportation processes (Shotwell, 1955). These are clearly extremes in a continuum, because the distance of transport must influence the quality of preservation.

The elements of a corpse may not remain closely associated and are unlikely to do so if it remains exposed on the surface for long. Scavengers may remove elements or consume some and damage others, in the process scattering elements over a wide area (Sutcliffe, 1970). Weathering (differential heating/cooling, freeze/thaw, chemical attack, eolian abrasion) may alter a bone's surface and destroy minor features; most important, it can split sutures and crack shafts of longbones. The extent of weathering can be used, in retrospect, to judge the length of exposure time (Behrensmeyer, 1978). Flowing water, whether in sheets on featureless surfaces or as streams or rivers, collects and sorts unburied skeletal elements for

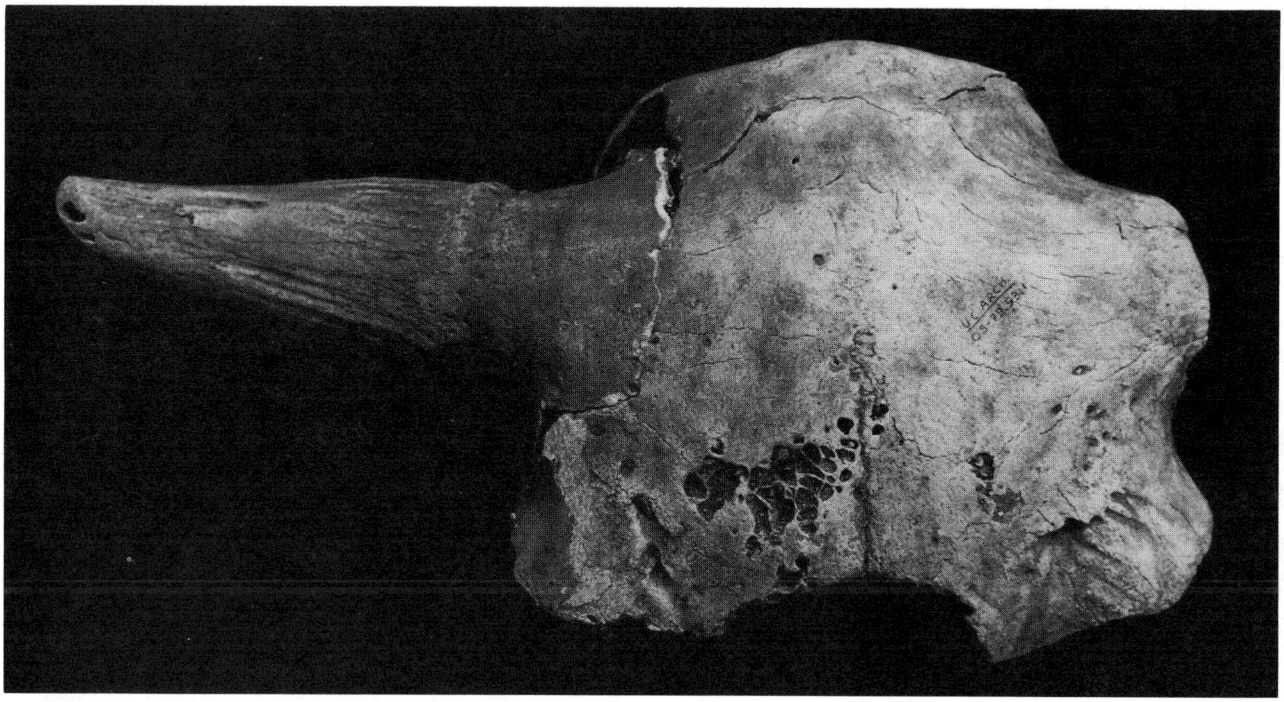

Figure 2 *Fossil bison skullroof (frontlet) with right horncore, old adult individual, Bighill Creek Formation, Bow River Valley, Alberta* (ca. *11,000 years B.P.). Fluvial transport has abraded the end of the right horncore, broken off the left horncore, and thinned bone on the frontal surface to expose sinuses. Comparative measurements of such waterworn specimens are untrustworthy and statistical treatment is unreliable based on specimens from such environments.*

deposition. Sorting occurs in the same way that other sedimentary particles are segregated; bones in such an environment can be viewed as nothing more than unusually shaped grains of sediment (Figure 2). Sheets of water from a sudden intense thunderstorm can carry large bones and even small boulders across a gently sloping surface and leave them in local hollows as collections of isolated elements, without any articular relations and representing a mixed thanatocoenosis. Once buried, these accumulations occur in depressions on the ancient surface.

It is not possible in this space to review the entire field of vertebrate taphonomy. Readers are referred to other reviews (Shipman, 1981; Behrensmeyer, 1984; M.V.H. Wilson, 1988) for more information. As an example, we will consider the effects of fluvial transport (Figures 3 and 4). Rivers receive carcasses of drowned animals or can entrain carcasses from the floodplain during floods or through lateral cutting on meanders. These remains can be transported far downstream, through floating of bloated carcasses or through more complex processes (M.V.H. Wilson, 1988). The skeleton will become mechanically disarticulated by current action unless, through desiccation, the skeleton has become "armoured" within its hide. Small bones similar in width and length behave as small pebbles and roll along the river bottom, bouncing in the current and being carried great distances. Large bones, in which there are fat-filled cavities, have a density significantly less than that of water, and tend to roll about their long axes, which become oriented at right angles to the current. As long as the bottom is even and the current moderately strong, such bones will move downstream. Bones or elements that have large air-filled cavities, such as sinuses in the skull, or a cavity that may be filled with either fat or gases from decomposition, may float on the surface and may therefore be carried farther than those that move along the river bottom. Portions of the structure that hang down from the centre of buoyancy, such as horns, antlers or facial bones, will drag or hit against the bottom or any projecting rocks, and may be broken away (Figure 2). Thus crania are often separated from facial regions and the ornamentations are damaged or absent, especially in the ungulates. Irregular structures, such as pelvic girdles, broken skulls with antlers or horns, or mandibles, have little flotation ability because they lack large cavities. Some bones are too heavy and large for a stream to roll or move by force, and their projections lodge in the bottom sediments. However, human skulls tend to behave like air-filled balls and float for long distances (Boaz and Behrensmeyer, 1976). Elements that lodge on the bottom are transported the least distance and are frequently buried by bottom sediments almost as soon as they are disarticulated from the carcass. Because bone is less dense than the minerals of the associated sediment grains, sorted bones will generally be larger than the sedimentary matrix in which they are buried. Given knowledge of the densities of bone and sediment grains, hydraulic equivalencies can be computed and associations broadly predicted. A seeming exception to the rule of relative size is the occurrence of tiny bones in pebble or cobble gravels, but this in fact parallels the processes producing bimodality in conglomerates.

The sorting and redistribution of skeletal elements downstream in a river bed is a study in itself (*e.g.*, Boaz and Behrensmeyer, 1976; Frison and Todd, 1986; Shipman, 1981). Small, dense pieces (such as isolated teeth or fragments of broken longbone shaft) and large, awkward elements (such as antlers or pelves) move only short distances before they are deposited. Teeth and small bone fragments may, under the right circumstances, travel farther, in which case they can become so abraded as to be indistinguishable in shape from the associated pebbles. Small, but light, bones such as carpals and tarsals move well and can become similarly abraded, but most often are trapped in sand or fine gravel. Longbones with resistant shafts and extensive lumina or fat-filled cavities roll along the bottom, while skulls with sinuses or braincases filled with gases from decomposition float easily. Thus, it is almost inconceivable that many elements of a skeleton will remain close together after significant transport. Most elements that have been transported will show signs of

Figure 3 *Pleistocene horse* (Equus *sp.) rib in bedding plane or paleosurface of a river bar. Note that the rib is associated with a change in depositional texture marking the surface of a paleo-bar form. Sangamonian or Mid-Wisconsinan deposits at Mitchell Bluff, South Saskatchewan River Valley, Medicine Hat, Alberta.*

Figure 4 *Pleistocene horse* (Equus *sp.) toe bone (proximal phalanx) in sand deposits. Note that the bone is larger than the surrounding matrix grains demonstrating its lesser density. Sangamonian or Mid-Wisconsinan deposits at Mitchell Bluff, South Saskatchewan River Valley, Medicine Hat, Alberta.*

abrasion or breakage from contact with the substrate (Figure 5). After they become lodged in the bottom, exposed areas are still subject to damage from contact with other transported grains, ranging from fractures by cobble impact to "sand-blasting" by finer grains. The ends, or epiphyses, of longbones are formed of softer cancellous bone than the shafts, or diaphyses, which are constructed of dense compact bone; after extensive transport, this differential resistance to erosion means that a bone will have lost its significant articular characters. Even lightly abraded bones will give lower than expected articular measurements, while some will be reduced to cylinders lacking articular ends. Such abraded cylinders are also produced by scavengers chewing on the soft articular ends and "rasping" out the softer internal structure with their roughened tongues (Binford, 1981) (Figure 6), so varied taphonomic pathways can lead to similar ends. Microscopic analysis of surface characteristics may be necessary to resolve such ambiguities.

Figure 5 *Extinct horse* (Equus ?conversidens) *metapodial or cannon bone in cobble gravels. Note ends worn to show cancellous inner structure and broken shaft from loading by gravels. Kansan age, Low Bluff, South Saskatchewan River Valley, Medicine Hat, Alberta.*

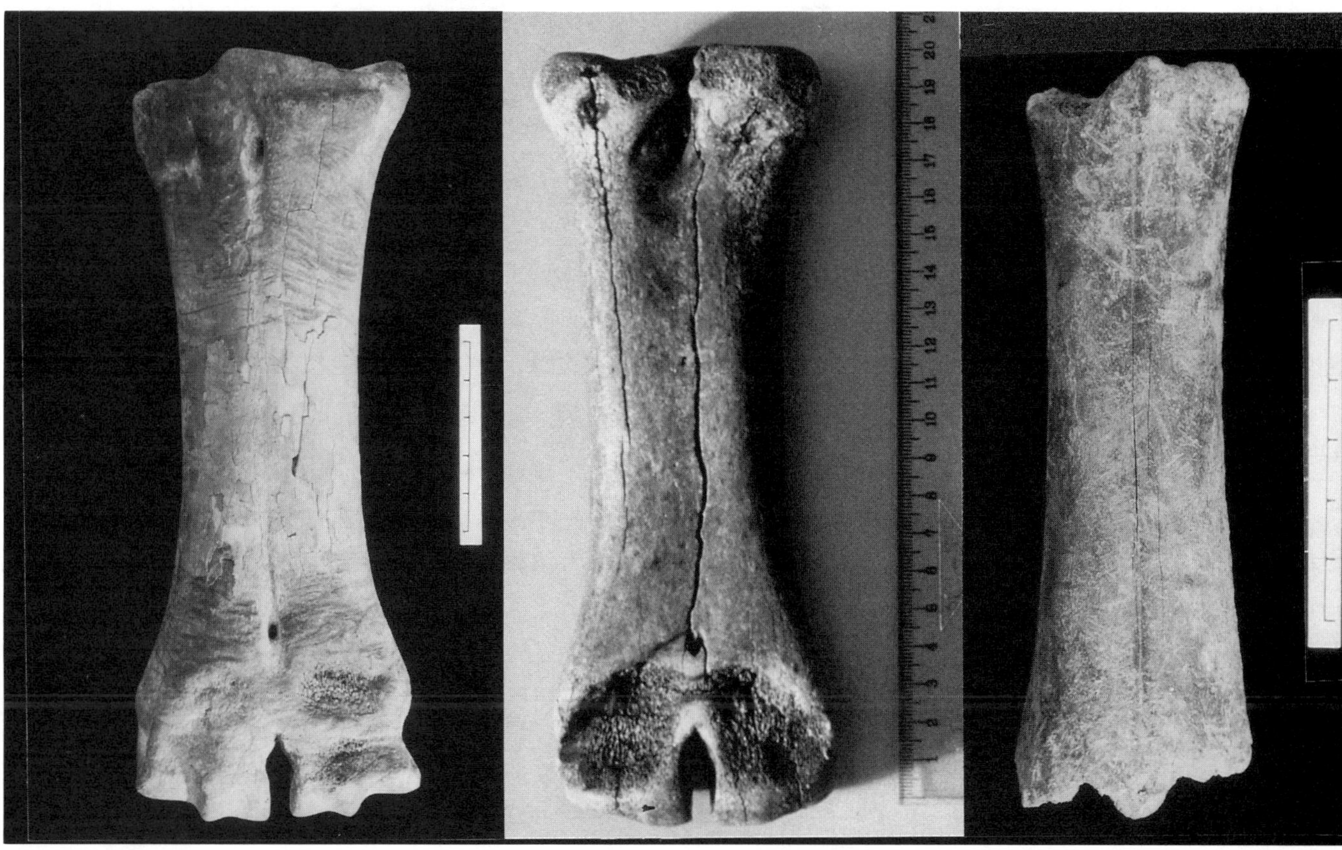

Figure 6 *Fossil bison metacarpals (or front limb cannon bones) showing action of scavenging canids (dogs or wolves).* ***Left****: anterior surface of mature metacarpal showing extensive scoring by teeth;* ***centre****: (posterior surface), exposure of cancellous bone at both ends of bone;* ***right****: anterior surface of immature metacarpal reduced to cylinder by removal of articular ends' cancellous tissue. Left and right, Hitching Post Ranch site, Alberta, about 3500 years old. Centre, Late Pleistocene gravels, Peace River area, Alberta. Note differing scales.*

Figure 7 *Old Crow River Locality 11A, Yukon Territory. Pointbar with bones and wood derived from reworking of cutbanks. Specimens have been collected from surface of sands and fine gravels composing the bar.*

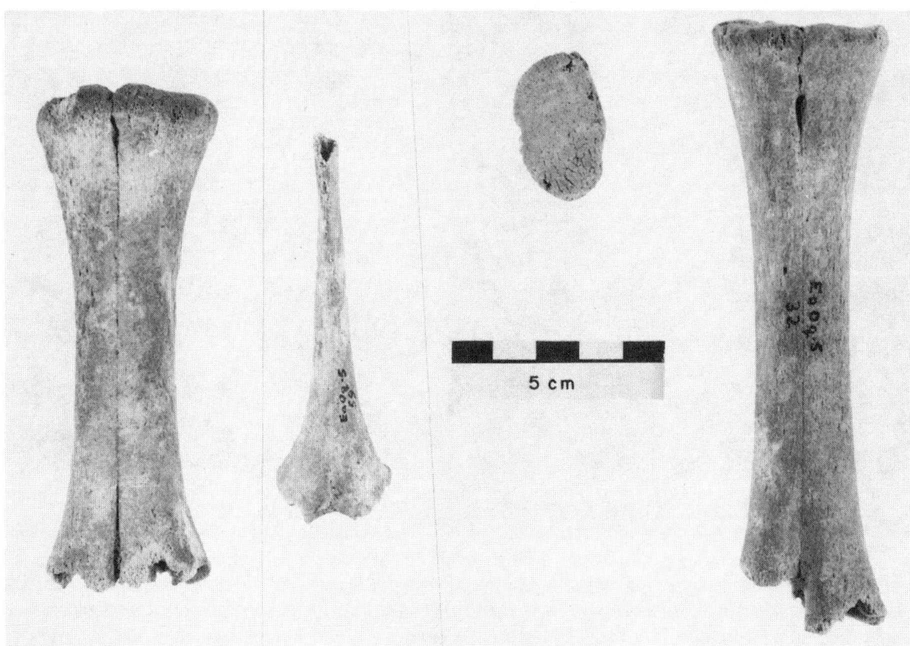

Figure 8 *Fossil foetal or neonatal bison bones, Reservoir Gully site, Medicine Hat, Alberta* (ca. *1000 years B.P.).* **Left to right:** *metacarpal, vertebral spine, tarsal (ectocuneiform) and metatarsal (cannon bone of hind limb). These are from primary deposition in an archaeological site; such fragile elements would likely be destroyed in fluvial transport.*

Bones are deposited whenever the current slacks, reducing transporting ability. Slower currents occur near the bottom and sides of a river due to frictional drag. Thus, a bone that is travelling over a smooth surface in mid-stream will move farther and faster than one near the stream's margin. The current is also slack or weak on the inside of a meander curve, so that bones and other sediment tend to be deposited on point bar surfaces (Figure 7). Fossil bones in the Old Crow Basin represent complex depositional histories (Morlan, 1980). Today, the Old Crow River is reworking its old sediments through lateral channel migration and, because its exit from the basin is still lowering, is winnowing the bones from the basin fill, transporting them along its bottom and, when in flood, carrying them up and onto point bars on major river bends, where they are available for recovery by scientists.

Taphonomic factors can influence the identification of paleospecies. In a group with features (*e.g.*, horn-cores) that continue to change with age, it becomes necessary to compare samples of similar population composition. In bison, for example, a gravel pit "population", reflecting alluvial transport of animals accumulated through attritional mortality, would ideally, and because of the nature of the attritional mortality profile, be rich in juvenile and very old individuals (Reher, 1970; Voorhies, 1969). Because juvenile skulls are fragile and more susceptible to damage in fluvial transport, the "population" will be dominated by old individuals (Figure 2). A population of comparable antiquity from a drive-kill site, such as a buffalo jump or trap, would be rich in animals from the intervening age groups, would be dominated by females, and would contain numerous immature calves and even foetal bones (Figure 8). Adult males from the bison kill would typically exhibit more slender horn cores than would males from the alluvial gravels, because the average individual age of the former would be significantly younger (Wilson, 1978). This is one of several factors that led to oversplitting of fossil bison species during the past 100 years. Reliable comparisons can only be made of faunas or samples that have similar taphonomic histories or that have been corrected to compensate for known differences.

Detailed study of the depositional environment and reconstruction of the taphonomic history of a bone occurrence are necessary prerequisites to paleoenvironmental inferences. As noted earlier, some occurrences may be obviously disharmonious, but others may be more intriguing than suspicious. A mountain goat *(Oreamnos americanus)* from a prairie point bar deposit might elicit discussion of changing ecological preferences if a researcher does not take time to consider the implications of its depositional context.

Fossil Material.

Fossils have traditionally been considered to be the hard parts remaining from an animal after death and decomposition of the soft parts, but there are other possibilities. Hornsheaths, eggshells or feathers may be preserved or traces may remain that also attest to the existence of an organism (*e.g.*, burrows, footprints, nests or tooth marks). Such shed elements as antlers and deciduous teeth resemble trace fossils in not requiring the death of the organism to enter the geological record. In general, robust elements preserve better than others and are less distorted or broken. Large elements are differentially preserved because of their strength and because, even when fragmented, their parts may be preserved as well as or better than smaller and lighter whole elements. Dense cortical bone preserves better than lighter structures, and cancellous articular bone preserves less well and with more distortion than heavier dense shaft bone. Thus, fossil vertebrate evidence is biased toward the remains of large robust animals or elements in any environment where there has been a reasonably high level of transporting energy (for example, a beach or river bed), while small, lightly built animals are best known from areas of quiet deposition (for example, backswamps, lakes or swales). Delicate structures may also be preserved where an animal has died in primary position, as in the case of a rodent or frog hibernating in its burrow. The bias toward large animals is compounded by their greater visibility in the field and such factors as fluvial sorting (Dodson, 1973), but such a bias can be overcome through planned use of special recovery techniques (see below).

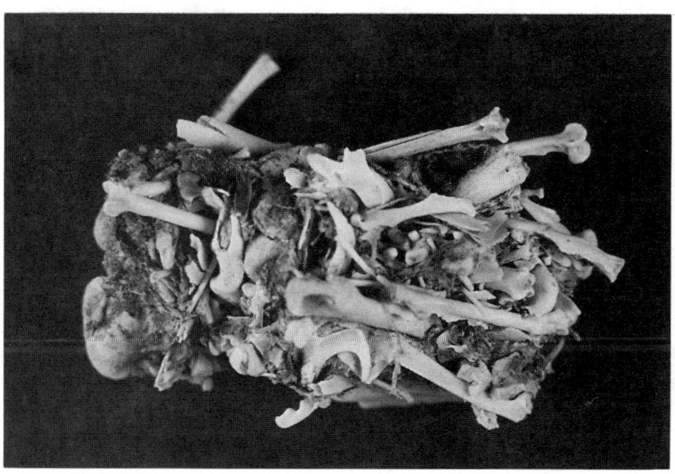

Figure 9 *Owl pellet containing rodent and bird bones, bound together with matted hair. Note unbroken longbones, vertebrae, and parts of skull. Length of pellet is 4.5 cm. (Photograph by M.B. Smith).*

Figure 10 *Waterworn partial mammoth* (Mammuthus primigenius) *tooth. Note cobble-like form with transverse grinding ridges worn smooth within general outline and broken right margin. Late Pleistocene (probably Late Wisconsinan) gravels at West Peace River, Alberta. This specimen is in the Peace River Centennial Museum.*

PRESERVED ELEMENTS FOR MAJOR VERTEBRATE GROUPS

Fishes.

Typical material can include complete skeletons, whether articulated or disturbed but having all parts associated, as flattened whole animals on bedding planes of fine-grained sediments. Good examples are the well-preserved skeletons in calcareous nodules from Champlain Sea deposits at Green Creek near Ottawa (McAllister *et al.*, 1981). Burrows or nests, as for aestivating lungfish *(Ceratodus)*, may be preserved. A desiccating pond or mass kills associated with volcanic ashfalls or anoxic conditions may give rise to the preservation of remains of many individuals. Isolated bones or clusters of bones such as heavy spines, vertebrae or disarticulated head or gill arch elements, scales and teeth (*e.g.*, shark, pike, sturgeon) may accumulate in small current eddies, either in freshwater streams or lakes, or beaches, or in backwaters and swales.

Amphibians.

Remains of amphibians may include complete or disarticulated skeletons of adults or larvae, often flattened on bedding planes. Isolated bones, usually limb-bones, vertebrae or disarticulated skull elements, including tooth-bearing elements, may be encountered. Fine-grained sediments may preserve footprints. Sediments of desiccating ponds or winter hibernacula, the latter ranging from burrows in soft sediment to cavities in coarse rubble, provide another opportunity for the preservation of amphibian remains (Bayrock, 1964).

Reptiles.

Material includes bones and teeth, occasionally skulls with many elements associated, and (rarely) complete skeletons or major portions of a skeleton. Isolated bones usually include disarticulated skull elements, longbones or vertebrae, scutes or carapace fragments (*e.g.*, turtles), and isolated shed teeth in groups that replace teeth (*e.g.*, crocodilians). Egg-shell fragments, gizzard stones (gastroliths), burrows or nests, tooth marks, or footprints record past reptilian activity. Quaternary reptilian remains are often small isolated elements because modern reptiles tend to be small, except for Crocodilia and the largest Chelonia, Lacertilia and Squamata. Like the remains of amphibians, reptilian remains may occur in aggregations deposited in eddies of small streams and in assemblages of articulated or partially articulated elements from winter hibernacula, usually cavities in coarse rubble or fractured bedrock.

Birds.

Material includes isolated bones and, rarely, whole skulls, limbs or skeletons (*e.g.*, Steadman and Miller, 1987; Harington, 1990), as their light bones are easily crushed or fragmented. Egg shell fragments and occasionally whole eggs, and rarely nests or burrows, may be found in fossil sequences. Regurgitated stomach debris (owl-pellets)(Figure 9), sometimes in massive form if beneath a roost, or debris beneath a raptor's nest, reflect diet of the roosting predator. Further analyses of bones contained in the regurgitated debris are informative as to microfauna in the area, and may record the presence of smaller birds. Feathers and their impressions or footprints, generally of wading forms, may also be preserved.

Figure 11 *Modern woodrat (=packrat) midden, Three Forks area, Montana. The midden is composed mainly of juniper (*Juniperus *sp.) and other shrub twigs and small branches, with mammal, bird and reptile bones, and pieces of paper and plastic.*

Figure 12 *South Saskatchewan River, north of Medicine Hat, Alberta, showing slumping exposing tills, sands, and gravels to left (south) at Mitchell Bluff and pointbar formation to right (north) downstream from Island Bluff in distance.*

Mammals.

Material includes either whole or fragmentary bones, such as skulls, limb girdles, partial vertebral columns and limbs, depending upon the energy of the depositional environment, and may be well water-rounded in high energy deposits. Teeth often preserve better than other bones (Figure 10), but can be fragmented due to cracking and splitting under alternating temperature or water stresses and drying. Whole animals or complete skeletons are rare unless they are aquatic forms and were deposited in water. They may be entombed in permafrost terrain as in the case of the Beresovka and other Siberian mammoths *(Mammuthus primigenius)* (Heintz and Garutt, 1965; Sokolov *et al.*, 1986), or woolly rhinoceros *(Coelodonta antiquitatis)* (Vereschagin and Baryshnikov, 1962), or buried in burrows (*e.g.*, hyena, *Crocuta* sp. (Sutcliffe, 1970)). Mummified remains occur in arid environments: *e.g.*, the camel *Camelops hesternus* in the American Southwest (Martin and Guilday, 1967; Romer, 1929), the marsupial wolf *Thylacinus cynocephalus* in Western Australia (Merrilees, 1970), and the aodad or Barbary sheep *Ammotragus lervia* in Egypt (Osborn and Helmy, 1980). Partially articulated or associated skeletons may be found in archaeological sites, particularly at kill sites. Partial skeletons may exist in pellets of raptorial birds or in burrows, nests or middens. The nests themselves may be rich in plant remains that are valuable in paleoenvironmental reconstruction, as in Arctic ground squirrel nests (Harington, 1984a) or woodrat middens in arid areas (Figure 11) (van Devender, 1987; Wells, 1976). Tooth marks caused by carnivore chewing (Haynes, 1980) or rodent gnawing, as well as footprints, and feces as coprolites provide additional evidence of past mammal activity. Under the right circumstances (minimal transportation and rapid burial, as, for example, at an archaeological kill site) even the bones of unborn (foetal) animals may be preserved (Wilson, 1974)(Figure 8).

Figure 13 *Extinct bison* (Bison latifrons) *skull in stratified sands and minor gravels of Mitchell Bluff, South Saskatchewan River, Medicine Hat, Alberta, Sangamonian or Middle Wisconsinan age. Back of skull is to left, with occipital condyle and auditory bulla visible. Top of skull is sheared through the sagittal plane. Nasals, teeth, maxillae, and premaxillae were removed before deposition and right side of skull was removed with slumping of slope.*

Figure 14 *Extinct camel* (Camelops hesternus) *cannon bone or metacarpal in sandy silt at Lindoe Bluff, South Saskatchewan River, Medicine Hat, Alberta. Possibly Illinoian age. Note horizontal attitude and arcuate rotational breaks from slumping matrix.*

COLLECTING AND LABORATORY TECHNIQUES

Locating Fossils.

Ground that has been exposed by gravel-pit operations, road making or other construction activities may yield fossils that have suffered little weathering after being exposed to wind, rain, sun and temperature changes. Other erosional environments lacking trees or lower vegetation, such as river cuts, arid or desert environments, beaches and cliffs, may also yield fossils (Figure 12). However, these exposures, especially those in deserts and savannahs, usually have a rigorous climate and rapid weathering of the fossils may take place. Fossils may be found *in situ* in vertical exposures where they protrude from the matrix (Figure 13), may be exposed on the erosional surfaces of beaches or river benches, or may be exposed and often weathered and fragmented by the actions of the weather and of tree or shrub roots when there is intermittent vegetation and a deflated surface.

The best matrices in which to locate fossils are alluvial sands, gravels, silts or clays, or eolian sand or loess, where rapid burial occurs before severe weathering or breakage and dispersal (Figures 14 and 15) However, the major drawback to most of these matrices is the high probability of disarticulation and dispersal by scavengers or currents. Likely exposures may be located from a distance through fieldglasses. Amateurs, most often people working close to the ground (*e.g.*, farmers, gravel pit operators, construction workers), may inform scientists of the presence of fossils in places known to them. Examination of the exposure must be on site, by foot or on hands and knees, and in a thorough and methodical manner. Prospecting potential sites is best done when the sun is not too high, or the temperature too hot, as the recognition of small fossils is less easy when conditions are trying. Even the location of large fossils under such conditions is less effective, and a return to the same site when conditions are better usually will reveal specimens that have been overlooked.

Vertebrate fossils are often found in gravel pits in Canada, especially those in the Prairies or southern Ontario. Marine mammal remains are abundant in Quaternary deposits on raised beaches in Arctic Canada (Harington, 1990). Glacial outwash gravels are usually barren of fossils, and so those of the Yukon, northern Prairies or Ontario and Quebec seldom produce vertebrates. Most fossils found in gravel pits are water-worn or crushed or both, due to the ball-mill effect of rolling cobbles and their weight in compaction. Fossils found in bogs or silts or sands may be better preserved, but may be less obvious as the matrix obscures them and is not always clean. Sands or minor gravels provide an ideal preservational environment in that they are not deposited by a high-energy flow, do not contain the larger cobbles that may damage the bones, are not usually acidic so as to denature the bones or teeth, and are easy to excavate either by hand or mechanically. Sand lenses in gravel pits often contain the best pre-

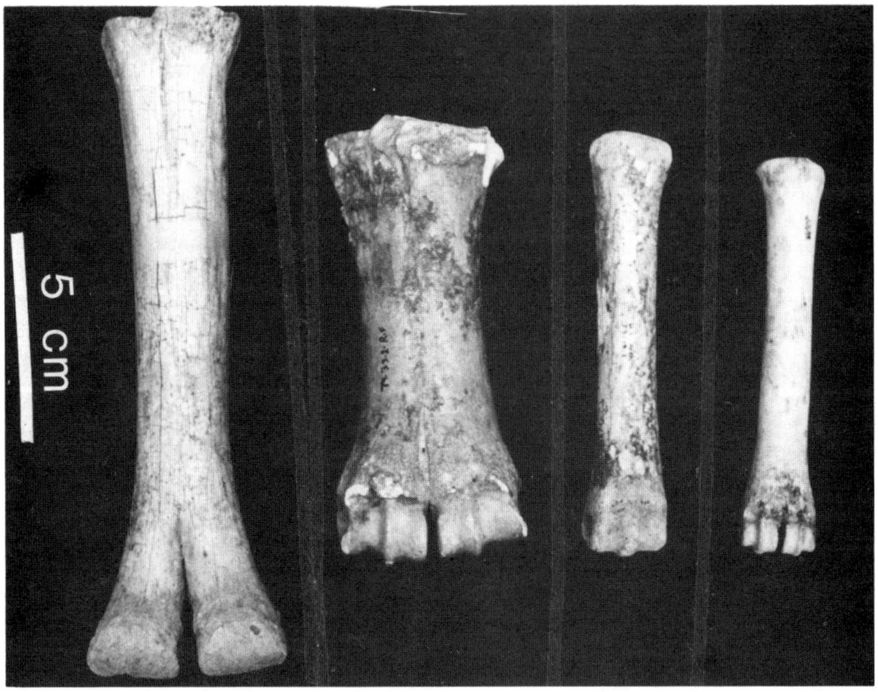

Figure 15 *Large vertebrate bones from a Late Quaternary assemblage. The specimens are all metacarpals (or front limb cannon bones) from extinct large hoofed mammals, collected from the Bighill Creek Formation, Bow River Valley, Alberta, (*ca. *11,000 years B.P.). From left to right: camel* (Camelops hesternus)*, bison* (Bison b. antiquus)*, small horse* (Equus conversidens) *and large mountain sheep* (Ovis canadensis catclawensis).

Figure 16 *Excavation at Mitchell Bluff, South Saskatchewan River Valley, Medicine Hat, Alberta. Excavated face is topped by till and stratified sands and minor gravels below are of Mid-Wisconsinan or Sangamonian age. Sandy soil has been bulldozed to the right and as there are bones within the sands, the slope has been walked to search for them. The pattern of footprints illustrates the procedure described in the text.*

served vertebrate fossils, often associated with mollusc shells and other invertebrate remains.

Fossils can also occur within concretions, as at Green's Creek, near Ottawa, Ontario. Here a variety of fish, small invertebrates and even some terrestrial vertebrates are found together with leaves and seeds within concretions (McAllister *et al.*, 1981, 1988). A fine collection of these concretions is kept in the Division of Palaeobiology, Canadian Museum of Nature (formerly National Museum of Natural History) in Ottawa.

If a slope is examined and a broken piece of a fossil is found at the foot of the slope, the areas below and above the first-found piece should be carefully examined to locate either other broken pieces that could extend the specimen or the source in the vertical face from which either the whole specimen or the pieces that lay on the slope may have been derived. This process applies to talus slopes in quarries, at the foots of cliffs or man-made cuttings, and similar natural scarps such as Scarborough Bluffs, Toronto, or the high cutbanks along many prairie rivers. Where the talus slope is extensive, it should be carefully walked along the base, where the rocks are large, but where large bones may "float" on the surface and be easily seen. Next, the prospector should walk the slope along horizontal lines at about one-metre vertical intervals, while searching the slope surface nearest to his or her face. When a specimen is located, the prospector must not act precipitately and move over areas of the slope that have not been prospected. Fragments of fossils may lie just under the surface of the talus slope and a prudent prospector will winnow through these surficial layers by hand or with small hand tools seeking other fragments that may be hidden.

When the slope has been completely prospected, a series of footprints may be seen in lines along the contours of the slope, and all areas will have been examined (Figure 16). If there is a vertical scarp above the talus, as on a cliff or in a quarry, it should be carefully examined from a short distance, preferably not more than a metre, and usually two passes are required to examine both the basal section of the face and that higher up. The prospector will thus pass backward and forward across the upper level of the talus while both the top level of the talus and the scarp face are fully examined.

Macrovertebrates.

The recovery of elements of large vertebrate fossils, those that have survived mainly intact, and are entombed within a soft matrix, requires removal of the surrounding matrix and exposure of the fossil. This may be done using hand tools such as stiff putty knives, pointing trowels, awls, and paint bushes, or even small spades or shovels, with larger spades, shovels and picks used to remove excessive overburden. Once a fossil is located, it should be carefully exposed by first removing any overhanging or overlying overburden, working down through the matrix until it is grossly exposed, and then removing the final matrix that closely adheres to the bones with small handtools and brushes. Specimens preserved in permafrost are less accessible and, rather, may need to be melted out of the permafrost before removal. Only when a specimen has been completely isolated from the matrix may an attempt to lift it be contemplated. Examination and experience will tell whether the bone is cracked or weak, and whether it may be lifted intact without added support. If it may be lifted as it is preserved, it is often prudent to place the fossil in a sealed plastic bag, so that it will not dry out too quickly, and will be less likely to crack or warp as it dessicates. The field number and any notes relevant to the fossil, such as orientation, level or association with other fossils should be noted. The fossil should now be double-bagged, with a separate label with the essential data placed between the two bags; double bagging prevents the labels from becoming water-logged or destroyed by mildew or mold.

Should a fossil be poorly preserved or so large that it might break when lifted, it should be jacketed. This is provided by a plaster jacket moulded around the fossil and the immediately surrounding matrix (Figure 17). The fossil is exposed as described above, but does not have to be exposed in all detail. It is exposed on a pedestal that is undercut all around, somewhat resembling a mushroom. A separator is laid over the fossil, usually consisting of a number of layers of wet tissue, paper towelling, or the like, so that the plaster jacket will not adhere to the fossil. The separating material should be carefully shaped around the fossil and packed into depressions to help stabilize it within the jacket for transport. Strips of burlap or sacking, about four to six inches wide, are soaked in a thin mix of plaster of Paris and laid over the separator so as to overlap each other and form a web-like support. Under very hot field conditions, rock salt

Figure 17 *Field jacketing of fossil bone with burlap and plaster. In this small-scale example, a thin slurry of plaster has been mixed in a bowl and the burlap soaked in it and wrapped around the exposed fossil. Note the undercut edge of the plastered specimen and that the burlap has been tucked under it. The jacket is being shaped and smoothed around the specimen, and will be labelled with field number and "TOP".*

may be added to the plaster and water to adjust the setting characteristics of the plaster. The ends are tucked round the margins and into the undercut edge. Should the fossil be massive or very dense, then supporting rods, either wood or metal, may be cut to shape, fitted to the uneven contours, or bent to fit, and smoothed in place with some of the remaining wet plaster. Before the internal plaster and the supporting pieces are too dry, a second layer of plaster-soaked strips is arranged over the supports and into the undercut margin. The cast is then marked with the specimen number, possibly the locality if it is not inherent in the field number, possibly the date, and the fact that this side is the TOP. This last is important, as it will tell the preparator which side to open first! The whole is now allowed to dry for at least 24 hours and sometimes longer, until the plaster is firm.

When the first half-jacket is dry, rods, chisel, spades, or crowbars are driven below the fossil through the matrix pedestal, and the matrix with the fossil lifted from its bed and turned over so that it rests on the plaster top. The inverted matrix base below the specimen is now pared down until either the fossil is exposed or the excavator believes that the fossil might be damaged by removal of too much matrix and loss of strength. One is always weighing the advantages of strength against the disadvantages of weight, especially when the fossil may have to be carried some distance. The exposed base is now wrapped in another double set of plaster-soaked strips, sometimes with additional reinforcing rods, as was the top, and allowed to dry.

Microvertebrates.
Remnants of small vertebrates such as rodents, rabbits, small birds, lizards, snakes and fish are often scattered as small isolated elements within a granular matrix such as sand. These items are almost impossible to locate and recover individually in the field and alternative methods have been devised. The matrix, which may be easily friable, as sand, or may be more massive, as clay or marl, is gently shaken, if dry and friable, or washed, if resistant to disintegration unless wet, through a sieve, allowing the fine parts of the matrix to pass through the mesh and retaining the sandy grains or small fossils. The size of the mesh may be varied to catch the size of items desired, but the finer the mesh, the slower is the sieving process. Whichever method is used, care and patience are required to recover the smaller fossils which may be a few millimetres in diameter and as few as two or three in a cubic metre of matrix!

Frames with a cross-handle on top and mesh on the bottom are made of wood, so that they may float. Robust timber is required. Plywood splits easily and often delaminates with continued use in wet washing. The handles are placed diagonally across the frame to give rigidity and the mesh is fastened to the bottom with nails and square battens. The mesh at the bottom may have to take considerable weight for a brief time, and steel mesh rather than copper, aluminum or nylon is preferred. If very heavy loads are anticipated, then battens should be fixed below the mesh and screwed into the sides from below.

The matrix to be washed is conveyed to a source of water, often a stream or river, and portions placed in each frame. If the matrix is solid, the frames with matrix may be placed in the edge of a stream to soak and disintegrate. Large-scale screening operations can introduce much sediment into a stream and may violate local pollution laws, so one should be familiar with appropriate regulations. It should be mentioned that, as the sediment blocks disintegrate, it is not unusual for the fines to slip through the mesh or float out in the current if the frames are submerged, and for the load in a frame to be so reduced that the frame may float away with the remaining concentrate. Thus, it is best to link the frames together with rope. If the fines do not pass easily through the mesh, they may be assisted by moving the frame gently up and down in the water, or gently massaging the matrix against the screen. This should be done carefully as rocks or pointed fossils may penetrate the screen and, once there is a hole in the mesh, it is impossible to estimate how many microfossils may pass through the hole in a stream of water. After the fine silty portion of the matrix has been washed away, the concentrated small grains are laid out on sheets to dry before being bagged. While drying, it is imperative to ensure that the drying concentrate is not blown away by gusts of wind. Heavy rocks around the edge of the sheet or baulks of timber along the edges and between the drying samples both hold the sheets in place and reduce the effect of the wind in removing the dried concentrate. The concentrate is then taken back to the laboratory for examination under a binocular microscope.

The sorts of fossils that may be recovered from such washing may be isolated small mammal teeth, small vertebrae of fish or mammals, partial long-bones or spines of small mammals and fish, and possibly invertebrate exoskeletal fragments, mollusc shells, and plant remains in some deposits.

Dry screening may also be carried out, either using hand-held screens such as the traditional circular ones once common in farms for sifting grain, or the more modern ones pioneered by archaeologists. It is possible to use a series of progressively finer screens to sort the items by size and to retain the smallest remains without their being mixed with the larger items that might crush them. Often the screen is suspended from a frame, like a small child's swing, and swung to and fro by handles. The matrix may also be gently tossed within the frame, or the materials massaged through the mesh. If the mesh is fine, additional support may be given by backing the fine mesh with a coarser mesh, such as chicken wire. The concentrate is bagged for eventual examination in the laboratory.

LABORATORY TECHNIQUES

As soon as a fossil is safely in the laboratory, the task of preparing it so that it will not deteriorate further, and so that it will yield its biological information to researchers begins. As the fossils come in from the field, they should be checked against the field numbers, and provisionally accessioned. Should any be missing, it is best to know this as soon as possible as it may be possible to retrieve any that are missing or at least save one the trouble of looking for it years later when memories are faint.

Preparation.
Basic preparation involves cleaning and restoring the fossil, followed by its stabilization and preservation. Cleaning is accomplished by the use of hand tools such as dental picks, small stiff and soft brushes (*e.g.*, old toothbrushes), acid development if the matrix is amenable to acid removal without harming the fossil, and cleaning with ultrasound. The acids used are dilute weak acids. Glacial acetic acid diluted to 10% by volume in tap

water will remove a carbonate matrix or encrustation without weakening any of the phosphate bonds in bones or teeth. At all times the fossils should be prepared so that they do not become too fragile, consistent with their use and storage conditions.

The fossil is cleaned as far as possible. If it is a single element such as an isolated bone or tooth that has not been damaged, it may only require some scraping and brushing before the stabilizer is applied. Stabilizer should be used only if absolutely necessary as stabilizer will limit future use of the fossils for dating or any further analysis. If a fossil is broken into a number of pieces, it may be possible to glue the parts back together to restore its original shape and size. However, when gluing broken fossils back together, the greatest care must be exercised to see that the joints are clean and fit snugly with the minimum of glue to prevent any distortion. It is an axiom of modern museum conservation that specimens should not be irreversibly restored (*e.g.*, using epoxy). A polyvinyl acetate plastic dissolved in acetone or amyl acetate is reasonably suitable and commercially available as household cement. Even when a specimen is not entire, it is frequently worthwhile to glue as much of it back together as possible for convenience of storage as well as to provide the greatest amount of evidence. Fossils that are in more than one piece often get separated. Should the specimen be a complex, such as a skull, it may be broken into many pieces by crushing. If it had to be retrieved without being jacketed, it may be that each piece has first to be cleaned and given a light dose of preservative. If it was collected as a unit within a plaster jacket, the jacket should be laid on sturdy table, propped up with sandbags, with its TOP uppermost. The field notebook should be consulted to see if there were any noteworthy aspects of the preservation or jacketing that suggest that the opening should be carried out in any particular way. With these strictures in mind, the plaster is cut around the rim of the upper surface and the "lid" lifted off. If the jacket has been properly made, the lid should separate easily from the fossil through the separating tissue layers. The fossil is then carefully exposed as the matrix is removed piecemeal. If the fossil is fragmented, but its pieces fit well together, it may be feasible to reconstruct units of the specimen as one proceeds to remove the matrix, but care must be taken to see that distortion does not occur if the eventual units are to be reassembled. As preparation proceeds, the sides of the jacket are removed, and the bone(s) or elements recovered, stabilized, and reassembled.

Preservation.
Bones that have retrieved without any jacketing should be dried out slowly, if wet, and preserved. Quick drying will result in drying the ouside more quickly than the inside, thus setting up internal stresses, and will result in longitudinal cracks in dense shafts and distortion of flat bones or thin areas. Exposure for 30 to 45 minutes daily and then rebagging will allow drying to occur over four to six weeks, and the loss of water at each exposure will be sufficiently small that the moisture content will equalize itself within the fossil during the bagged periods. When a fossil is completely dry, it may be very light and fragile. It is then advisable to soak into it applications of a thinned stabilizer such as alcohol-thinned PVA glue, acetone-thinned household cement, or one of the proprietory brands such as "Gelva". It is always good to seek the advice of an experienced conservator, if in doubt. The preservative should not be too thick as it will not sink into the pores of the bone and may lift the surface as it hardens, thus damaging the fossil. The first and second application should seem to have little effect on the porous bone, but gradually a fine sheen will be seen on the surface and the solution will sink into the bone more slowly. This should signal the preparator that the bone's pores are filling with the solution and that a final coating with a non-hygroscopic solution is all that is required. Water-based stabilizers are not recommended as they expand and contract with variations in the atmospheric humidity, thus working the fossil and eventually causing its structural failure. At this point, it may be convenient to inscribe the catalogue number on the specimen. Numbers can be written in permanent ink on light bone or, if the bone is dark, a white panel is painted on in a paint that is not miscible with the stabilizer and then the number is inscribed. "Magic markers" should be avoided because many have light-sensitive ink that fades through time.

Concentrate Sorting.
The dry fossil concentrate is placed in trays, usually fairly small to fit under a binocular microscope, and the concentrate is distributed in a thin layer. This is then examined under the binocular microscope, with either a small probe used to winnow the fossils from the rock debris or a pair of fine forceps for removing the fossils as they are found. The fossils are retained in vials for later identification and mounting.

Each fossil is identified as far as possible and, if it is important, catalogued. Often, bulk catalogue numbers are given to the many fragments that are retrieved, but are not noteworthy. An important microscopic fossil, such as a rodent tooth, may be mounted on the end of a pin with a small blob of glue. Each pin is then stuck into the inside end of a cork, which is stopped into a small plastic or glass vial. The vial can be labelled with a label either on the inside or the outside, and the cork may have a short identification number on it so that specimens and their identities are not confused. The vials may be stored in boxes or drawers for convenience. An alternative is the use of "card vials" which allows storage of fossils (Fitzgerald, 1986).

Labelling and Cataloguing.
Each specimen should be accompanied by a label that gives field number and final accession number (which should be written on the fossil if possible), its identification, collector, date and place of collection, rock unit or stratigraphic level, and other important data. This information should also be kept in a ledger or master catalogue, where all specimens are entered by number, with all pertinent data, and any publications or exceptional aspects also noted. By going to the master catalogue, one should be able to learn all that is known of a fossil's history without actually examining it.

Data pertaining to a fossil, such as black-and-white or colour photographs, tables of measurements, line illustrations, maps, sectional diagrams, records of radiocarbon dates performed on it or on associated materials, publications referring to it, or similar evidence resulting from its use after preparation, should be kept on file and noted in the master catalogue as being available. Likewise, a record of the geological evidence should be available, including the field notes, and any important rock samples, and taphonomic evidence as to its depositional history, disease, or the effect of other animals' chewing or burrowing on it. Because other workers will use fossils and come to diverging conclusions, it is often difficult to keep the later reference data complete or up-to-date.

APPLICATIONS

Paleoenvironmental Reconstructions.

Vertebrates reflect an impressive array of parameters in reconstruction of past environments (*e.g.*, Davis, 1987; Graham and Semken, 1987; Harington, 1984a,b), and differ dramatically in the degree to which they can be used to reconstruct paleoenvironments. Terrestrial vertebrates range from large and highly mobile forms (*e.g.*, large ungulates) to small species of limited mobility (*e.g.*, rodents or salamanders). The former are useful in reconstructing communities on a regional scale, such as grassland *versus* forest, even though they may have the ability to move across ecotones in the course of a year. Many are, in fact, migratory, with the wildebeest *(Connochaetes taurinus)* of East Africa being one of the best-known examples (Vesey-Fitzgerald, 1960). Microvertebrates may provide specific microenvironmental information relating to substrate, cover type, and even particular food plant species (Auffenberg and Milstead, 1965). Thus, in a paleoenvironmental study one cannot treat all species as of equal analytical weight. Efforts at quantitative analysis of bone assemblages are becoming more widely used (Graham, 1976; Grayson, 1984; R.A. Martin, 1984; Rhodes and Semken, 1986). It should be possible to apply transfer functions comparable to those used by palynologists in the not-too-distant future.

For a time, it appeared that maps of sympatry would be a central element in paleoecological studies. Maps of the modern distribution of still-extant species identified in a Late Quaternary fauna would be superimposed, and the area in which all or most overlapped (the area of sympatry) would be taken as the best modern analogue for past environmental conditions at the site. While this approach retains utility, it embodies several assumptions: for example, differential dispersal rates of colonizing species are ignored, modern ranges are assumed to (but frequently, in reality, do not) document physiological limits, and possible microevolutionary change in habitat preference is discounted. Graham and Semken (1987), while reaffirming the power of this approach, warned against uncritical over-use of maps of sympatry, urging researchers to employ other methods as well and, in particular, to analyze the implications of a fauna species-by-species, integrating the environmental data afterward. Baseline species-by-species studies of the modern equivalents are an obvious prerequisite; in fact, they are seldom available in the detail desired. One exception might be the study of Champlain Sea fishes (McAllister *et al.*, 1988).

Substrate.

Certain species are useful indicators of specific ecological niches or local environmental parameters. Substrates, such as the character of geological deposits and bedrock, and the soils developed within them, are important constraints for burrowing species. In general, burrowing vertebrates reflect soft, uncompacted soils, often sands or loams. Some insectivores (*e.g.*, moles) and rodents (*e.g.*, pocket gophers) are more highly specialized and efficient burrowers than their relatives and can take advantage of more challenging substrates. Pocket gophers *(Geomys* and *Thomomys)* possess specialized front teeth and powerful clawed feet, and the ability to close the mouth behind the teeth. Like moles, they effectively extend their burrows in order to forage underground. Burrowing ground squirrels *(Spermophilus)* and prairie dogs *(Cynomys)* lack these specializations and, while they also use teeth and claws in burrowing, they do so less efficiently and do much of their foraging above ground. Substrate is also reflected in foot structures of larger forms, such as ungulates. For example, hoof structures of mountain goat *(Oreamnos americanus)* and caribou *(Rangifer tarandus)* differ dramatically: the former concentrates its weight upon a relatively small surface for rock climbing, while the broadly splayed large hooves of the latter are developed for locomotion on soft unconsolidated substrates. Similarly, the insulated and cushioned "hooves" (pads) of camels are ideal for locomotion on hot sand, but are relatively delicate and quickly wear down on angular rocky or gravelly surfaces. Furthermore, it is likely that direct loading and compaction by trampling of tundra habitats by mammoths and other large mammals could have caused dramatic changes in vegetation and associated microvertebrate faunas (Guthrie, 1982).

Paleoclimates.

Vertebrates are useful indicators in paleoclimatic reconstructions, through either their known direct requirements (temperature, availability of water) or indirect requirements (changes in vegetation or substrate). For example, thawing of permafrost in peatland terrain would render a formerly productive area physically impassible to most large ungulates, resulting in a lower diversity of species (Guthrie, 1990). Climatic warming, which might contribute to high biotic diversity and productivity, could, in this case, produce a substrate change that would offset any adaptive advantages conferred by temperature. Paleoclimatic information is not restricted to land vertebrates, as freshwater fishes are sensitive to changes in water temperature (Cross, 1970). Unfortunately, it is difficult to isolate the interrelated variations in temperature, atmospheric humidity, and seasonal extremes, but it may not always be necessary to do so. Many modern vertebrate adaptations depend upon the interaction of several factors. For example, an index of snow severity involving both depth and low temperatures has proved more informative than either parameter alone in understanding year-to-year variations in populations of white-tailed deer *(Odocoileus virginianus*; Ozoga and Gysel, 1972; Verme, 1968). The past record of this species in areas near the margin of its range could be used, in part, as a proxy record of snow severity (Wilson and Hills, 1984). Such an approach may not hold for all species, as the rapidity with which species respond to environmental changes can vary according to variations in species mobility. Even though a vast area of suitable range may open up, such as through deglaciation, a given species can only move into this territory at a rate governed by its dispersal ability (Hoffman and Jones, 1970). Consequently, there can be considerable lag between time of climatic change, habitat availability and occupation of new territory.

Bones and teeth provide a record of physiological changes as a result of fluctuating climatic factors during the life of an individual animal. Paleopathological studies can therefore reveal environmental as well as behavioural information. Dental pathologies (growth arrests, or hypoplasias) in *Bison*, for example, record seasonal stresses such as winter severity and summer droughts. The time of stress is detectable on the basis of the location of the hypoplasia on the tooth crown (M.C. Wilson, 1988). Climatic conditions associated with composition of vegetation are reflected in the carbon isotope composition of bone collagen in herbivores (Chisholm *et al.*, 1986; Land *et al.*, 1980: Tieszen *et al.*, 1979). A temporal record of environmental change can be reconstructed in a stratified site where a single species is well represented (*e.g.*, McKinnon, 1986). Changes through time in rates of bison tooth attrition also reflect changes in quality of forage (M.C. Wilson, 1988). Changes in tooth wear over large geographic distances may reflect either temporal or environmental differences, or both, and may be impossible to distinguish in the record. Climatic stresses can also cause disruptions in population structure, producing unusual recruitment or survivorship curves (M.C. Wilson, 1974; Reher, 1978; Reher and Frison, 1980).

Vegetation Communities.
Dietary specializations in many species are known and are reflected in dental structure (*e.g.*, high-crowned teeth in grazing forms) and can in many cases be inferred for fossil species (Janis, 1984; Solounias and Dawson-Saunders, 1987). The high-crowned/low-crowned dental conformation for the grazer/browser relationship is not infallible, however, and must be used with caution (Graham and Semken, 1987). Direct recovery of vegetable food remains from crevices in enamel folds or tooth fossettes or otherwise in close association with a fossil is possible on occasion and allows more sophisticated dietary analyses (Akersten *et al.*, 1988; Voorhies and Thomasson, 1979), as does analysis of dessicated dung (coprolites) from exceptional environments such as dry caves (Mead *et al.*, 1986a,b). Dental microwear patterns can also indicate the type of diet that an animal eats (Walker *et al.*, 1978). Vegetation also provides protection for many vertebrates, either in the form of cover to hide in, or as a broken background to camouflage an otherwise distinct outline. Reconstruction of protective cover types depends upon analogies from modern forms, and evolutionary behavioural changes pose a problem with this approach. The habitat requirements for many species can be very specific, especially for arboreal forms such as tree squirrels (*Sciurus* and allies) and most primates, which can provide a reliable indication of past vegetation or cover types. Some species can be highly destructive of vegetation, influencing the distribution of other species; a good example is destruction and trampling of trees by elephants in Africa today.

There are times when vegetational reconstruction on the basis of the bones of animals is successful (Graham, 1976; Land *et al.*, 1980; Guthrie, 1982; Rhodes and Semken, 1986). In fact, vertebrate remains can at times be better than pollen in vegetation reconstructions, because vertebrate remains can be identified to species level with considerable accuracy, and, with the exception of some fluvial sequences, vertebrates are often of more local origin as opposed to pollen which can be derived from great distances (Guthrie, 1990). Many pollen records are incomplete from the northern Great Plains because of the paucity of permanent lakes during the early postglacial period, but abundant vertebrate remains from alluvial gravels deposited during this time are dominantly grazing forms, indicating an open steppe or steppe-tundra (Churcher, 1968, 1975; Wilson and Churcher, 1978, 1984). Pollen evidence is beginning to confirm this (Barnosky, 1989; Barnosky *et al.*, 1987; MacDonald, 1989).

Clinal Variation and Microevolution.
Many living species are polytypic, varying clinally in size, colour, and bodily proportions over geographic distances. Populations that differ significantly in such characters are regarded as taxonomic subspecies, with arbitrary lines drawn between them to segment the clines. Given that neighbouring subspecies overlap morphologically, especially in the area of the imposed boundary, taxonomists can only identify individuals as to subspecies with respect to the collecting locality and expected range boundaries. Joysey (1972) has argued persuasively that Late Quaternary chronicles should be segmented for fossil remains by establishing arbitrary temporal boundaries analogous to the geographical boundaries within neospecies. This has the salutary effect of making subspecies three-dimensional (both areally and temporally distributed) and more amenable to incorporation into micro-evolutionary schemes. Other paleontologists have considered that subspecies based on morphological details are impossible to recognize in the fossil record as they are overshadowed by sexual, age and individual variation, because of the general paucity of materials or small sample sizes, or unreliable stratigraphic correlations.

Clinal variation throughout time (Figure 18) has been documented for such disparate groups as rodents and lagomorphs (Davis, 1987; Purdue, 1980), carnivores (Klein and Scott, 1989), proboscideans (Churcher, 1986; Maglio, 1973) and artiodactyls such as bison (Wilson, 1974, 1980) and white-tailed deer (Purdue, 1989; Purdue and Styles, 1986). Direct environmental influences can be implicated for some modern clines; for example, Purdue (1989) related size variation in squirrels (*Sciurus* spp.) and cottontail rabbits (*Sylvilagus floridanus*) to an east-west moisture gradient. Range boundaries frequently represent not sudden environmental changes, but the attainment of a threshold value on an environmental gradient (*e.g.*, mean winter temperature). Temporal variations in the location of the threshold isopleth are reflected in extensions and contractions of range as, for example, along the northern boundary of the range of the pronghorn antelope (*Antilocapra americana*) where occasional severe winters cause dramatic population reductions and range contraction, but during the intervening mild years the range can expand dramatically into marginal areas. Certain sequences suggesting morphological change through time at one locality may be related directly to geographic displacement or clines, as in the case of *Sciurus* spp. and *Sylvilagus floridanus* (Purdue, 1980) and short-tailed shrews, *Blarina* spp. (Graham and Semken, 1976; Jones *et al.*, 1984), and thus need not mark true evolutionary changes (Koch, 1986). Displacement of the cline in the opposite direction would bring back the original morph (body type) to the locality in question, and apparent trends in one area could be balanced by conservatism of morphology elsewhere. However, some chronoclines seem to represent true pervasive evolutionary change, as in *Bison* (Wilson, 1974, 1978, 1980), muskrat, *Ondatra* spp. (L.D. Martin, 1984; Nelson and Semken, 1970), and mammoths, *Mammuthus* spp. (Churcher, 1986; Maglio, 1973). Geist (1971) argued for a strong phenotypic component in this, linked to dispersal history and dietary trends, in which case the "evolution" would be reversible under appropriate conditions, but McDonald (1981) argued that changes in bison chronology were beyond simple phenotypic variation. Given the evidence for modern environmental influences in fostering clinal variation, there is much potential in using chronoclinal shifts to infer changing paleoenvironments.

When both spatial and temporal variation are documented for extinct forms, as for the elk-moose *Cervalces* (Figure 19), inference of the causative factors is difficult in the absence of direct analogues in modern forms. However, examination of paleoenvironmental associations can reveal correlations of value. Morphological change is only one possible response to environmental change. An animal's body shape may be conserved if a species migrates to an area that approximates its former environ-

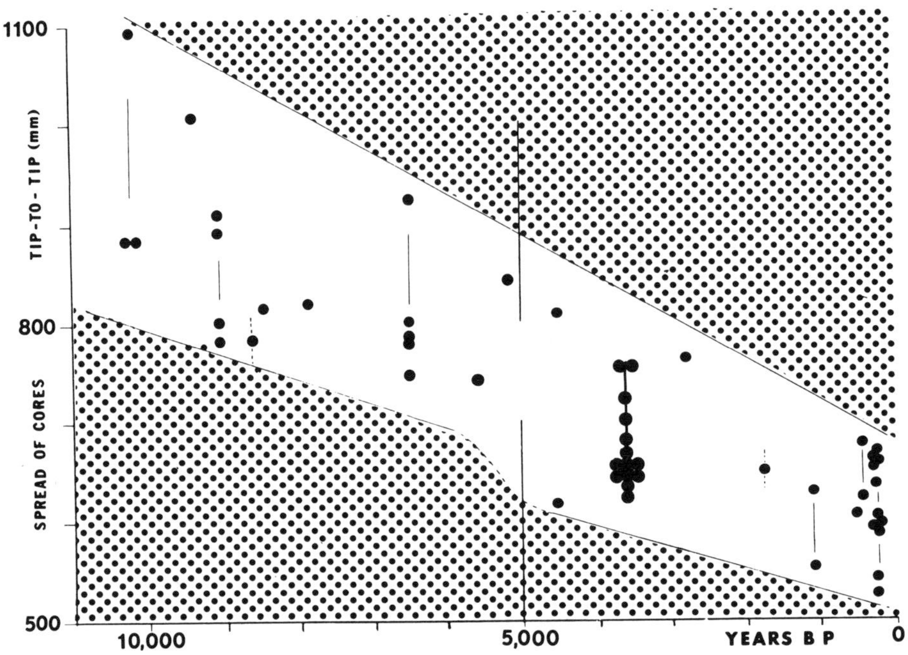

Figure 18 *Chronoclinal variation in* Bison *horncore spread (tip-to-tip) on the northern Great Plains, from 11,000 years B.P. to present. Adapted from Wilson (1980) with additional unpublished data.*

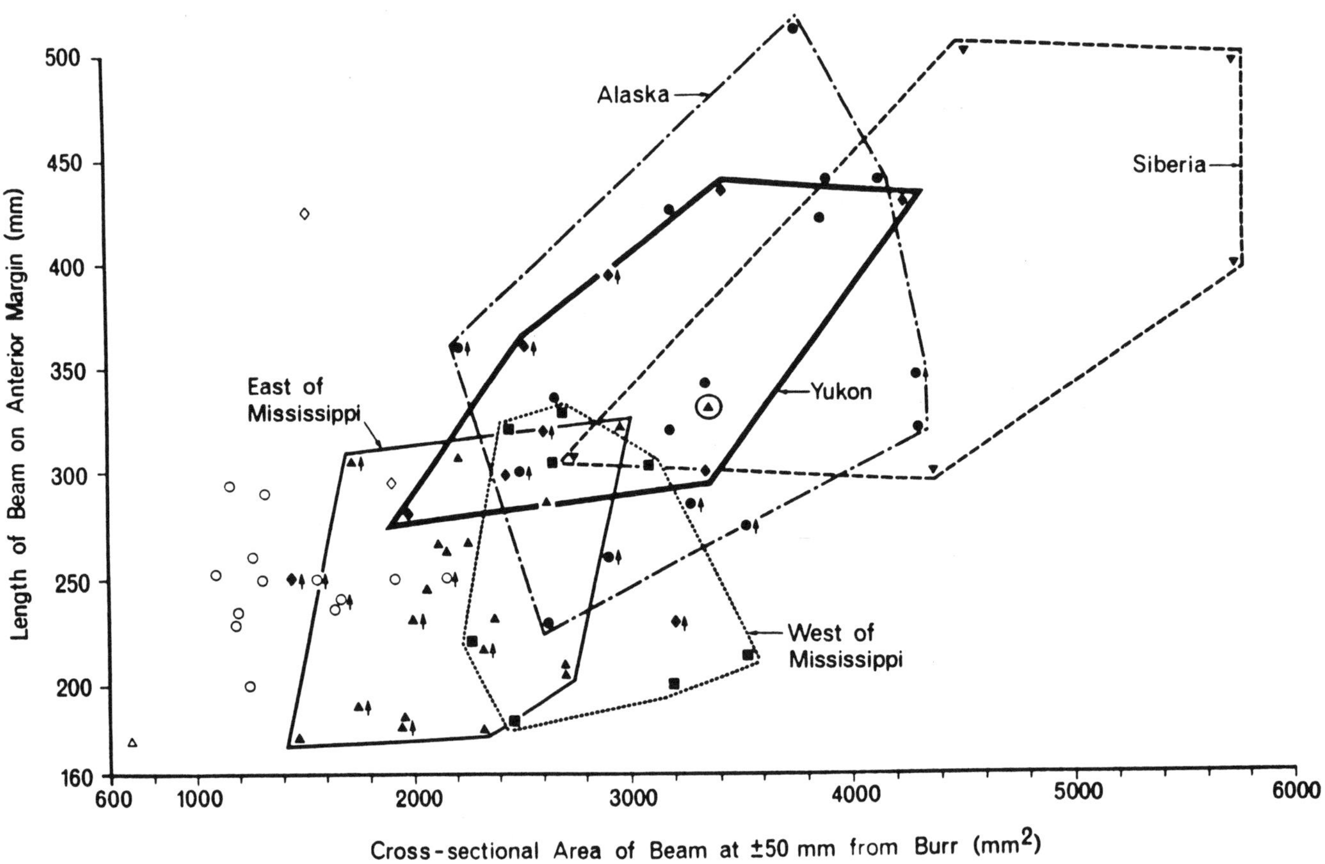

Figure 19 *Geographic clinal variation in antler beam of the extinct elk-moose* (Cervalces). *Antlers derived from presumed adult males are indicated by solid symbols, those from young males by open circles. Arrows indicate plots for which length of beam is minimal. Polygons outline plots from geographic areas. (Taken from Churcher and Pinsof, 1987).*

ment. Migration, then, is another possible response to environmental change.

Extinctions.

An important area in which paleoecological studies of vertebrates are pre-eminent is in the search for causes of Quaternary extinctions, which differentially affected megavertebrates (Martin and Klein, 1984). Such ecological factors as habitat destruction, lessened availability of nutritious vegetation, and disruption of feeding strategies probably were involved; these could have resulted from changes in seasonal climate patterns changing the length and severity of seasonal minima (Graham, 1986). These considerations do not negate the possibility that human hunters administered the *coup de grâce* to many species. Authors debating the case for human agency must clearly distinguish the *process* of extinction from the *event* of extinction: the killers of the last few mammoths did not necessarily play a role in bringing them to the brink of extermination.

Local extirpation or outright extinction are the extreme results of environmental change that exceeds the ability of a species to adapt. It should also be remembered that modern ranges do not necessarily reflect the physiological limits of species, but can be influenced by inter-specific competitive exclusion, predation (especially by humans), or other factors. Even today the ranges of some species are poorly known. Careful baseline studies of modern distribution and physiological adaptations are required for more detailed paleoecological reconstructions. If the modern ranges are themselves undergoing change (*e.g.*, some species are still dispersing into territory opened up by deglaciation millenia ago), the establishment of a

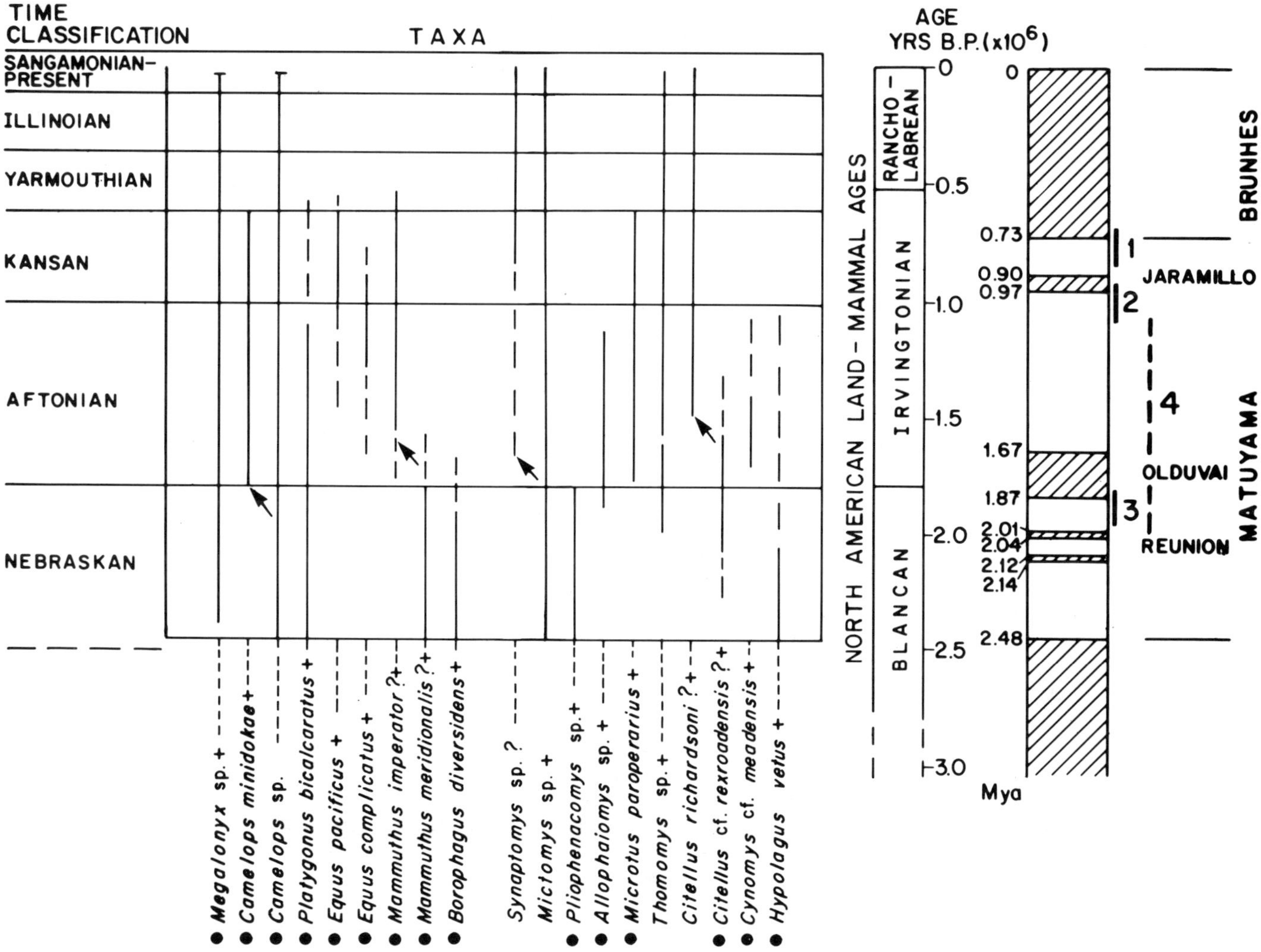

Figure 20 *An example of the taxonomic data available for faunal dating of a site, in this case the Wellsch Valley site near Stewart Valley, Saskatchewan. Taxa with dots before their names are extinct, though they may have relatives that survive to the present. Those with pluses (+) after their names are present in the Wellsch Valley record; diagonal arrows (→) indicate replacement or derivation of taxa, an important consideration if identification is difficult. Age ranges of taxa as documented at other sites are shown as solid lines. Land-mammal ages and magnetic polarity events are shown at right; normal polarity is shaded and reversed polarity is white. Lines 1 to 4 on right show possible ages for Jaw Face section at Wellsch Valley based on combined faunal, paleomagnetic and stratigraphic considerations; alternatives 1 and 4 are preferred over 2 and 3. The potential temporal range for the Wellsch Valley Local Fauna appears to fit best within the Aftonian, with the caveat that the utility of early glacial/interglacial temporal units is itself in question (see text, p. 60).* Borophagus *and* Pliophenacomys *suggest an older age for the fauna or later persistence of these taxa than formerly documented. (Modified after Barendregt* et al., *in preparation).*

modern "baseline" can be difficult.

Biostratigraphy.

A major change in this century has been the establishment of a time-scale, to which sites' faunas or individual taxa can be related. Such chronological control has allowed more complete paleoenvironmental reconstructions to be made with multifaceted analysis of fossil remains. Current emphasis, given an ability to date materials and sites directly by radiometric, paleomagnetic, or other means, allows the fine-resolution study of the rapidity of environmental change and microevolutionary change, millennium by millennium, with standard errors of less than 200 years for most of the Holocene. It is now possible to document with great precision chronoclinal trends as well as events of dispersal and extinction; given dated sequences from disparate areas we can now search for "critical dates" of change that reflect widespread paleoclimatic or other paleoenvironmenatal changes (*e.g.*, Bryson *et al.*, 1970). However, the lack of a reliable chronological control covering most of the Quaternary is still one of the greatest problems for vertebrate paleontologists. Only when there are better dating techniques and agreement among geologists and their interpretations, will it be possible to refine the vertebrate time-scale.

Some 80 years ago, many Quaternary geologists identified strata by their contained index fossils (Savage, 1951). There are, at least in terms of the degree of time resolution demanded by Quaternary scientists, no worldwide terrestrial index fossils. An additional problem is that vertebrates have always been less abundant than invertebrates. Absence of a given species could as easily reflect sampling biases or accidents of preservation as it could any environmental, physiological, competitive or extinction factor. Thus, with terrestrial fossils, it became necessary to have a suite of index species before an age could be suggested. This procedure was codified in a series of "Land Mammal Ages" for the Tertiary and Quaternary, in which the indicator species were included. For example, the presence of mammoths (genus *Mammuthus*) in North America indicates an age of about 2.3 million years ago or less (Churcher, 1986; Harington, 1984a), whereas the presence of bison (*Bison* spp.) indicates an age of 1.2 million years or less in northern North America and considerably less (<400,000 years ago) in the south (Graham and Semken, 1987; Harington, 1984a,b). Both of these genera originated in Eurasia and arrived in North America by migration, providing convenient sudden appearances in the record. Alternatively, horses (genus *Equus*) evolved in North America. Fossils of equid teeth are intergradational between the ancestral genus *Piohippus* and its descendant *Equus* poses a problem in assigning stratigraphic ages. Only broad age limits of perhaps a few hundred thousand years can be used.

At present, even Land Mammal Ages are considered insufficiently reliable or too coarse in temporal resolution; if a deposit yields an extensive fauna it is designated as a local fauna and is then matched taxon by taxon with temporal distributions established on evidence from other sites, as has been done for several parts of North America (Churcher, 1984; Lundelius *et al.*, 1987; Harington, 1978; Kurtén and Anderson, 1980) (Figure 20). The best possible fit is made and any discrepancies are checked for errors in identification. If the discrepancies are confirmed, it can then be considered that the species in question were present earlier or later than previously recorded, or that they changed their geographic distribution through time (so that their previously established temporal range remains valid for one area but not for another; in effect, being time transgressive). The latter is known to have happened in Recent faunas either through human introduction of exotics, as in the case of opossums *(Didelphis virginianus)* released in California, or through colonization without human intervention, as in the northward range extension of opossums into Ontario and British Columbia, of fishes in the Great Lakes (Stewart and Lindsay, 1983), or the colonization of South America and then North America by cattle egrets *(Bubulcus ibis)* from Africa.

Whenever there has been a major faunal change, the recovered faunas are often confusing or contradictory, because there can be different "faunal facies" in deposits that are apparently correlative or geographically contiguous. Thus, where contacts between riverine, riparian forest, grassland, and rocky hillslope or mountain environments are narrow or marked, only the more mobile and ecologically tolerant species will appear in more than one habitat and thus in the representative deposits. Animals with limited mobility, either because of small size or because of structural limitations, will be better indicators of local environments. However, a mechanism of transport and deposition that samples several environments (*e.g.*, a river passing through different vegetational communities) can accumulate mixed "faunas" that did not co-exist in life (Shotwell, 1955). Animals that undertake seasonal migration may reside sequentially in different environments and may therefore exhibit specific adaptations for each; environmental changes with metamorphosis in amphibians are an obvious example, but other groups (*e.g.*, birds) require specific habitats during the period in which young are dependent upon their parents for protection and feeding. Seemingly productive vegetational communities might become unattractive in winter because of snow cover, causing occupants to move to otherwise less favorable communities and resulting in seasonal mixing of species that occupy distinct areas in summer.

Many authors have described Pleistocene faunas that exhibit unusual associations of forms that seemingly represent markedly different, and hence contradictory, environments. These "disharmonious faunas" could, in part, represent artificial associations, like those described above, but are frequent enough to have elicited hypotheses that the environments in which they lived were not like any of the present day. An example of such a disharmonious fauna from a northern Alabama cave includes northern species (caribou, *Rangifer tarandus*; marten, *Martes americana*; fisher, *Martes pennanti*; and brown bear, *Ursus americanus*), western species (prairie chicken, *Tympanuchus cupido*; and thirteen-lined ground squirrel, *Spermophilus tridecemlineatus*), and southern species (extinct armadillo, *Dasypus bellus*; extinct tapir, *Tapirus* cf. *veroensis*; jaguar, *Panthera onca*; and extinct peccary, *Platygonus compressus)* (Churcher *et al.*, 1988). Even discounting the extinct species, this assemblage has no modern counterpart. Such an unusual fauna, if not taphonomically mixed, could signal a vegetational community no longer represented, seasonal alternation of discrete fauna, or any number of other possibilities (see

Graham, 1986; Graham and Semken, 1987). Most importantly in terms of the present paper, it reinforces the need for studies of entire assemblages, rather than interpretation based on a single "indicator" species.

Zooarchaeology.

An almost separate subdiscipline of Quaternary paleontology is the burgeoning field of faunal interpretation of remains from archaeological sites, known by the names of "zooarchaeology", "archaeozoology" or "faunal analysis". It is beyond the scope of this paper to discuss this important area of research in detail. The indisputable association of artifacts with extinct faunas, and later with hominids that could be seen to differ from modern humans, validated the concept that humans had coexisted with extinct mammals and had an evolutionary history that included a "human" (toolmaker) stage of more than one million years' duration. This period is often known as the "Anthropogene" in Europe and the Soviet Union in recognition of sequences of fossil hominids and of stone tool industries that established that humans had evolved in a manner similar to that of the rest of the biotic world. In North America the involvement of Barnum Brown in the excavations at Folsom, New Mexico firmly cemented the link between archaeology and vertebrate paleontology that continues to this day. Analysis of the animal remains associated with human remains or occupational sites has developed dramatically in the past thirty years, building especially upon the pioneering work of vertebrate paleontologist T.E. White (1952-55). It boasts distinctive methodologies (*e.g.*, Grayson, 1984; Klein and Cruz-Uribe, 1984) and extensive cross-links with taphonomy (*e.g.*, Binford, 1981; Morlan, 1980), and tool analysis (Loy, 1983), subdivisions for each of the taxonomic groups and biotic provinces and, in some cases, for culturally mediated divisions (*e.g.*, fish, bird, or mammal studies; Old World *versus* New World; Neolithic *versus* Paleolithic or Industrial; Classical Greek *versus* Roman, Egyptian or Mesopotamian; sedentary *versus* transhumance).

CONCLUDING REMARKS

Clearly, the study of Quaternary vertebrate faunas has played, and will continue to play, an important part in Quaternary ecology. Fruitful areas of future research will include refinements in the use of vertebrates as biostratigraphic tools, more attention paid to the record of microvertebrate remains, and co-ordinated studies of all biotic remains from single sites containing vertebrate fossils for more complete paleoenvironmental reconstructions.

REFERENCES

Agassiz, L., 1833-1844, Recherches sur les poissons fossiles, 5 vols: Neuchatel, 1420 p. With supplement, Monographie des poissons fossiles du vieux grès rouge ou systéme Devonien [Old Red Sandstone] des Iles Britanniques et de Russie, 171 p.

Agassiz, L., 1850, Lake Superior. Its physical character, vegetation, and animals: Gould, Kendall and Lincoln, Boston, 428 p.

Akersten, W.A., Foppe, T.M. and Jefferson, G.T., 1988, New source of dietary data for extinct herbivores: Quaternary Research, v. 30, p. 92-97.

Auffenberg, W. and Milstead, W.W., 1965, Reptiles in the Quaternary of North America, *in* Wright, H.E., Jr. and Frey, D.G., eds., The Quaternary of the United States: Princeton University Press, Princeton, New Jersey, p. 557-568.

Barendregt, R.W., Thomas, F.F., Irving, E., Stalker, A. MacS. and Churcher, C.S., in preparation, Stratigraphy and paleomagnetism of the Jawface Section, Wellsch Valley Site, Saskatchewan.

Barnosky, C.W., 1989, Postglacial vegetation and climate in the northwestern Great Plains of Montana: Quaternary Research, v. 31, p. 57-73.

Barnosky, C.W., Anderson, P.M. and Bartlein, P.J., 1987, The northwestern U.S. during deglaciation: vegetational history and paleoclimatic implications, *in* Ruddiman, W.F. and Wright, H.E., Jr., eds., North America and Adjacent Oceans during the Last Deglaciation: Decade of North American Geology Series, v. K-3, Geological Society of America, Boulder, Colorado, p. 289-321.

Bayrock, L.A., 1964, Fossil *Scaphiopus* and *Bufo* in Alberta: Journal of Paleontology, v. 38, p. 1111-1112.

Behrensmeyer, A.K., 1978, Taphonomic and ecologic information from bone weathering: Paleobiology, v. 4, no. 2, p. 150-162.

Behrensmeyer, A.K., 1984, Taphonomy and the fossil record: American Scientist, v. 72, p. 558-566.

Billings, E., 1870, On bones of a whale lately discovered in Cornwall, Ontario.: Canadian Naturalist and Quarterly Journal of Science, New Series, v. 5, p. 438-439.

Binford, L.R., 1981, Bones: Ancient men and modern myths: Academic Press, New York, 320 p.

Boaz, N.T. and Behrensmeyer, A.K., 1976, Hominid taphonomy: transport of human skeletal parts in an artificial fluviatile environment: American Journal of Physical Anthropology, v. 45, p. 53-60.

Bryson, R.A., Baerreis, D.A. and Wendland, W.M., 1970, The character of late-glacial and post-glacial climatic changes, *in* Dort, W., Jr. and Jones, J.K., Jr., eds., Pleistocene and Recent Environments of the Central Great Plains: University of Kansas, Department of Geology, Special Publication No. 3, p. 53-74.

Buckland, W., 1824, Notice on the *Megalosaurus*, or great fossil lizard of Stonesfield: Transactions of the Geological Society of London, Series 2, v. 1, p. 390-396.

Buckland, W., 1836, Geology and mineralogy considered with reference to natural theology, 2 vols.: Bridgewater Treatises No. 6, London.

Chisholm, B.S., Driver, J., Dube, S. and Schwarcz, H.P., 1986, Assessment of prehistoric bison foraging and movement patterns via stable-carbon isotope analysis: Plains Anthropologist, v. 31, p. 193-205.

Churcher, C.S., 1968, Pleistocene ungulates from the Bow River gravels at Cochrane, Alberta: Canadian Journal of Earth Sciences, v. 5, p. 1467-1488.

Churcher, C.S., 1975, Additional evidence of Pleistocene ungulates from the Bow River gravels at Cochrane, Alberta: Canadian Journal of Earth Sciences, v. 12, p. 68-76.

Churcher, C.S., 1984, Faunal correlations of Pleistocene deposits in western Canada, *in* Mahaney, W.C., ed., Correlation of Quaternary Chronologies: GeoBooks, Norwich, United Kingdom, p. 145-158.

Churcher, C.S., 1986, A mammoth measure of time: molar compression in *Mammuthus* from the Old Crow Basin, Yukon: Current Research in the Pleistocene, v. 3, p. 61-64.

Churcher, C.S., Parmalee, P.W., Bell, G.L. and Lamb, J.P., 1988, Caribou from the Late Pleistocene of northwestern Alabama: Canadian Journal of Earth Sciences, v. 67, p. 1210-1216.

Churcher, C.S. and Pinsof, J.D., 1987, Variation in the antlers of North American *Cervalces* (Mammalia; Cervidae): review of new and previously recorded specimens: Journal of Vertebrate Paleontology, v. 7, p. 373-397.

Coleman, A.P., 1901, Glacial and interglacial beds near Toronto: Journal of Geology, v. 9, p. 285-310.

Cross, F.B., 1970, Fishes as indicators of Pleistocene and Recent environments in the Central Great Plains, *in* Dort, W., Jr. and Jones, J.K., Jr., eds., Pleistocene and Recent Environments of the Central Great Plains: University of Kansas, Department of Geology, Special Publication No. 3, p. 241-257.

Cuvier, G., 1812, Recherches sur les ossemens fossiles, où l'on rétablit les caractères de plusiers animaux dont les revolutiones du globe on détruit les espéces, 4 vols., Paris.

Davis, L.C., 1987, The mammalian record for Late Pleistocene and Holocene environmental changes in the Central Great Plains of the United States, *in* Graham, R.W., Semken, H.A., Jr. and Graham, M.A., eds., Late Quaternary Mammalian Biogeography of the Great Plains and Prairies: Illinois State Museum Scientific Papers, v. 22, p. 88-143.

Dawkins, W.B., 1878, Monograph of the British Pleistocene Mammalia, a preliminary treatise on the relation of the Pleistocene Mammalia to those now living in Europe: Palaeontographical Society Monographs, pt. A, p. i-xxxviii.

Dawson, J.W., 1878, Note on a fossil seal from the *Leda* clay of the Ottawa valley: Canadian Naturalist, New Series, v. 8, p. 340-341.

Dawson, J.W., 1883, On portions of the skeleton of a whale from gravel on the line of the Canadian Pacific Railway, near Smith's Falls, Ontario: Canadian Naturalist, New Series, v. 10, p. 385-387.

Dawson, J.W., 1891, Note on a fossil fish and marine worm found in the Pleistocene nodules of Green's Creek on the Ottawa: Canadian Record of Science, v. 4, p. 86-88.

Dawson, J.W., 1893, The Canadian Ice Age: W.V. Dawson, Montreal, 301 p.

Dodson, P., 1973, The signficance of small bones in paleoecological interpretation: University of Wyoming Contributions to Geology, v. 12, p. 15-19.

Efremov, I.A., 1940, Taphonomy: a new branch of paleontology: Pan-American Geologist, v. 74, p. 81-93.

Fitzgerald, G.R., 1986, The card-vial system for the storage of small vertebrate fossils, *in* Waddington, J. and Rudkin, D.M., eds., Workshop on Care and Maintenance of Natural History Collections: Royal Ontario Museum, Life Sciences Miscellaneous Publications, p. 81-83.

Flint, R.F., 1971, Glacial and Quaternary Geology: John Wiley and Sons, New York, 553 p.

Frison, G.C. and Todd, L.C., 1986, The Colby Mammoth Site: Taphonomy and archeology of a Clovis kill in northern Wyoming: University of New Mexico Press, Albuquerque, 238 p.

Geist, V., 1971, The relation of social evolution and dispersal in ungulates during the Pleistocene, with emphasis on the Old World deer and the genus *Bison*: Quaternary Research, v. 1, p. 283-315.

Glob, P.V., 1969, The Bog people: Iron Age man preserved; translated by R. Bruce-Mitford: Cornell University Press, Ithaca, New York, 200 p.

Graham, R.W., 1976, Late Wisconsin mammalian faunas and environmental gradients of the eastern United States: Paleobiology, v. 2, p. 343-350.

Graham, R.W., 1986, Plant-animal interactions and Pleistocene extinctions, *in* Elliott, D.K., ed., Dynamics of Extinction: John Wiley and Sons, New York, p. 131-154.

Graham, R.W. and Semken, H.A., Jr., 1976, Paleoecological significance of the short-tailed shrew *(Blarina)*, with a systematic discussion of *Blarina ozarkensis*: Journal of Mammalogy, v. 57, p. 433-449.

Graham, R.W. and Semken, H.A., Jr., 1987, Philosophy and procedures of paleoenvironmental studies of Quaternary mammalian faunas, *in* Graham, R.W., Semken, H.A., Jr. and Graham, M.A., eds., Late Quaternary Mammalian Biogeography of the Great Plains and Prairies: Illinois State Museum, Scientific Papers, v. 22, p. 1-17.

Grayson, D.K., 1984, Quantitative Zooarchaeology: Academic Press, Orlando, 202 p.

Guthrie, R.D., 1982, Mammals of the mammoth steppe as paleoenvironmental indicators, *in* Hopkins, D.M., Matthews, J.V., Jr., Schweger, C.E. and Young, S.B., eds., Paleoecology of Beringia: Academic Press, New York, p. 307-326.

Guthrie, R.D., 1990, Frozen Fauna of the Mammoth Steppe: the story of Blue Babe: University of Chicago Press, Chicago, 323 p.

Harington, C.R., 1978, Quaternary vertebrate faunas of Canada and Alaska and their suggested chronological sequence: National Museum of Natural Sciences, Ottawa, Syllogeus 15, 105 p.

Harington, C.R., 1984a, Mammoths, bison and time in North America, *in* Mahaney, W.C., ed., Quaternary Dating Methods: Elsevier, Amsterdam, p. 299-309.

Harington, C.R., 1984b, Quaternary marine and land mammals and their paleoenvironmental implications — some examples from northern North America, *in* Genoways, H.H. and Dawson, M.R., eds., Contributions to Quaternary Vertebrate Paleontology: A Volume in Memorial to John E. Guilday: Carnegie Museum of Natural History, Special Publication no. 8, p. 511-525.

Harington, C.R., 1990, ed., Canada's Missing Dimension: Science and history in the Canadian Arctic Islands: Canadian Museum of Nature, Ottawa, Vol. 1.

Harington, C.R. and Occhietti, S., 1988, Inventaire systematique et paléoécologie des mammifères marins de la mer de Champlain (fin de Wisconsinien) et de ses voies d'accès: Géographie physique et Quaternaire, v. 42, p. 45-64.

Haynes, G., 1980, Evidence of carnivore gnawing on Pleistocene and Recent mammalian bones: Paleobiology, v. 6, p. 341-351.

Heintz, A. and Garutt, 1965, Determination of the absolute age of the fossil remains of mammoth and woolly rhinoceros from the permafrost of Siberia by the help of radio carbon (C^{14}): Norsk Geologisk Tidsskrift, v. 45, p. 73-79.

Hibbard, C.W., 1958, Summary of North American Pleistocene mammalian faunas: Papers of the Michigan Academy of Science, Arts and Letters, v. 43, p. 3-32.

Hibbard, C.W., 1970, Pleistocene mammalian local faunas from the Great Plains and Central Lowland provinces of the United States, *in* Dort, W., Jr. and Jones, J.K., Jr., eds., Pleistocene and Recent Environments of the Central Great Plains: University of Kansas, Department of Geology, Special Publication no. 3, p. 395-433.

Hibbard, C.W., Ray, D.E., Savage, D.E., Taylor, D.W. and Guilday, J.E., 1965, Quaternary mammals of North America, *in* Wright, H.E., Jr. and Frey, D.G., eds., The Quaternary of the United States: Princeton University Press, Princeton, New Jersey, p. 509-525.

Hoffman, R.S. and Jones, J.K., Jr., 1970, Influence of Late-Glacial and Post-Glacial events on the distribution of Recent mammals on the northern Great Plains, *in* Dort, W., Jr. and Jones, J.K., Jr., eds., Pleistocene and Recent Environments of the Central Great Plains: Department of Geology, University of Kansas, Special Publication No. 3, p. 355-394.

Hutton, J., 1795, Theory of the earth, with proofs and illustrations: Edinburgh.

Janis, C.M., 1984, The use of fossil ungulate communities as indicators of climate and environment, *in* Brenchley, P., ed., Fossils and Climate: John Wiley and Sons, New York, p. 85-104.

Jefferson, T., 1799, A memoir on the discovery of certain bones of a quadruped of the clawed kind in the western parts of Virginia: Transactions of the American Philosophical Society, v. 4, p. 246-260.

Jones, C.A., Choate, J.R. and Genoways, H.H., 1984, Phylogeny and paleobiogeography of short-tailed shrews (genus *Blarina*), *in* Genoways, H.H. and Dawson, M.R., eds., Contributions in Quaternary Vertebrate Paleontology: A Volume in Memorial to John E. Guilday: Carnegie Museum of Natural History, Special Publication No. 8, p. 56-148.

Joysey, K.A., 1972, The fossil species in space and time: some problems of evolutionary interpretation among Pleistocene mammals, *in* Joysey, K.A. and Kemp, T.S., eds., Studies in Vertebrate Evolution: Winchester Press, New York, p. 267-280.

Klein, R.G. and Cruz-Uribe, K., 1984, The Analysis of Animal Bones from Archaeological Sites: University of Chicago Press, Chicago, 266 p.

Klein, R.G. and Scott, K., 1989, Glacial/Interglacial size variation in fossil spotted hyenas *(Crocuta crocuta)* from Britain: Quaternary Research, v. 32, p. 88-95.

Koch, P.L., 1986, Clinal geographic variation in mammals: implications for the study of chronoclines: Paleobiology, v. 12, p. 269-281.

Kurtén, B., 1968, Pleistocene Mammals of Europe: Aldine Publishing Co, Chicago, 317 p.

Kurtén, B. and Anderson, E., 1980, Pleistocene Mammals of North America: Columbia University Press, New York, 422 p.

Land, L.S., Lundelius, E.L., Jr. and Valestro, S., 1980, Isotopic ecology of deer bones: Palaeogeography, Palaeoclimatology, Palaeoecology, v. 32, p. 143-152.

Leidy, J., 1856, Note on the remains of a species of seal, from the Post-Pleistocene deposit of the Ottawa River: Proceedings of the Academy of Natural Sciences of Philadelphia, v. 8, p. 90-91.

Loy, T.H., 1983, Prehistoric blood residues: detection on tool surfaces and identification of species of origin: Science, v. 220, p. 1269-1271.

Lundelius, E.L., Jr., Graham, R.W., Anderson, E., Guilday, J., Holman, J.A., Steadman, D.W. and Webb, S.D., 1983, Terrestrial vertebrate faunas, *in* Porter, S.C., ed., Late Quaternary Environments of the United States, v. 1, The Late Pleistocene: University of Minnesota Press, Minneapolis, Minnesota, p. 311-353.

Lundelius, E.L., Jr., Churcher, C.S., Downs, T., Lindsay, E.H., Schultz, G.E., Semken, H.A., Webb, S.D. and Zakrzewski, R.J., 1987, The North American Quaternary sequence, *in* Woodburne, M.O., ed., Cenozoic Mammals of North America: University of California Press, Berkeley, p. 211-235.

Lyell, C., 1845, Travels in North America in the years 1841-1842, with geological observations in the United States, Canada, and Nova Scotia: 2 vols., Wiley and Putnam, New York.

MacDonald, G.M., 1989, Postglacial palaeoecology of the subalpine forest-grassland ecotone of southwestern Alberta: new insights on vegetation and climate change in the Canadian Rocky Mountains and adjacent foothills: Palaeogeography, Palaeoclimatology, Palaeoecology, v. 73, p. 155-173.

Maglio, V., 1973, Origin and evolution of the Elephantidae: Transactions of the American Philosophical Society, v. 63, p. 1-149.

Mantell, G.A., 1825, Notice on the *Iguanodon*, a newly discovered fossil reptile, from the sandstone of Tilgate Forest, in Sussex: Philosophical Transactions of the Royal Society, v. 115, p. 179-186.

Martin, L.D., 1984, Phyletic trends and evolutionary rates, *in* Genoways, H.H. and Dawson, M.R., eds., Contributions in Quaternary Vertebrate Paleontology: A Volume in Memorial to John E. Guilday: Carnegie Museum of Natural History, Special Publication No. 8, p. 526-538.

Martin, L.D. and Martin, J., 1987, Equability in the Late Pleistocene: Kansas Geological Survey Guidebook No. 5, p. 123-127.

Martin, P.S. and Guilday, J.E., 1967, A bestiary for Pleistocene biologists, *in* Martin, P.S. and Wright, H.E., Jr., eds., Pleistocene Extinctions: the search for a cause: Yale University Press, New Haven, p. 1-62.

Martin, P.S. and Klein, R.G., eds., 1984, Quaternary Extinctions: A prehistoric revolution: University of Arizona Press, Tucson, 892 p.

Martin, R.A., 1984, The evolution of cotton rat body mass, *in* Genoways, H.H. and Dawson, M.R., eds., Contributions in Quaternary Vertebrate Paleontology: A Volume in Memorial to John E. Guilday: Carnegie Museum of Natural History, Special Publication No. 8, p. 179-183.

McAllister, D.E., Cumbaa, S.L. and Harington, C.R., 1981, Pleistocene fishes *(Coregonus, Osmerus, Microgadus, Gasterosteus)* from Green Creek, Ontario, Canada: Canadian Journal of Earth Sciences, v. 18, p. 1356-1364.

McAllister, D.E., Harington, C.R., Cumbaa, S.L. and Renaud, C.B., 1988, Paleoenvironmental and biogeographic analyses of fossil fishes in peri-Champlain Sea deposits in eastern Canada, *in* Gadd, N.R., ed., The Late Quaternary Development of the Champlain Sea Basin: Geological Association of Canada, Special Paper 35, p. 241-258.

McDonald, J.N., 1981, North American Bison: their classification and evolution: University of California Press, Berkeley, 316 p.

McKinnon, N.A., 1986, Paleoenvironments and Cultural Dynamics at Head-Smashed-In Buffalo Jump, Alberta: The carbon isotope record: M.A. thesis, University of Calgary, Calgary, Alberta, 219 p.

Mead, J.I., O'Rourke, M.K. and Foppe, T.M., 1986a, Dung and diet of the extinct Harrington's mountain goat *(Oreamnos harringtoni)*: Journal of Mammalogy, v. 67, p. 284-293.

Mead, J.I., Agenbroad, L., Davis, O.K. and Martin, P.S., 1986b, Dung of *Mammuthus* in the arid Southwest, North America: Quaternary Research, v. 25, p. 121-127.

Merrilees, D., 1970, A check on the radiocarbon dating of desiccated thylacine (marsupial "wolf") and dog tissue from Thylacine Hole, Nullarbor Region, Western Australia: Helictite, v. 8, p. 39-42.

Morlan, R.E., 1980, Taphonomy and Archaeology in the Upper Pleistocene of Northern Yukon Territory: a glimpse of the peopling of the New World: National Museum of Man Mercury Series, Archaeological Survey of Canada, Ottawa, Paper 94, 380 p.

Nelson, R.S. and Semken, H.A., Jr., 1970, Paleoecological and stratigraphic signficance of the muskrat in Pleistocene deposits: Geological Society of America, Bulletin, v. 81, p. 3733-3738.

Osborn, D.J. and Helmy, I., 1980, The Contemporary Land Mammals of Egypt (including Sinai): Fieldiana (Zoology), New Series, No. 5, 579 p.

Osborn, H.F., 1936-1942, Proboscidea: a monograph of the discovery, evolution, migration, and extinction of the mastodonts and elephants of the world, 2 vols.: American Museum of Natural History, New York, 1675 p.

Owen, R., 1846, A History of British Fossil Mammals and Birds: John van Voorst, London, 560 p.

Owen, R., 1849-1884, A History of British Fossil Reptiles: reprinted from publications of the Palaeontographical Society, etc., London, 4 vols.

Ozoga, J.J. and Gysel, L.W., 1972, Response of white-tailed deer to winter weather: Journal of Wildlife Management, v. 36, p. 892-896.

Peale, R., 1803a, An historical disquisition on the mammoth, or great American incognitum, an extinct, immense, carnivorous animal whose remains have been found in North America: London, 91 p.

Peale, R., 1803b, Account of some remains of a species of gigantic oxen found in America and other parts of the world: Philosophical Magazine, v. 15, p. 325-327.

Purdue, J.R., 1980, Clinal variation of some mammals during the Holocene in Missouri: Quaternary Research, v. 13, p. 242-258.

Purdue, J.R., 1989, Changes during the Holocene in the size of white-tailed deer *(Odocoileus virginianus)* from central Illinois: Quaternary Research, v. 32, p. 307-316.

Purdue, J.R. and Styles, B.W., 1986, Dynamics of mammalian distribution in the Holocene of Illinois: Illinois State Museum, Reports of Investigations No. 41, 63 p.

Reher, C.A., 1970, Population dynamics of the Glenrock bison population, *in* Frison, G.C., ed., The Glenrock Buffalo Jump, 48CO304: Late Prehistoric Period buffalo procurement and butchering on the northwestern plains: Plains Anthropologist, v. 15, no. 50, Memoir No. 7, p. 51-55.

Reher, C.A., 1978, Buffalo population and other deterministic factors in a model of adaptive process on the shortgrass plains, *in* Davis, L.B. and Wilson, M.C., eds., Bison Procurement and Utilization: a symposium, Plains Anthropologist, v. 23, no. 82, Memoir No. 14, p. 23-39.

Reher, C.A. and Frison, G.C., 1980, The Vore Site, 48CK302, a stratified buffalo jump in the Wyoming Black Hills: Plains Anthropologist, v. 25, Memoir No. 16, 190 p.

Reynolds, S.H., 1939, A monograph on the British Pleistocene Mammalia, Volume III, Part VI. The Bovidae: Palaeontographical Society, London, p. 1-65.

Rhodes, R.S., II and Semken, H.A., Jr., 1986, Quaternary biostratigraphy and paleoecology of fossil mammals from the Loess Hills region of western Iowa: Proceedings of the Iowa Academy of Science, v. 93, p. 94-130.

Richmond, G.M. and Fullerton, D.S., 1986, Introduction to Quaternary glaciations in the United States of America, *in* Sibrava, V., Bowen, D.Q. and Richmond, G.M., eds., Quaternary Glaciations in the Northern Hemisphere: Quaternary Science Reviews, v. 5, p. 3-10.

Romer, A.S., 1929, A fresh skull of an extinct American camel: Journal of Geology, v. 37, p. 261-267.

Savage, D.E., 1951, Late Cenozoic vertebrates of the San Francisco Bay region: University of California Publications, Bulletin of the Department of Geological Sciences, v. 28 p. 215-314.

Scott, W.B., 1913, A history of land mammals in the western hemisphere: Macmillan, New York, 693 p.

Semken, H.A., Jr., 1983, Holocene mammalian biogeography and climatic change in the eastern and central United States, *in* Wright, H.E., Jr., ed., Late Quaternary Environments of the United States, v. 2, The Holocene: University of Minnesota Press, Minneapolis, Minnesota, p. 182-207.

Shipman, P., 1981, Life History of a fossil, An introduction to taphonomy and paleoecology: Harvard University Press, Cambridge, 222 p.

Shotwell, J.A., 1955, An approach to the paleoecology of mammals: Ecology, v. 36, p. 327-337.

Simpson, G.G., 1951, Horses: Oxford University Press, New York, 247 p.

Smith, W., 1817, Stratigraphical System of organized fossils: London, 118 p.

Sokolov, V.Y., Dubrovo, I.A., Lebedkina, N.S. and Ponomarenko, N.G., eds., 1986, Yuribeyskiy momont [The Yuribey mammoth]: Iztadel'stvo "Nauka", Moscow, 160 p.

Solounias, N. and Dawson-Sanders, B., 1987, Dietary adaptations and paleoecology of the Late Miocene ruminants from Pikermi and Samos in Greece: Palaeogeography, Palaeoclimatology, Palaeoecology, v. 65, p. 149-172.

Steadman, D.W., and Miller, N.G., 1987, California condor associated with spruce-pine woodland in the Late Pleistocene of New York: Quaternary Research, v. 28, p. 415-426.

Stewart, K.W. and Lindsay, C.C., 1983, Postglacial dispersal of lower vertebrates into the Lake Agassiz region, *in* Teller, J.T. and Clayton, L., eds., Glacial Lake Agassiz: Geological Association of Canada, Special Paper 26, p. 391-419.

Sutcliffe, A.J., 1970, Spotted hyaena: crusher, gnawer, digester and collector of bones: Nature, v. 227, p. 1110-1113.

Tieszen, L.L., Hein, D., Qvortrup, S.A., Troughton, J.H. and Imbamba, S.K., 1979, Use of $\delta^{13}C$ values to determine vegetation selectivity in East African herbivores: Oecologia, v. 37, p. 351-359.

van Devender, T.R., 1987, Paleoclimatic implications for the monsoonal Sonoran Desert of a packrat midden vegetation sequence from the Puerto Blanco Mountains, southwestern Arizona: Quaternary Research, v. 27, p. 51-72.

Vereshchagin, N.K. and Baryshnikov, G.F., 1982, Paleoecology of the mammoth fauna in the Eurasian Arctic, *in* Hopkins, D.M., Matthews, J.V., Jr., Schweger, C.E. and Young, S.B., eds., Paleoecology of Beringia: Academic Press, New York, p. 267-279.

Verme, L.J., 1968, An index of winter weather severity for northern deer: Journal of Wildlife Management, v. 32, p. 566-574.

Vesey-Fitzgerald, D.F., 1960, Grazing succession among East African game animals: Journal of Mammalogy, v. 41, p. 161-172.

Voorhies, M.R., 1969, Taphonomy and population dynamics of an Early Pliocene Vertebrate Fauna, Knox County, Nebraska: University of Wyoming, Contributions to Geology, Special Paper, v. 1, 69 p.

Voorhies, M.R. and Thomasson, J.R., 1979, Fossil grass anthoecia within Miocene rhinoceros skeletons: diet in an extinct species: Science, v. 206, p. 331-333.

Walker, A., Hoeck, H. and Perez, L., 1978, Microwear of mammalian teeth as indicators of diet: Science, v. 201, p. 908-910.

Wells, P.V., 1976, Macrofossil analysis of wood rat *(Neotoma)* middens as a key to the Quaternary vegetational history of arid America: Quaternary Research, v. 6, p. 233-248.

White, T.E., 1952-1955, Observation on the butchering techniques of some aboriginal peoples, nos. 1 to 9: American Antiquity, v. 17, p. 337-338; v. 19, p. 160-161; v. 19, p. 254-264; and v. 21, p. 170-178.

Wilson, M.C., 1974, The Casper Local Fauna and its fossil bison, *in* Frison, G.C., ed., The Casper Site: a Hell Gap bison kill on the High Plains: Academic Press, New York, p. 125-171.

Wilson, M.C., 1978, Archaeological kill site populations and the Holocene evolution of the genus *Bison*, *in* Davis, L.B. and Wilson, M.C., eds., Bison Procurement and Utilization: A symposium: Plains Anthropolgist, v. 23, Memoir No. 14, p. 9-22.

Wilson, M.C., 1980, Morphological dating of Late Quaternary bison on the northern plains: Canadian Journal of Anthropolgy, v. 1, p. 81-85.

Wilson, M.C., 1988, Bison dentitions from the Henry Smith Site, Montana: evidence for seasonality and paleoenvironments at an Avonlea bison kill, *in* Davis, L.B., ed., Avonlea Yesterday and Today: Archaeology and prehistory: Saskatchewan Archaeological Society, Saskatoon, p. 203-225.

Wilson, M.C. and Churcher, C.S., 1978, Late Pleistocene *Camelops* from the Gallelli Pit, Calgary, Alberta: morphology and geologic setting: Canadian Journal of Earth Sciences, v. 15, p. 729-740.

Wilson, M.C. and Churcher, C.S., 1984, The Late Pleistocene Bighill Creek Formation and its equivalents in Alberta: correlative potential and vertebrate palaeofauna, *in* Mahaney, W.C., ed., Correlation of Quaternary Chronologies: Geo Books, Norwich, United Kingdom, p. 159-175.

Wilson, M.C. and Hills, L.V., 1984, Holocene white-tailed deer *(Odocoileus virginianus)* from the foothills northwest of Calgary, Alberta, Canada: Canadian Journal of Earth Sciences, v. 21, p. 365-371.

Wilson, M.V.H., 1988, Taphonomic processes: Information loss and information gain: Geoscience Canada, v. 15, p. 131-148.

Wright, G.F., 1890, The ice age in North America and its bearings upon the antiquity of man: D. Appleton and Company, New York, 622 p.

Originally published in
Geoscience Canada Volume 17 Number 2
(June 1990)

Other Fossils

Barry G. Warner
Department of Earth Sciences
and Quaternary Sciences Institute
University of Waterloo
Waterloo, Ontario N2L 3G1

SUMMARY

There are several plant and animal fossil types that are commonly encountered during routine analysis of other more conventional fossil groups. These less familiar or diagnostic remains can be important paleoecological indicators, often complementing paleoenvironmental reconstructions based on other more widely used fossil groups. The composition of the fossil sediment or peat itself contains important paleoecological information, in addition to the fossils contained within its matrix. Analysis of peat components can refine species identifications and, in general, yield useful information on peat-forming plant communities.

INTRODUCTION

As surprising as it may seem, the vast majority of terrestrial and freshwater plants and animals leave remains that may become fossilized in Quaternary deposits. Almost all non-vascular and vascular plant species produce some part that may be fossilized. In the case of animals, the fossils can be egg capsules, protective shells of resting bodies, fragments of larvae or early instars, or some part of the adult animal. It is probably not an exaggeration to state that almost any organism, given the right set of circumstances, may become fossilized. Some of the rarest fossils have been found entombed in calcareous concretions, amber, permafrost, tar seeps, and deep peat deposits. The problems, however, lie in recovering these fossils from the enclosing matrix, and the ability of a Quaternary paleontologist to identify these less familiar or diagnostic remains. The lack of suitable taxonomic aids, and inadequate information for interpreting the paleoecological significance further hamper routine use of these less conventional fossil types in paleoecological reconstructions.

Although the authors of the papers in this series have attempted to provide an overview of the most widely used groups in Quaternary ecology, this paper discusses the status and paleoecological significance of a number of additional plant and animal remains that rarely are reported in most publications. I do not intend this to be a complete list or detailed account of all fossil types reported here, but it is especially important to mention those only recently discovered in North American deposits or those types having the best potential in paleoecological studies given further work. The following papers provide additional information on the Quaternary record of the fossil groups considered in this paper: Frey (1964), Grosse-Brauckmann (1986), and Gray (1988), and the papers by Van Geel and co-workers (*i.e.*, Van Geel, 1978; Van Geel *et al.*, 1980, 1982; Pals *et al.*, 1980).

Numerous workers have noted that the next phase in Quaternary ecological research will involve multi-component studies of single cores or sites in an attempt to reconstruct whole ecosystems as completely as possible (West, 1985). Great strides have been made in recent years to improve processing techniques and data presentation, and hence the detail and accuracy of paleoenvironmental interpretations. The next most important contributions to Quaternary ecology should involve development of new fossil groups into useful paleoecological indicators. I echo some of Frey's (1964; p. 4-5) introductory comments made 25 years ago; the situation has not changed much in the intervening period: "the time is propitous for coordinated research by geologists, palynologists, paleobotanists, paleolimnologists, and zoologists on strategic sections... to attain that maturity of insight and understanding necessary for a realistic comprehension of the Quaternary and its biological consequences...". Furthermore, "Sometimes even these approximate identifications [to at least genus] are very useful in interpreting past ecological conditions... ".

LABORATORY PROCEDURES

Collection of field samples is the same as for any routine paleontological study. Extraction of many of the fossils encountered in Quaternary deposits, and of those included in this paper, need not involve specialized laboratory procedures. Procedures used for pollen analysis, extraction of macroscopic remains, or the observation of fossil remains in wet mounts of fresh sediment are probably sufficient for the identification and tabulation of most fossils, and any detailed analysis should probably employ all three treatments. Stains such as safranin-O or gentian violet may accentuate certain fossils in temporary slide mounts or permanent preparations of processed concentrate. As more is learned about the nature and occurrence of each fossil group, it might be expected that more specialized laboratory procedures might be developed for efficient, accurate, and quantitative analysis. For the time being, it seems appropriate to analyze all microfossil types composed of chitin and other inert polymers in pollen preparations, siliceous types in the residue used for diatom analysis, and other more fragile microfossils in water mounts.

PLANTS

Polypodiaceae.

Spore-cases, or sporangia, are tiny stalked capsules containing the spores on the fronds of ferns. Sporangia are encircled by a jointed ring of cells which shrinks upon ripening and drying, splitting the sides of the sporangium to release the spores into the air. The ripened empty sporangia may be released from the frond either before or after death only to be incorporated into the underlying limnic or peat sediments (Figure 1).

There is not much morphological variation within sporangia of the Polypodiaceae to permit confident identification beyond the family level. The sporangia are not likely to be transported far beyond the plant, thereby providing evidence for local spore sources and presence of the plant. Sporangia are most likely recovered from peat or peaty gyttja where such wetland species as *Dryopteris thelypteris* or *Onoclea sensibilis* ferns may have lived. Sporangia have been found in abundance in interglacial deposits in Idaho (Bright, 1982). They were interpreted to represent a marsh where *Cystopteris fragilis* was abundant locally.

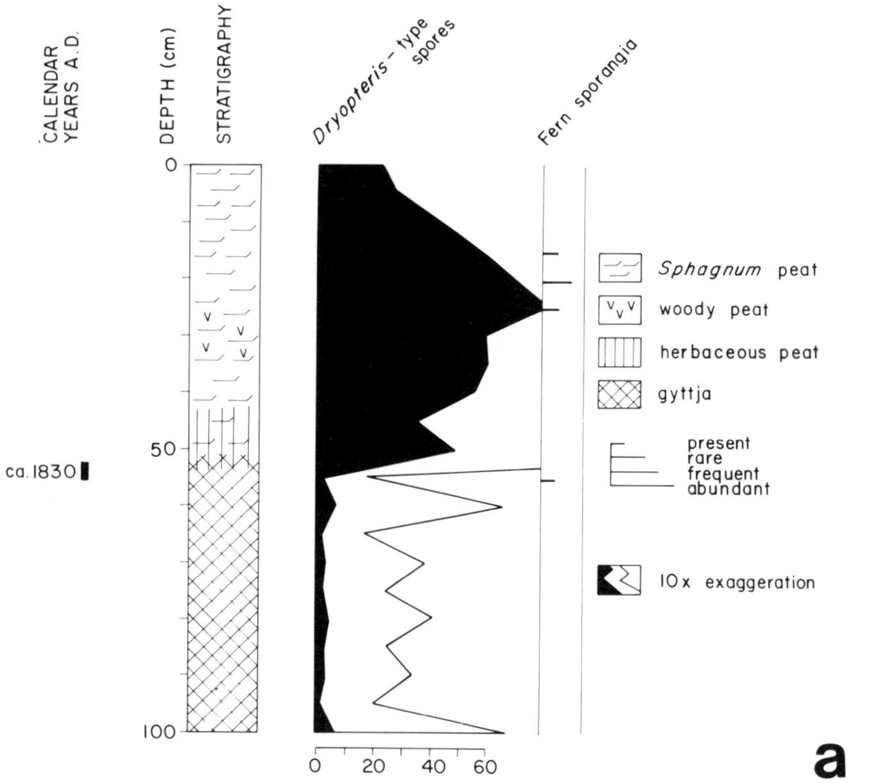

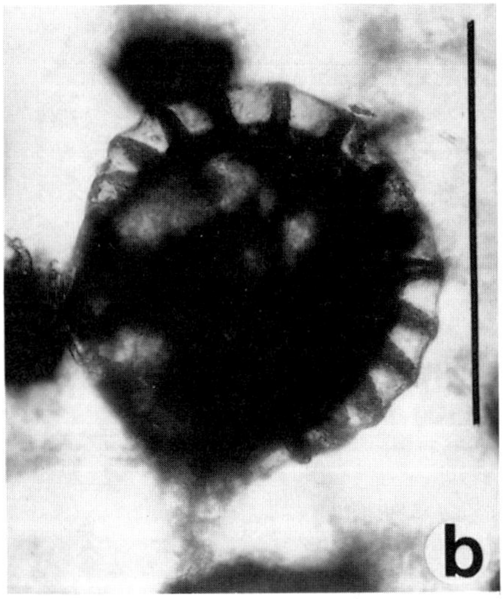

Figure 1 **(a)** *Comparison of profiles of* Dryopteris-*type spores (presented as percent of total pollen sum) and fern sporangia (presented as relative abundance) illustrating the local spore sources in a floating bog near Woodstock, Ontario.*
(b) *Sporangium of Polypodiaceae. Note the peripheral ring of cells and the tiny stalk. Scale bar equals 100 μm.*

Gramineae.
Among the suite of mineralized biological structures encountered in fossil deposits, cytoliths constitute one group of cellular structures or mineralized cells that may include phytoliths (Bombin, 1984). Phytoliths are opaline silica or biosilica bodies of diverse shapes that are produced by a variety of vascular plants. They occur in the epidermis (including hair cells, hair bases, and stomata), mesophyll, sclerenchyma, and vascular tissue (Figure 2). Despite their occurrence throughout both monocotyledons and dicotyledons, phytoliths have been studied mostly from grasses, sedges, and the palms. Grass phytoliths, perhaps, are the best known, largely as a result of a short-sighted belief among early workers that their phytoliths were the most abundant and diagnostic (Piperno, 1985). Phytoliths are abundant and well preserved in geological and archaeological deposits, and are morphologically diagnostic (Rovner, 1971; Carbone, 1977; Bombin, 1984; Piperno, 1985, 1988; Powers and Gilbertson, 1987). The concentration of phytoliths in some fossil deposits may reach up to 1.5–2.3 mg/cm^3.

Piperno (1985) provides the following comments on fossil phytolith classification. Phytoliths of silicified hair cells are tapering and triangular along the shaft, the base of which is flat, spheroidal or ellipsoidal. The hair cells may be segmented or non-segmented. Both the hair cells and hair bases can be characterized further by shapes, sizes, and

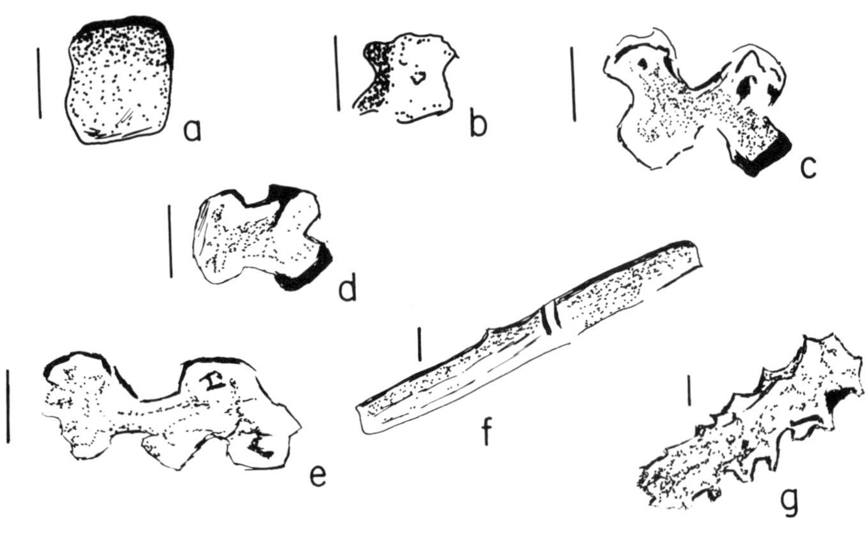

Figure 2 *Representative types of phytoliths. Scale bars equal 50 μm.*

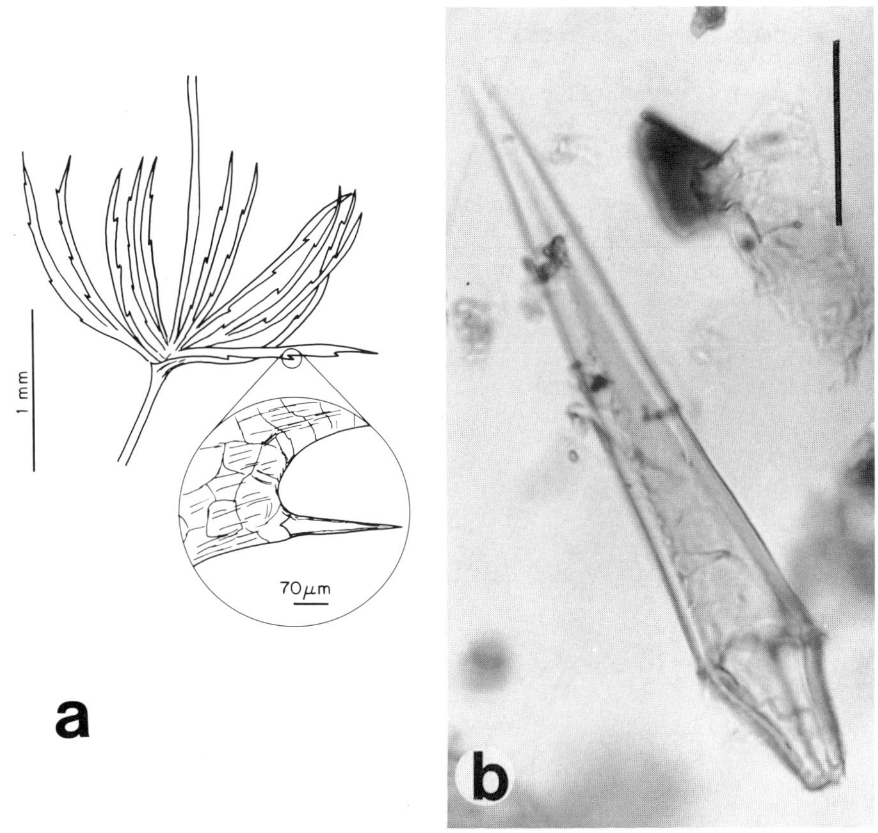

Figure 3 **(a)** *Sketch of forked segments of* Ceratophyllum demersum *showing the position of marginal leaf spines.* **(b)** *Fossil leaf spine from a Holocene marl deposit in northern Ontario. Scale bar equals 50 μm.*

Table 1 **Summary of water chemistry values for *Ceratophyllum demersum* in North America.**

Location and Reference	pH	Alkalinity (mg·L⁻¹CaCO$_3$)	Conductivity (μmhos·cm⁻¹@25°C)
North Carolina (Beals, 1977)	6.3 – 8.3	–	–
NE USA (Hellquist and Crow, 1980)	5.7 – 7.8	5.5 – 151.0	–
Central North America (Pip, 1988)	–	40.0 – 558.0	–
Alberta (Doherty and LaRoi, 1973)	5.7 – 10.5	70.0 – 260.0	–
Minnesota (Moyle, 1945)	6.3 – 9.0	23.0 – 376.0	–
Minnesota and North Dakota (Birks, 1973)	–	–	180.0 – 300.0
Southern Ontario (Wile and McCombie, 1972)	7.8 – 8.4	172.0 – 204.0	–
Southern Ontario (Crowder *et al.*, 1977)	–	–	124.0 – 316.0
Alaska/NWT (Holmquist, 1971)	7.7 – 9.2	77.0 – 142.0	–
Central British Columbia (Reynolds and Reynolds, 1975)	7.8 – 9.7	–	600.0 – 830.0

unique surface patterns. Phytoliths within epidermal cells occur in a variety of shapes, from conical and spherical to non-spherical shapes. Sclereids (situated within the thick and lignified sclerenchyma tissue) and tracheids (occurring within the vascular tissue of xylem) may be silicified also. Good introductions to phytolith classification and analysis are Bombin (1984) and Piperno (1985; 1988).

As phytoliths are an important component in atmospheric dust, their use in paleoecology is much under-used. Phytolith analysis is potentially as important as pollen or diatom analysis in terms of abundance in and ease of recovery from sediments, and in importance as a source of paleoecological information (Bombin, 1984).

In a detailed study of phytoliths from Alaska and the Yukon, Bombin (1984) found that phytoliths were abundant and widespread in non-fluvial deposits, including dung deposits from Beringia, and their stratigraphic changes correlated well with changes in other more conventional fossil indicators such as pollen and diatoms.

In addition to phytoliths, grasses possess distinctive leaf cuticular epidermal features (hairs, papillae, cell shapes, and stomata). Both fossil grass phytoliths and cuticle have been most widely

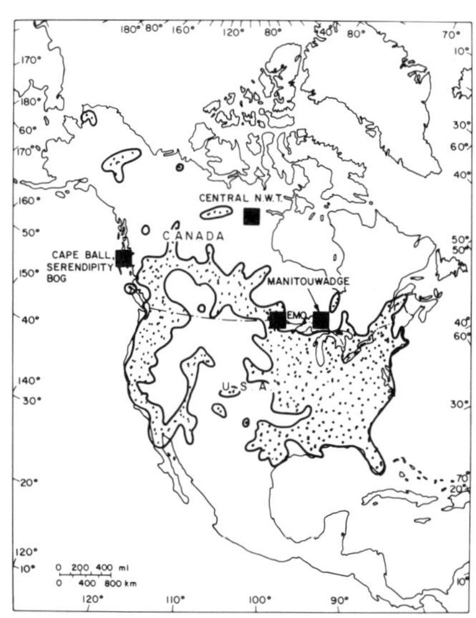

Figure 4 *Modern distribution of* Ceratophyllum demersum *(after Riley and MacKay, 1980; and simplified after Les, 1986). Occurrences of fossil leaf spines are shown with squares (Terasmae and Craig, 1958; Warner, 1984b, and unpublished data).*

used in tropical and other arid regions where grasses are diverse and a major component of the vegetation (Palmer, 1976). Warner *et al.* (1987) found fossil leaves of wild rice *(Zizania palustris)*, in mid-Holocene deposits in north-central Ontario which confirmed the natural range limit of the plant, disproving beliefs that the plant was a recent anthropogenic introduction to the local region.

Ceratophyllaceae.

Ceratophyllum is a perennial, free-floating, submerged aquatic plant lacking roots. Two species are known in Canada, *C. demersum* and *C. echinatum*; the former is the most widespread (Les, 1986). The principal leaves are in forked segments with distinct serrations and marginal leaf spines (Figure 3a). The marginal teeth are readily preserved in limnic sediments and can be encountered in pollen preparations.

The leaf spines vary from 190 to 220 μm in length (Figure 3b). They are straight, tapering to a sharp point, and thick walled. The base comes to a pyramidal point that fits into the main leaf segment of the plant. The leaf spines are probably siliceous in composition, in which case they might be regarded as a type of phytolith.

Ceratophyllum leaf spines are useful fossil indicators in Quaternary ecology because they are preserved better and produced in greater abundance than are the pollen or seeds, particularly in instances where only vegetative plants may have been present. Leaf spines of *Ceratophyllum* are known from a few localities in Canada (Figure 4); four localities lie beyond or near the modern distributional range limit. The paleoecological significance and possible reasons for these distributional shifts are probably interdependant and complex. *C. demersum* grows in habitats

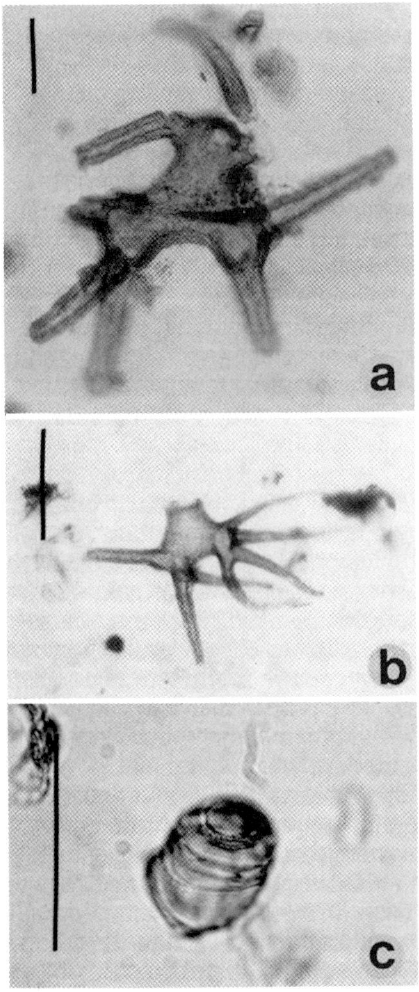

Figure 5 (a,b) *Fossil Nymphaeaceae sclerids.* **(c)** *Fossil leaf hair basal cells of Nymphaeaceae. Scale bars equal 100 μm for a, b, and 50 μm for c.*

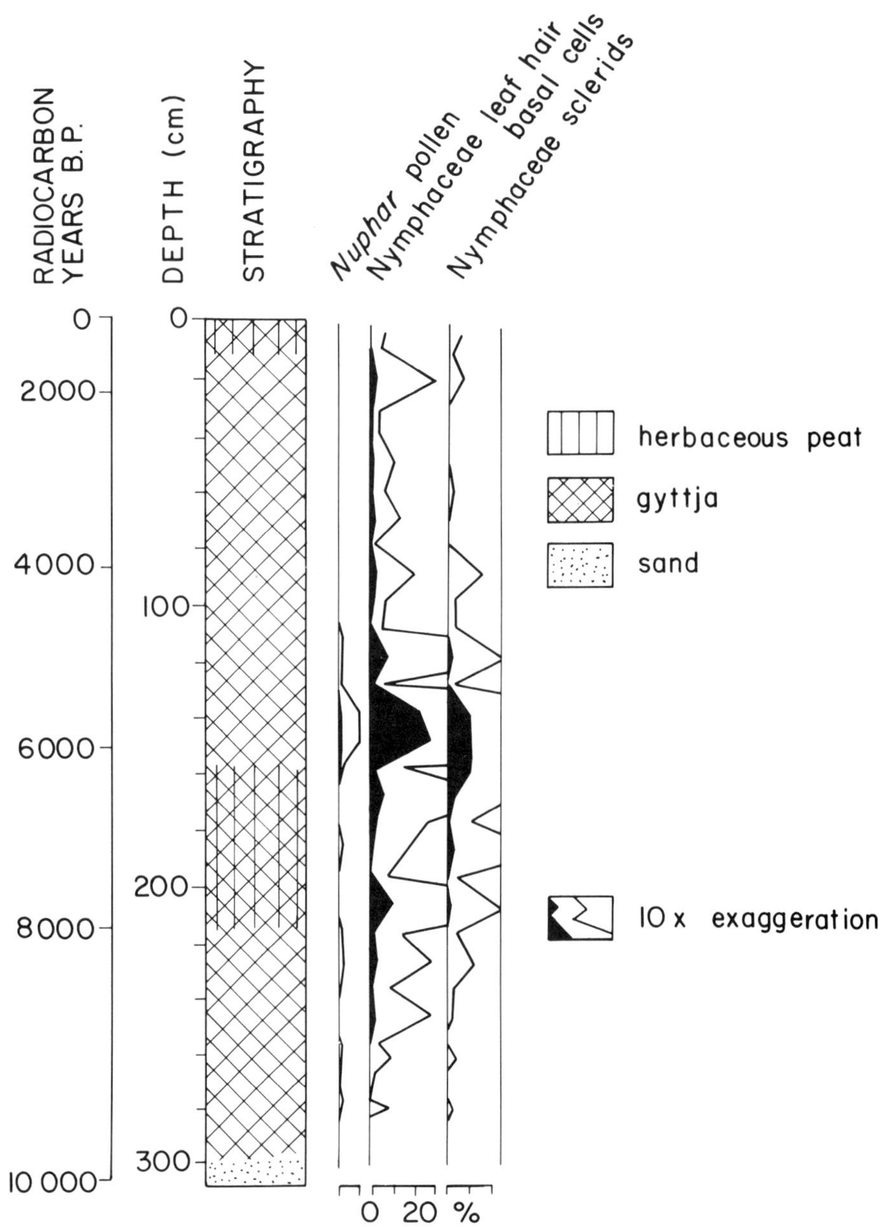

Figure 6 *Profile of fossil Nymphaeaceae pollen, leaf hair basal cells, and sclerids from Boulton Lake, Graham Island, British Columbia (Warner, 1984b). Pollen is presented as percentage of total pollen including aquatic taxa, and the basal cells and sclerids are presented as percentage of total pollen and total of basal cells and sclerids. Note the peak of all fossil types at about 140 cm depth.*

with high pH and alkalinities, and in waters where dissolved free CO_2 is limited or totally depleted (Table 1). The capacity of *C. demersum* to take up low amounts of CO_2 or to utilize bicarbonate ions in photosynthesis probably contributed to its success and more widespread distribution in postglacial waterbodies that probably had low amounts of free CO_2. Some workers have suggested that distribution beyond its northern range limit may be due to more favourable climatic conditions (Terasmae and Craig, 1958; Petersen *et al.*, 1983). It is possible that its more widespread distribution was controlled by its photosynthesizing capacity, the hydrochemical and physical characteristics of the site, and dispersibility of diaspores, rather than past climatic conditions (Warner, 1984a). Furthermore, its establishment and survival can be temporary, and the apparent range expansions in the past may really be of little paleoenvironmental significance.

Nymphaeaceae.

Sclerenchyma tissue in the petiole and peduncle of *Nuphar* and *Nymphaea* contain specialized stellate cells referred to as stone-cells, or sclereids (Ogden, 1974). They are thick walled, consisting of lignin and hardened cellulose, and are present singly or in clusters (Figures 5a and b). The sclereids are microscopic, ranging in size from 100 to 400 μm. The complex polymers comprising the lignin and cellulose make the sclereids much more resistant to decay than the enclosing sclerenchyma. The sclereids survive acid treatments in pollen analysis, and often can be viewed in the slide preparations.

Nuphar and *Nymphaea* have cells at the base of leaf hairs composed of suberin. The fatty acids of the suberin are derived from respiratory processes and presumably aid in maintaining buoyancy of the large, floating leaves of the plant. The cells are thin walled, clear, and circular in cross-section, varying from 35 to 42 μm in width (Figure 7b). They may be composed of many individual round, stacked cells. The suberized basal cells are readily incorporated and fossilized in the sediment following the death of the plant. The suberin resists acid treatments, and can be viewed in pollen slide preparations.

Both sclereids and the suberized basal cells of the Nymphaeaceae leaf

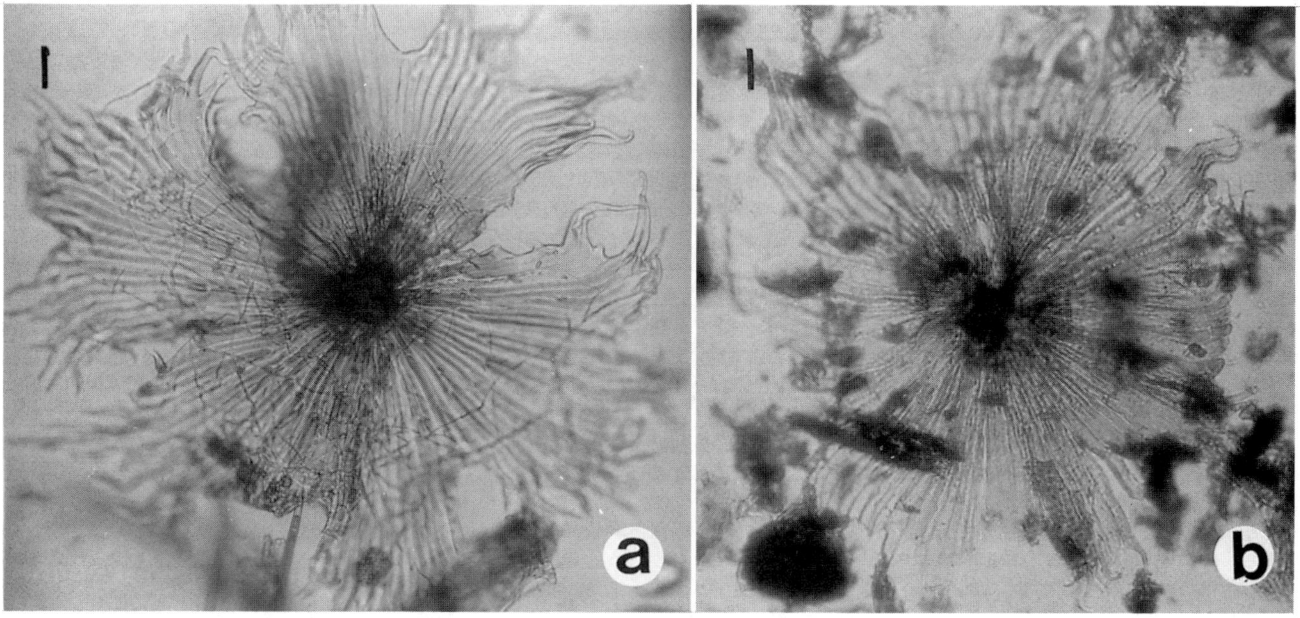

Figure 7 **(a)** *Modern and* **(b)** *fossil peltate trichomes of* Shepherdia canadensis. *Most fossils occur as fragments of the complete trichome. Scale bars equal 100 μm.*

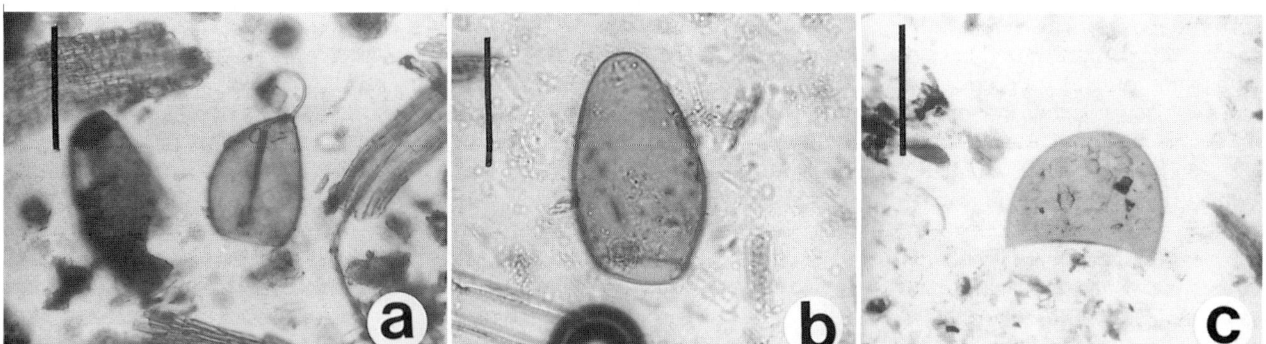

Figure 8 *Neorhabdocoel egg capsules.* **(a)** Gyratrix-*type;* **(b)** Microdalyellia-*type;* **(c)** *spherical-shaped capsule. Scale bars equal 50 μm.*

hairs are most common in gyttja and peaty gyttja sediments, although they may be found in deposits of aquatic environments that may have supported Nymphaeaceae.

A good example of Nymphaeaceae sclereids and leaf hair cells is from a lake on Graham Island, British Columbia (Warner, 1984b). Fossil pollen analysis shows that the sclereids and leaf hair cells are from *Nuphar lutea* ssp. *polysepala* (Figure 6). It is interesting to note that in samples where no pollen was found, sclereids and leaf hair cells were abundant, contributing to a much more complete paleoenvironmental representation of the plant.

Elaeagnaceae.

The shrubs, *Shepherdia canadensis* and *S. argentea*, possess deciduous leaves, with rusty-coloured scales, and silvery down on the underside. These rusty-coloured scales are umbrella shaped, peltate trichomes that preserve and can be recovered with the pollen. The trichomes may be up to 0.5 mm in diameter, with a radiating pattern of linear cells (Figure 7).

The trichomes are usually found with notable proportions of *S. canadensis* pollen (Mathewes, 1973; Warner, unpublished), particularly from late-glacial sites where this shrub was especially important in the local vegetation. Perhaps careful attention to fossil trichomes in the prairie region of Canada where *Shepherdia* attains its greatest representation in the vegetation will reveal a long and more complete fossil record for the shrubs.

ANIMALS

Flatworms.

Flatworms of the Order Neorhabdocoela (Class Turbellaria, Phylum Platyhelminthes) produce egg capsules that are incorporated and preserved in fossil freshwater deposits. The neorhabdocoeles reproduce sexually, and lay single eggs, some of which are encapsulated in a stalked coccoon for attachment to some substrate. Two types of eggs are laid: thin-shelled "summer" eggs which hatch promptly, and thick-shelled eggs which may lie dormant for long periods (Pennak, 1978). The latter types are those that are most likely to be recovered as fossils. The egg capsules vary from 80 to 600 μm in size. They are elliptical to spherical in shape, consisting of the main body and an operculum. The operculum is usually missing in the fossils (Harmsworth, 1968). There are three main types of capsules. The first type is the *Gyratrix*-type which is stalked, shaped like a wine glass, dark coloured, and may or may not be striated (Figure 8a). The second type, referred to as the *Microdalyellia*-type, is U-shaped and dark in colour (Figure 8b). The third type is a spherical-shaped capsule, much lighter in colour than the other two types (Figure 8c).

It is difficult to assess the paleoecological significance of the fossils as little is known about both the modern ecology and paleoecology of neorhabdocoeles. However, it would seem that neorhabdocoel egg capsules hold much promise largely because the animals are cosmopolitan in distribution, and different species occupy entire lake basins in a variety of microhabitats. The capsules can be abundant and well preserved, they are readily recovered from the sediment, and can even be observed in fossil pollen preparations (Harmsworth, 1968; Pennak, 1978). In one of the few systematic studies of fossil capsules, Sreenivasa (1973) found them to exceed 1500 per gram (ash weight) of lake sediment at a site in southern Ontario. Their peak abundances corresponded to eutrophic phases in the lake, but whether neorhabodocoel capsules can be used as eutrophic indicators remains to be established by comparison with modern faunal surveys.

Rotifers.

Rotifers, classified in the Phylum Rotifera, are common within zooplankton communities. They are cosmopolitan in distribution, living in a vari-

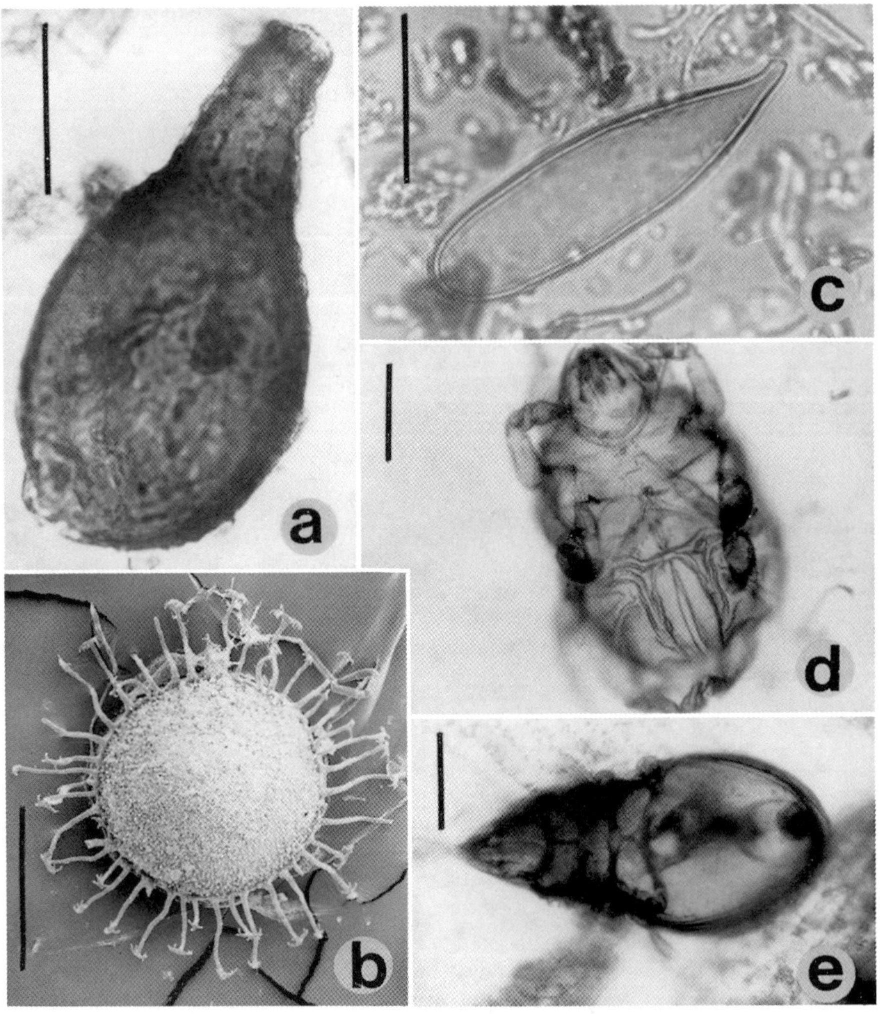

Figure 9 **(a)** *Fossil* Habrotrocha angusticollis *(Bdelloidea: Rotifera) capsule,* **(b)** *fossil* Cristatella mucedo *statoblast;* **(c)** *fossil copepod spermatophore;* **(d,e)** *representative examples of fossil oribatid mites. Scale bars equal 50μm for a,b; 25μm for c; and 100 μm for d,e.*

ety of aquatic and semi-aquatic habitats in lakes, puddles, damp soil, and wet mosses.

Members of the Order Bdelloidea occupy submerged, emergent, and mossy habitats that may be occasionally wet. The bdelloids are colourless, transparent animals possessing highly contractile bodies, some of which are protected in a shell. The dark, proteinaceous shells of one member of the Bdelloidea, *Habrotrocha angusticollis*, are particularly common in fossil peat and gyttja deposits (Figure 9a; Warner and Chengalath, 1988).

The fossil shells may be pale, to yellow-brown, to dark brown in transmitted light. They are flask-shaped and elongate, with a long neck and obovate body. The width across the body measures 100 μm, and around 56 μm at the base of the body. *H. angusticollis* shells are diagnostic because they are smooth, and do not contain extraneous material such as fungal hyphae, plant detritus, or mineral particles. They are striated, and have an obvious thickened lip around the anterior end (Figure 9a). The fossil shells resist destruction by acids used in fossil pollen preparations, and may be observed in wet mounts of untreated peat samples also.

The distribution and ecology of *H. angusticollis* in North America are not well known. This makes it difficult to assess its paleoecological significance. The bdelloids as a group are most characteristic of *Sphagnum* and other mossy habitats. They are associated with mosses in *Sphagnum* bogs, marshes, and with floating mats and other emergent vegetation along the shores of lakes and ponds. Shells of *H. angusticollis* have been found in *Sphagnum* bogs in Ontario throughout a range of soil moisture regimes from 82 to 98% (Warner and Chengalath, 1988).

Warner and Chengalath (1988) examined fossil *H. angusticollis* in two *Sphagnum* bogs in northern Ontario, and concluded that *H. angusticollis* flourished at a time when there was probably more open water with wet mosses and emergent vegetation than exists at the sites today.

Resting eggs of rotifers may become fossilized, but their morphology is not sufficiently diagnostic to confidently differentiate them from resting eggs of some other freshwater invertebrate groups.

Bryozoans.

Freshwater bryozoans are colonial, sessile animals forming large, encrusting masses on submerged stones, logs, twigs or any other solid substrate in shaded areas of permanent water bodies. Phylum Bryozoa contains mainly marine representatives, but there are at least 20 species confined to freshwater habitats in Canada and the United States (Pennak, 1978).

Individual bryozoans resemble some coelenterate polyps by being organized in cylindrical zooids with thin body walls. A mouth is situated on a large lophophore with ciliated tentacles on the anterior end of the animal. The lophophore, tentacles, and the base of the lophophore form a polypide, which comprises the living part of an individual bryozoan. The coelom is continuous from one zooid to another and interconnects the individuals into highly branched colonies.

Members of the Class Phylactolaemata produce asexual statoblasts. Statoblasts are internal buds which protect internal organelles in two tightly fitted convex sclerotized valves. The statoblasts are released by the animal and can be carried by water currents, wind, in mud on bird feet, or attached to animal bodies to establish new colonies. Statoblasts can be produced in such abundance as to form windrows along the shore of some waterbodies (Pennak, 1978). These factors make statoblasts conducive to preservation in limnic sediments.

There are three main types of statoblasts differentiated by their mode of production. All genera, except *Fredericella*, produce floatoblasts. Generally, they are released from the living zooid. Floatoblasts are equipped with an annulus of thin air cells or spines and hooked processes around the central thickened capsule (Figure 9b). Sessoblasts are produced by *Fredericella, Hyalinella, Plumatella* and *Stolella*. Usually they lack an annulus, and possess a peripheral girdle which attaches the sessoblast to the zooecium wall by cement. Lastly, only *Fredericella* produces somewhat more specialized statoblasts, called piptoblasts, which are neither cemented to the surface of the zooecium nor capable of floating. Usually piptoblasts sink to the bottom when the animal dies and the body decays.

Kuc (1973) recovered well-preserved *Cristatella mucedo* statoblasts in abundance from Early Pleistocene interglacial and interstadial deposits in the Arctic. These fossils are far north of the modern known distributional northern limit for the animal. He surmised that interglacial limnological conditions in the Arctic were much more oligotrophic than modern conditions in the region today.

In a detailed study of surficial sediments in 30 Florida lakes, Crisman *et al.* (1986) concluded that statoblasts are most abundant in lakes with an extensive littoral zone and a low phytoplankton biomass. This study emphasizes the potential for developing bryozoan statoblasts further as a new paleoecological tool, but problems exist such as the patchiness in space and time of modern faunas and the interpretation of their fossil counterparts. Of interest is Sreenivasa's (1973) interpretation of fossil statoblasts at Sunfish Lake. *Plumatella* statoblasts were found to be most abundant at a time when the littoral zone became larger compared to the open pelagic area. *Plumatella* seems to have taken advantage then of a larger area of submerged aquatic macrophytes, and an increase in water transparency and substrates for colonization.

Some other reports of fossil statoblasts in Canada include: Schofield and Robinson (1960), Matthews (1980), Klassen *et al.* (1983), Warner *et al.* (1984, 1987), and Garneau (1987). The oldest Quaternary records are from western Alaska (Matthews, 1974), the western Arctic (Kuc, 1973), and Toronto (Kerr-Lawson, 1985).

Earthworms.

Most everyone is familiar with earthworms found in terrestrial soils, but there are aquatic and amphibious species too, found in lakes, ponds, or marshes and in the wet soils at the edges of these water bodies. Although the soft-bodies of the adults are not conducive to preservation in fossil deposits, their chitinous-like coccoons are resistent to decay. The coccoons are smooth, spheroidal or ellipsoidal, with a spike, tube or tuft at opposite ends of the main structure. The size may reach 3 mm along the longest axis.

The first report in North America of fossil earthworm coccoons was found in a Holocene marl sequence in southern Ontario (Schwert, 1979). Subsequently, Warner *et al.* (1984) found them in

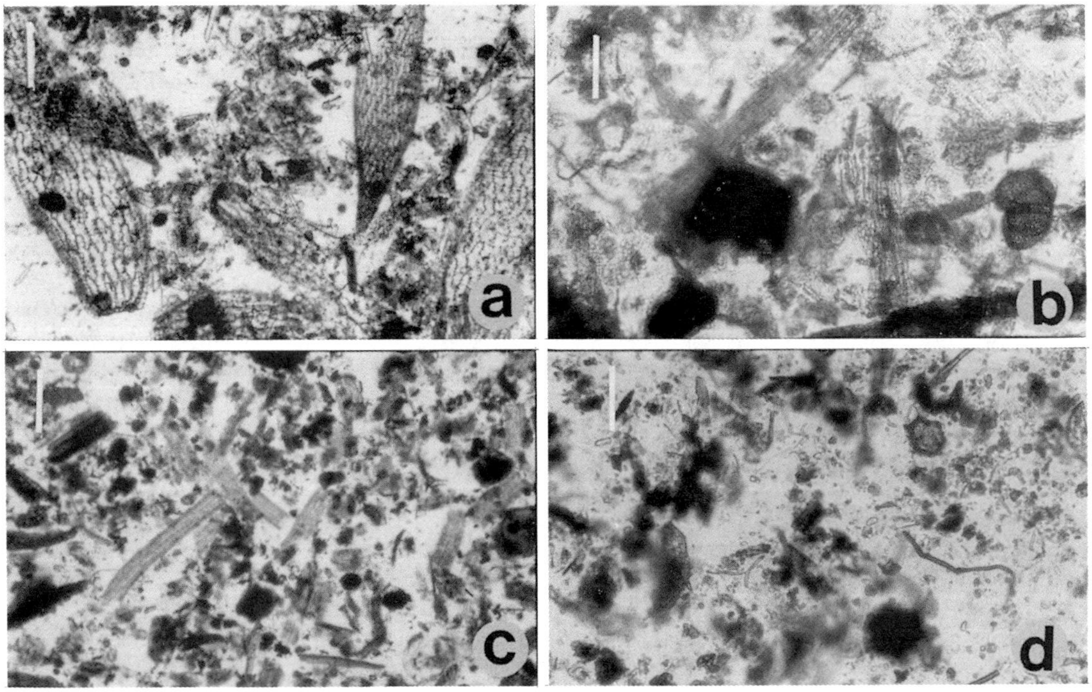
a
b
c
d

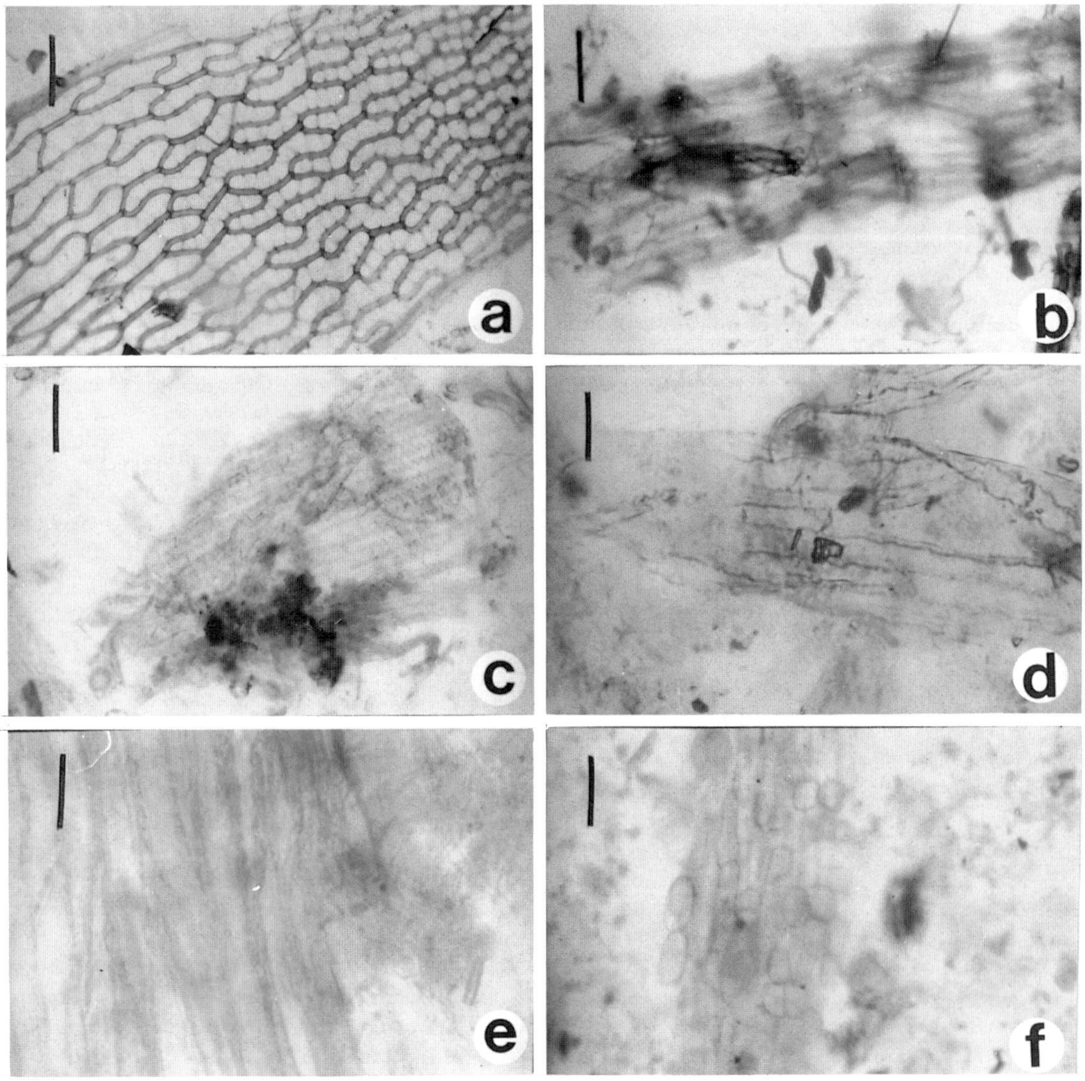
a
b
c
d
e
f

Holocene peats on Manitoulin Island. Most terrestrial earthworms are circumpolar or cosmopolitan in distribution. The occurrence of coccoons 7000 years old or older in Ontario dispels the long-held belief among zoologists that terrestrial earthworms in North America are European introductions (Schwert, 1979). More finds of fossil earthworm coccoons will illuminate the zoogeographic problems of distribution and dispersal of earthworms throughout the northern hemisphere.

Copepods.
Along with the cladocerans, ostracods, and rotifers, copepods are universally distributed in the planktonic, benthic, and littoral regions of freshwater bodies. Free-living forms include the suborders Calanoida, Cyclopoida, and Harpacticoida; they are not as morphologically modified and specialized as their parasitic relatives. The free-living copepods are the best known and most frequently collected. A single cyclopoid species commonly characterizes the limnetic plankton, occasionally with subordinate presence of a calanoid. Littoral communities, especially in aquatic vegetation, contain a variety of cyclopid and calanoid species but in lower densities.

Harpacticoids are confined to the benthos within lakes. In Europe, they have been found in unusual habitats such as in interstitial waters of sandy lake beaches, in springs, in caves, and in damp forest mosses (Pennak, 1978).

Breeding habits within the free-living copepods are similar. The male clasps the female with antennae and sometimes with a modified fifth leg to pair up the genital pores which are located on the genital (or last thoracic) segment. The male transfers the sperm to the female in small, pocket-like spermato-

Figure 10 ***(opposite page: upper)*** *Fossil peat types.* **(a)** Sphagnum *peat;* **(b)** *cyperaceous peat, with* Eriophorum *fragments;* **(c)** *brown moss; and detrital peat;* **(d)** *limnic peat, with fungal hyphae (arrow) and fine detritus. Scale bars equal 100 μm.*

Figure 11 ***(opposite page: lower)*** *Common plant fragment types encountered in peat deposits.* **(a)** Sphagnum *leaf;* **(b)** Sphagnum *stem;* **(c)** Eriophorum *leaf;* **(d)** Carex *leaf;* **(e)** Scheuzeria palustris *leaf;* **(f)** Carex limosa-*type rootlet. Scale bars equal 100 μm.*

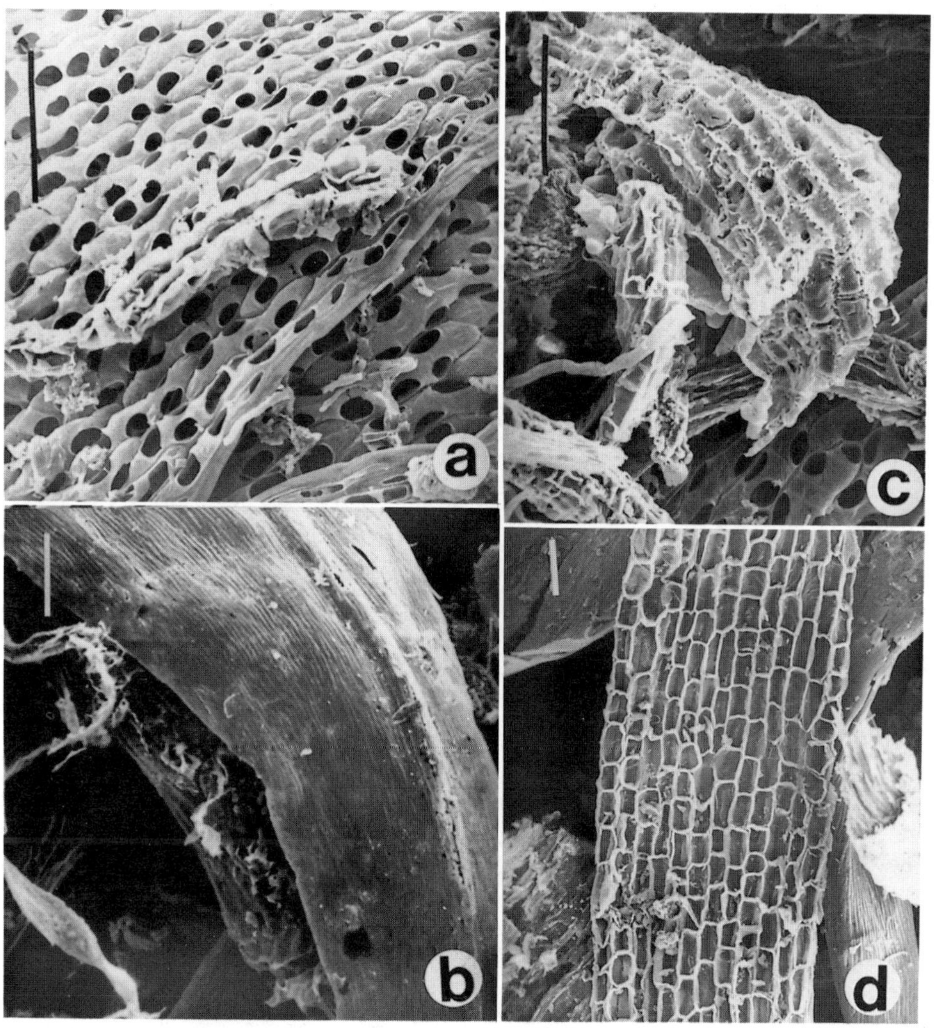

Figure 12 *Scanning electron micrographs of selected peat plant fragments types.* **(a)** Sphagnum *leaf;* **(b)** *brown moss leaf;* **(c)** *probably* Carex *leaf; and* **(d)** *Cyperaceae stem. Scale bars equal 100 μm.*

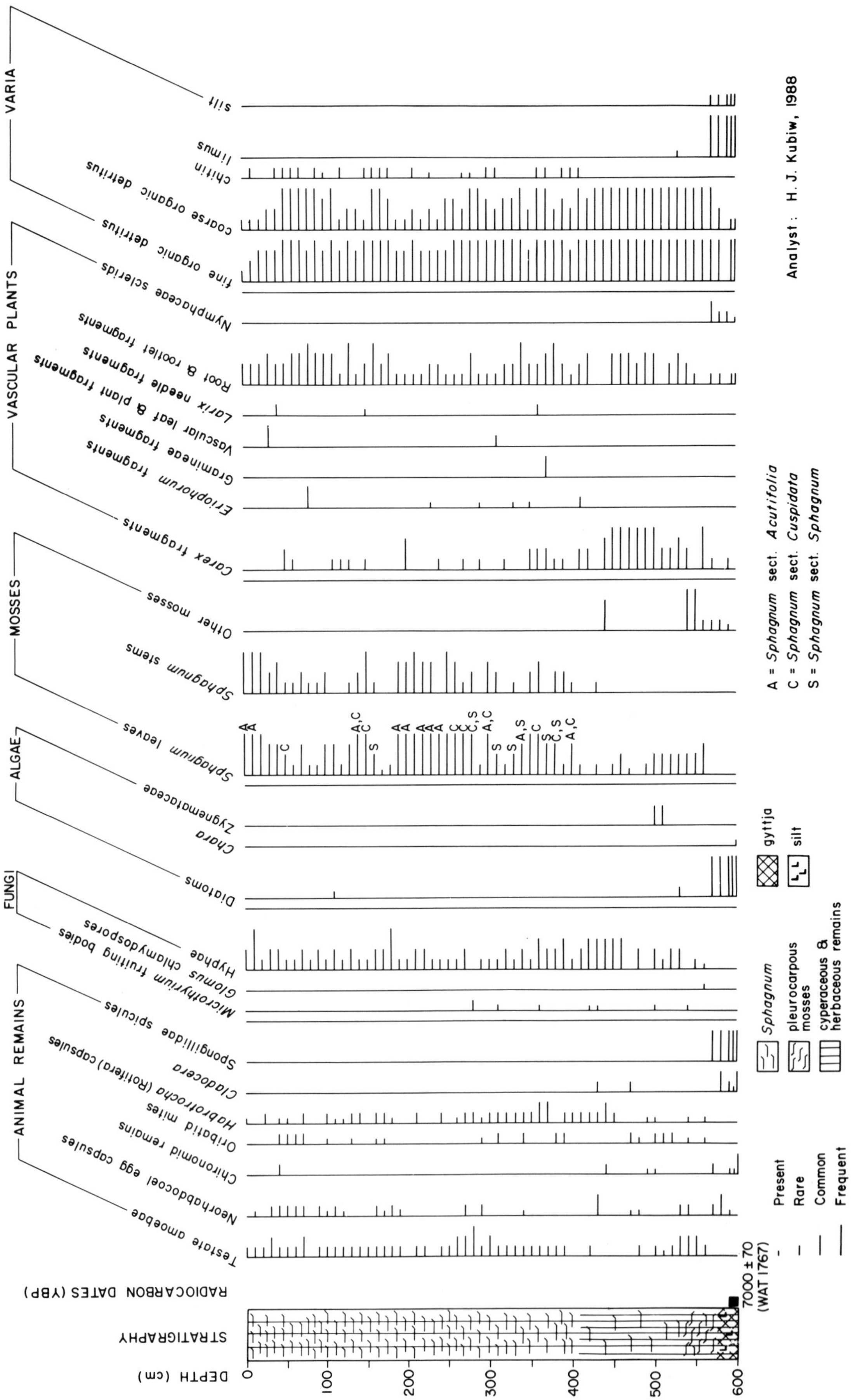

Figure 13 *A diagram of peat and sediment components from a Sphagnum bog near Thunder Bay, Ontario. (From Warner, 1989).*

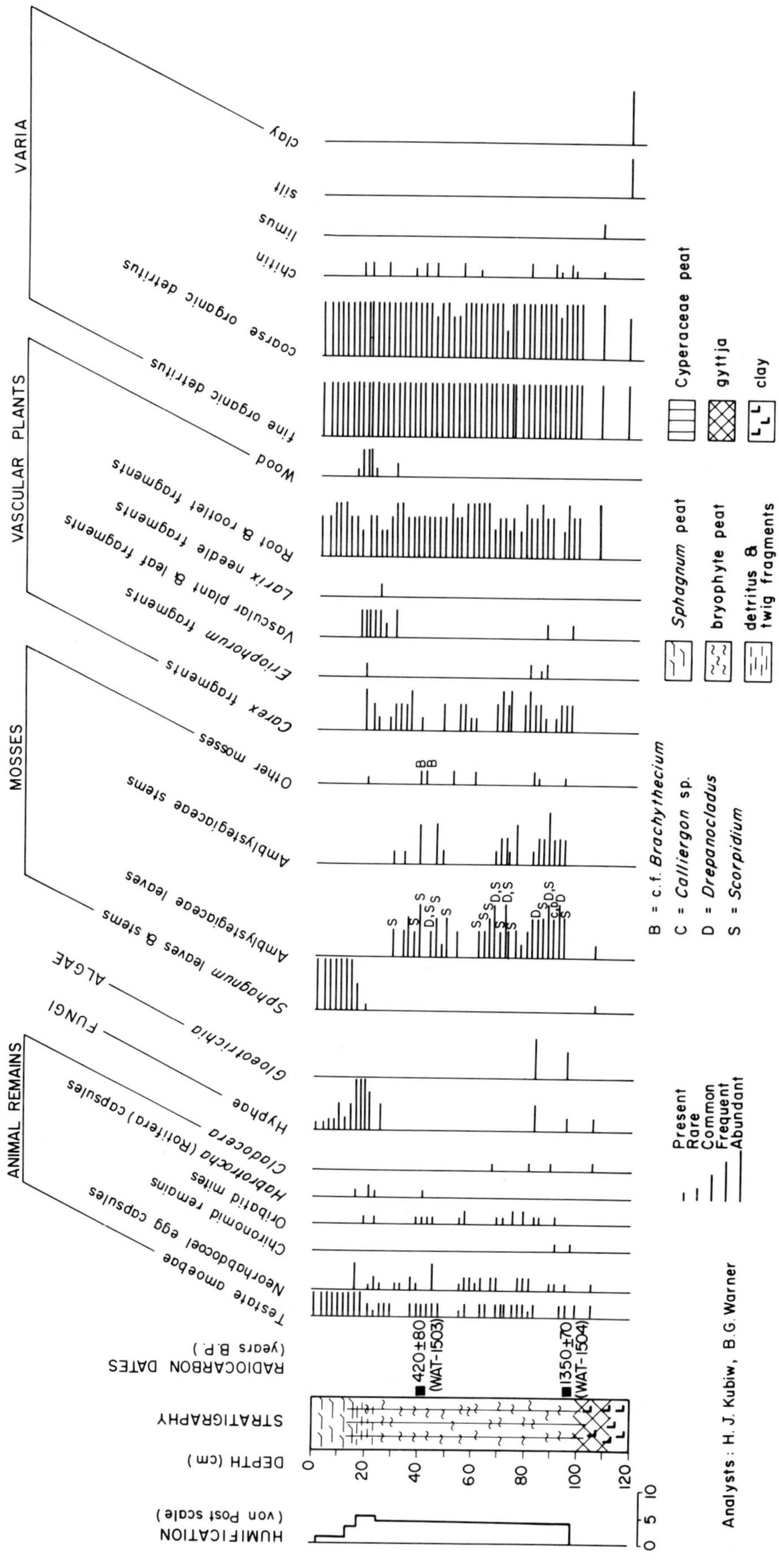

Figure 14 *Representative diagram of various peat and sediment components from a Sphagnum bog on the Albany River in northern Ontario. (From Warner, 1989).*

phores, usually with the aid of the legs during clasping. The female then stores the sperm in a special ventral area of her genital segment, into which she will eventually release eggs for fertilization. It is the male spermatophore that is preserved and fossilized in Quaternary limnic deposits.

The fossil spermatophores are usually translucent to pale yellow. They are tear-drop shaped with a short, coiled, tail-like appendage. The spermatophore is up to 50 μm in length. The entire tail segment is often broken off in fossil specimens (Figure 9c).

Copepods are soft-bodied individuals and undergo complete decomposition following death. Apart from resting eggs, which are not particularly diagnostic, spermatophores provide the best fossil record of copepods.

Mites.

Soil or moss mites, also referred to as oribatids, are a small group of arthropods living in terrestrial soils, peatlands, and fresh waters. Classified in Class Arachnida and Suborder Oribatida, the Subclass Acari are particularly abundant as fossils in lacustrine and peat deposits. Some fossil oribatids have been found in Mexican amber. Fossil numbers can reach 23 per gram of marl (Erickson, 1988).

The animals are small (usually less than 500 μm) and possess a coriaceous, sclerotized body that is readily preserved almost intact; such exceptional preservation provides all details of mouthparts and genital and anal areas necessary for identification (Frey, 1964; Erickson, 1988). Oribatids have a characteristic "tear-drop" shaped body, with fused cephalothorax and abdomen, six pairs of appendages, and are dark-coloured (Figures 9d and e). Ecologically, oribatids are most common in forest and peatland soils, litter, and mosses, and are less frequent in aquatic habitats (Pennak, 1978; Erickson, 1988). Aquatic species are most abundant in littoral submerged and emergent plant beds, and are less common on wave-washed lake shores, and in the deeper parts of lake bodies. Oribatid mites may be valuable indicators of change in peatlands due to drainage (Markkula, 1986).

Frey (1964), Krivolutsky and Druk (1986), Erickson (1988) and Schelvis and van Geel (1989) have reviewed the Quaternary record of oribatids; the latter paper is one of the most detailed studies in North America.

PEAT AND SEDIMENT COMPOSITION

The scientific definition of peat is partially decayed organic matter mainly of plant origin. Almost any plant community can form peat, however, only a few plants can grow in the wet conditions and saturated soils of peatlands. Peatlands store the plants that grew in them in the form of wood, stems, leaves, roots, rhizomes, fruits, seeds, and pollen.

Lakes, like peatlands, contain in their sediments fossil representatives of the plankton communites from both the littoral and pelagic parts of the basin that sink to the bottom sediments when they die. One main difference between lakes and peatlands is that the fossils in lake sediments may have moved from their original site of deposition either by within-basin processes or by external processes such as surface feeder streams or runoff. Such preferential sorting, long-distance transport, and reworking processes are negligible or minimized in peatlands.

Other contributions in this series, and the first part of this paper, have described some of the fossils in peatlands and their use in paleoecology. Not only are the fossils contained within them of importance, but the nature and composition of the sediments themselves contain a wealth of paleoecological information.

The unifying characteristic of aquatic and peatland plants is that their bulk is composed of aerenchyma and parenchyma tissue. These plant tissues form an open network of living cells separated by air spaces which supply the plant with oxygen and give it buoyancy to keep it afloat. Specific plants have their own individual systems to perform these functions by possessing their own intricate cellular designs. The fossil tissue patterns and structures can be identified under the microscope to recognize plant species and reconstruct past plant communities. These analyses are another area of paleoecological research, which when performed in conjunction with analysis of pollen, plant macrofossils, testate amoebae, and some other important animal types, provides a fairly complete representation of past peat-forming or lacustrine communities.

The guide by Lévesque *et al.* (1988) is a useful introduction to analysis of Canadian peats. Other keys such as Stoddard (1965) and Riegert and Singh (1982) may serve to identify plant fragments in certain types of fossil deposits. The following papers some aspects of peat analysis in more detail: Heikurainen and Huikari (1952), MacFarlane and Radforth (1968), Haihu and Etelämäki (1986), Svensson (1986), and Warner (1989).

Detailed examination of the character and composition of peat constituents can reveal information such as past hydrological and oxygenation conditions of the site by noting such features as the degree of humification and quality of preservation, or other details of the plant species acting as major peat-formers (Figures 10, 11, and 12). Mosses of the genus *Sphagnum*, commonly known as the "peat mosses", and mosses primarily of the family Amblystegiaceae, commonly known as "brown mosses", and vascular plants in the Cyperaceae are the main peat-formers in Canadian wetlands.

Although it is difficult to accurately quantify sedimentary and peat components, it is possible to estimate the relative proportion of the various identifiable remains. Profiles can be constructed in much the same manner as is done for other fossil groups, illustrating the various sedimentary components (Figures 13 and 14).

Clearly, much more work remains to be done on the lesser known fossil groups. Only with further information on the distribution and ecology of modern counterparts, will it be possible to improve the effectiveness of fossil representatives in the Quaternary fossil record.

ACKNOWLEDGEMENTS

R. Chengalath, R.P. Futyma, and P.J.H. Richard critically read early versions of this paper which contributed to its accuracy and clarity. I thank K.I. Hanf and H.J. Kubiw for providing data and technical assistance. N. Bahar drafted the figures and P. Fisher assisted with photography. On-going paleoecological research which contributed to this paper is supported by the Natural Sciences and Engineering Research Council of Canada for which I am grateful.

REFERENCES

Beals, E.O., 1977, A manual of marsh and aquatic vascular plants of North Carolina with habitat data: North Carolina Experiment Station, Raleigh, 298 p.

Birks, H.H., 1973, Modern macrofossil assemblages in lake sediments in Minnesota, *in* Birks, H.J.B. and West, R.G., eds., Quaternary Plant Ecology: Blackwell Scientific, Oxford, p. 173-188.

Bombin, M., 1984, On phytoliths, Late Quaternary paleoecology of Bergingia, and information evolutionary theory, Ph.D. Dissertation, University of Alberta, 164 p.

Bright, R.C., 1982, Paleontology of the lacustrine member of the American Falls Lake Beds: Idaho Bureau of Mines and Geology Bulletin, v. 26, p. 597-614.

Carbone, V.A., 1977, Phytoliths as paleoecological indicators: Annals of the New York Academy of Sciences, v. 288, p. 194-205.

Crisman, T.L., Crisman, U.A.M. and Binford, M.W., 1986, Interpretation of bryozoan microfossils in lacustrine sediment cores: Hydrobiologia, v. 143, p. 113-118.

Crowder, A.A., Bristow, J.M., King, M.R. and Vander Kloet, S., 1977, The aquatic macrophytes of some lakes in southeastern Ontario: Naturaliste canadien, v. 104, p. 457-464.

Doherty, M.J. and LaRoi, G.H., 1973, Some effects of fluctuating water levels and other factors on aquatic and semiaquatic vegetation in five water bodies of the Peace-Athabasca Delta, Alberta, *in* Reinelt, E.R., Laycock, A.H. and Schultz, W.M., eds., Proceedings of the Symposium on the Lakes of Western Canada: University of Alberta, Edmonton, p. 202-226.

Erickson, J.M., 1988, Fossil oribatid mites as tools for Quaternary paleoecologists: Preservation quality, quantities, and taphonomy: Bulletin of the Buffalo Society of Natural Sciences, v. 33, p. 207-226.

Frey, D.G., 1964, Remains of animals in Quaternary lake and bog sediments and their interpretation: Ergebnisse der Limnologie, v. 2, p. 1-114.

Garneau, M., 1987, Reconstitution paléoécologique d'une tourbière littorale de l'estuaire de Saint-Laurent: Analyse macrofossile et sporopollinique: Géographie physique et Quaternaire, v. 41, p. 109-125.

Gray, J., 1988, Evolution of the freshwater ecosystem: The fossil record: Palaeogeography, Palaeoclimatology, Palaeoecology, v. 62, p. 1-214.

Grosse-Brauckmann, G., 1986, Analysis of vegetative plant macrofossils, *in* Berglund, B.E., ed., Handbook of Holocene Palaeoecology and Palaeohydrology: J. Wiley and Sons Ltd., Chichester, p. 591-618.

Haihu, K. and Etelämäki, H., 1986, Two microscopic methods for the determination of peat types: Suo, v. 37, p. 29-33.

Harmsworth, R.V., 1968, The developmental history of Blelham Tarn (England) as shown by animal microfossils, with special reference to the Cladocera: Ecological Monographs, v. 38, p. 223-241.

Heikurainen, L. and Huikari, O., 1952, The microscopic determination of peat types: Communicationes Instituti Forestalis Fenniae, v. 40, p. 1-34.

Hellquist, C.B. and Crow, G.E., 1980, Aquatic vascular plants of New England: Part 1. Zosteraceae, Potamogetonaceae, Zannichelliaceae, and Najadaceae: New Hampshire Agriculture Experimental Station Bulletin, v. 515, p. 1-68.

Holmquist, C., 1971, Northerly localities for three aquatic plants, *Lemna trisulcata* L., *Ceratophyllum demersum* L., and *Myriophyllum spicatum* L.: Botaniska Notiser, v. 124, p. 335-342.

Kerr-Lawson, L.J., 1985, Gastropods and plant macrofossils from the Quaternary Don Formation (Sangamonian Interglacial), M.Sc. Thesis, University of Waterloo, 201 p.

Klassen, R.A., Matthews, J.V. and Philips, L.K., 1983, Taxa in lake sediments of the District of Keewatin: Geological Survey of Canada, Paper 83-1A, p. 357-361.

Krivolutsky, D.A. and Druk, A. Ya., 1986, Fossil oribatid mites: Annual Review of Entomology, v. 31, p. 533-545.

Kuc, M., 1973, Fossil statoblasts of *Cristatella mucedo* Cuvier in the Beaufort Formation and in interglacial and postglacial deposits of the Canadian Arctic: Geological Survey of Canada, Paper 72-28, p. 1-12.

Les, D.H., 1986, The phytogeography of *Ceratophyllum demersum* and *C. echinatum* (Ceratophyllaceae) in glaciated North America: Canadian Journal of Botany, v. 64, p. 598-509.

Lévesque, P.E.M., Dinel, H. and Larouche, A., 1988, Guide to the identification of plant macrofossils in Canadian peatlands: Research Branch, Agriculture Canada, Publication 1817, p. 1-65.

MacFarlane, I.C. and Radforth, N.W., 1968, Structure as a basis of peat classification: Proceedings of the Third International Peat Congress, Quebec, p. 91-97.

Markkula, I., 1986, Comparison of present and subfossil oribatid faunas in the surface peat of a drained pine mire: Annales Entomologici Fennici, v. 52, 39-41.

Mathewes, R.W., 1973, A palynological study of postglacial vegetation changes in the University Research Forest, southwestern British Columbia: Canadian Journal of Botany, v. 51, p. 2085-2103.

Matthews, J.V., 1974, Quaternary environments at Cape Deceit (Seward Peninsula, Alaska): Evolution of a tundra ecosystem: Geological Society of America, Bulletin, v. 85, p. 1353-1384.

Matthews, J.V., 1980, Paleoecology of John Klondike Bog, Fisherman Lake Region, southwest District of Mackenzie: Geological Survey of Canada, Paper 80-32, p. 1-12.

Moyle, J.B., 1945, Some chemical factors influencing the distribution of aquatic plants in Minnesota: American Midland Naturalist, v. 34, p. 402-420.

Ogden, E.C., 1974, Anatomical patterns of some aquatic vascular plants of New York: New York State Museum and Science Service, Bulletin, v. 24, p. 1-133.

Palmer, P.G., 1976, Grass cuticles: A new paleoecological tool for East African lake sediments: Canadian Journal of Botany, v. 54, p. 1725-1734.

Pals, J.P., van Geel, B. and Dafos, A., 1980, Palaeoecological studies in the Klokkeweel Bog near Hoogkarspel (Prov. of Noord-Holland): Review of Palaeobotany and Palynology, v. 30, p. 371-418.

Pennak, R.W., 1978, Fresh-water Invertebrates of the United States, Second Edition: J. Wiley and Sons, New York, 803 p.

Petersen, K.L., Mehringer, P.J. and Gustafson, G.E., 1983, Late-Glacial vegetation and climate at the Manis mastodon site, Olympic Peninsula, Washington: Quaternary Research, v. 20, p. 215-231.

Pip, E., 1988, Niche congruency of aquatic macrophytes in central North America with respect to five water chemistry parameters: Hydrobiologia, v. 162, p. 173-182.

Piperno, D.R., 1985, Phytolith analysis and tropical paleo-ecology: Production and taxonomic significance of siliceous forms in New World plant domesticates and wild species: Review of Palaeobotany and Palynology, v. 45, p. 185-228.

Piperno, D.R., 1988, Phytolith Analysis: An archaeological and geological perspective: Academic Press, Inc., San Diego, 280 p.

Powers, A.H. and Gilbertson, D.D., 1987, A simple preparation technique for the study of opal phytoliths from archaeological and Quaternary sediments: Journal of Archaeological Science, v. 14, p. 529-535.

Reynolds, J.D. and Reynolds, S.C.P., 1975, Aquatic angiosperms of some British Columbia lakes: Syesis, v. 8, p. 291-295.

Riegert, P.W. and Singh, N., 1982, A key to identify epidermal fragments of Saskatchewan rangeland plants: Canadian Journal of Botany, v. 60, p. 2461-2468.

Riley, J.L. and MacKay, S.L., 1980, The vegetation and phytogeography of coastal southwestern James Bay: Royal Ontario Museum, Life Sciences Contributions, v. 124, p. 1-81.

Rovner, I., 1971, Potential of opal phytoliths for use in paleoecological reconstruction: Quaternary Research, v. 1, p. 345-359.

Schelvis, J. and van Geel, B., 1989, A palaeoecological study of the mites (Acari) from a Lateglacial deposit at Usselo (The Netherlands): Boreas, v. 18, p. 237-243.

Schofield, W.B. and Robinson, H., 1960, Late-glacial and postglacial plant macrofossils from Gillis Lake, Richmond County, Nova Scotia: American Journal of Science, v. 258, p. 518-523.

Schwert, D.P., 1979, Description and significance of a fossil earthworm (Oligochaeta: Lubricidae) cocoon from postglacial sediments in southern Ontario: Canadian Journal of Zoology, v. 57, p. 1402-1405.

Sreenivasa, B.A., 1973, Paleoecological studies of Sunfish Lake and its environs, Ph.D. Dissertation, University of Waterloo, Ontario, 185 p.

Stoddard, E.M., 1965, Identifying plants by leaf epidermal characters: Connecticut Agricultural Experimental Station Circular, v. 227, p. 1-9.

Svensson, G., 1986, Recognition of peat-forming plant communities from their peat deposits in two south Swedish bog complexes: Vegetatio, v. 66, p. 95-108.

Terasmae, J. and Craig, B.G., 1958, Discovery of *Ceratophyllum demersum* L. in Northwest Territories: Canadian Journal of Botany, v. 36, p. 567-569.

van Geel, B., 1978, A paleoecological study of Holocene peat bog sections in Germany and the Netherlands: Review of Palaeobotany and Palynology, v. 25, p. 1-120.

van Geel, B., Bohncke, S.J.P. and Dee, H., 1980, A palaeoecological study of an Upper Late Glacial and Holocene sequence from "De Borchert", the Netherlands: Review of Palaeobotany and Palynology, v. 31, p. 367-448.

van Geel, B., Hallewas, D.P. and Pals, J.P., 1982, A Late Holocene deposit under the Westfriese Zeedijk near Enkhuizen (Prov. of Noord-Holland), the Netherlands: Review of Palaeobotany and Palynology, v. 38, p. 269-335.

Warner, B.G., 1984a, Commentary on plant indicator taxa in climatic reconstructions: Quaternary Research, v. 22, p. 266-268.

Warner, B.G., 1984b, Late Quaternary paleoecology of eastern Graham Island, Queen Charlotte Islands, British Columbia, Canada: Ph.D. Thesis, Simon Fraser University.

Warner, B.G., 1989, Geological and paleoecological aspects of *Sphagnum* bogs in Ontario: Proceedings of the Wetlands Conference, Federation of Ontario Naturalists, Toronto, p. 329-338.

Warner, B.G. and Chengalath, R., 1988, Fossil *Habrotrocha angusticollis* (Rotatoria: Bdelloidea) in North America: Journal of Paleolimnology, v. 1, p. 141-147.

Warner, B.G., Hebda, R.J. and Hann, B.J., 1984, Postglacial paleoecological history of a cedar swamp, Manitoulin Island, Ontario, Canada: Palaeogeography, Palaeoclimatology, Palaeoecology, v. 45, p. 301-345.

Warner, B.G., Karrow, P.F., Morgan, A.V. and Morgan, A., 1987, Plant and insect fossils from Nipissing sediments along the Goulais River, southern Lake Superior: Canadian Journal of Earth Sciences, v. 24, p. 1526-1536.

West, R.G., 1985, The future of Quaternary research: The Geologists' Association, Proceedings, v. 96, p. 193-197.

Wile, I. and McCombie, A.M., 1972, Growth of aquatic plants in southern Ontario impoundments in relation to phosphorous, nitrogen, and other factors: Ontario Ministry of the Environment, Toronto.

Originally published in
Geoscience Canada Volume 16 Number 4
(December 1989)

Index